Pontiac Fiero Automotive Repair Manual

by Mike Stubblefield and John H Haynes

Member of the Guild of Motoring Writers

Models covered:

All models, 2.5L four and 2.8L V6 engines
Manual and automatic transaxles
1984 through 1988

(79008-5E13)

Haynes Group Limited
Haynes North America, Inc.
www.haynes.com

Acknowledgements

Wiring diagrams originated exclusively for Haynes North America, Inc. by George Edward Brodd. Certain illustrations originated by Valley Forge Technical Information Services.

A book in the Haynes Automotive Repair Manual Series

ISBN-10: 1 85010 616 9
ISBN-13: 978 1 85010 616 6

Library of Congress Catalog Card Number 89-84581

Contents

Haynes mechanic, author and photographer with Pontiac Fiero

About this manual

Its purpose

The purpose of this manual is to help you get the best value from your vehicle. It can do so in several ways. It can help you decide what work must be done, even if you choose to have it done by a dealer service department or a repair shop; it provides information and procedures for routine maintenance and servicing; and it offers diagnostic and repair procedures to follow when trouble occurs.

We hope you use the manual to tackle the work yourself. For many simpler jobs, doing it yourself may be quicker than arranging an appointment to get the vehicle into a shop and making the trips to leave it and pick it up. More importantly, a lot of money can be saved by avoiding the expense the shop must pass on to you to cover its labor and overhead costs. An added benefit is the sense of satisfaction and accomplishment that you feel after doing the job yourself.

Using the manual

The manual is divided into Chapters. Each Chapter is divided into numbered Sections, which are headed in bold type between horizontal lines. Each Section consists of consecutively numbered paragraphs.

At the beginning of each numbered Section you will be referred to any illustrations which apply to the procedures in that Section. The reference numbers used in illustration captions pinpoint the pertinent Section and the Step within that Section. That is, illustration 3.2 means the illustration refers to Section 3 and Step (or paragraph) 2 within that Section.

Procedures, once described in the text, are not normally repeated. When it's necessary to refer to another Chapter, the reference will be given as Chapter and Section number. Cross references given without use of the word "Chapter" apply to Sections and/or paragraphs in the same Chapter. For example, "see Section 8" means in the same Chapter.

References to the left or right side of the vehicle assume you are sitting in the driver's seat, facing forward.

Even though we have prepared this manual with extreme care, neither the publisher nor the author can accept responsibility for any errors in, or omissions from, the information given.

NOTE

A **Note** provides information necessary to properly complete a procedure or information which will make the procedure easier to understand.

CAUTION

A **Caution** provides a special procedure or special steps which must be taken while completing the procedure where the Caution is found. Not heeding a Caution can result in damage to the assembly being worked on.

WARNING

A **Warning** provides a special procedure or special steps which must be taken while completing the procedure where the Warning is found. Not heeding a Warning can result in personal injury.

Introduction to the Pontiac Fiero

The Pontiac Fiero is the first American-built mid-engine sports car. Although it uses many existing GM components from other vehicles, the Fiero combines these parts in innovative ways. Moreover, one aspect of the Fiero's design that is shared with no other vehicle, American or foreign, is its "space frame" and chassis design with separate reinforced "Enduraflex" body panels. Corrosion free exterior panels are bolted to the space frame, making them easily removable if repair or service is required.

Two engines are available in the Fiero. The base engine is a 2.5 liter four-cylinder, mated to either a five-speed manual or three-speed automatic transaxle. In 1985, an optional 2.8 liter V6 engine, equipped with an automatic transaxle or a four-speed manual transaxle, was added to the line. In 1987, a five-speed manual transaxle became available as an option for the V6 engine. Some 1987 and all 1988 2.5L engines are equipped with a force balancer. This assembly is attached to the engine block at the bottom under the crankshaft and extending into the oil pan. This engine requires a cartridge type oil filter.

1988 models are also equipped with a redesigned front suspension. The upper control arm is shorter in length than the lower control arm. The shock absorbers and coil springs have been redesigned. The suspension components are not interchangeable between early and late models.

General dimensions

Overall length	
SE and GT models	165.8 in
all others	160.7 in
Overall width	68.9 in
Overall height	46.9 in
Wheelbase	93.4 in

Vehicle identification numbers

Modifications are a continuing and unpublicized process in automotive manufacturing. Because spare parts manuals and lists are compiled on a numerical basis, the individual vehicle numbers are essential to correctly identify the component required.

Body number plate

The body number plate is located in the front compartment on the left inner wheelhousing, just behind the left headlight. The body number plate identifies the model year, body color(s), division, body type, series, body style, assembly plant, body number, trim combination, seat option, paint type, time build code and roof option (see illustration). This plate is especially useful for matching the color and type of paint during repair work.

Vehicle Identification Number (VIN)

This very important identification number is located on a plate attached to the top left corner of the dashboard of the vehicle (see illustration). The VIN also appears on the Vehicle Certificate of Title and Registration. It contains valuable information such as the vehicle's manufacturing location and the date of its completion. The VIN label also contains information about the way in which the vehicle is equipped.

Engine identification number

The engine identification number is located on a flat space at the right (pulley) end of the block on 1984 engines (see illustration). On 1985 through 1988 four-cylinder models, the engine ID number is located on a flat space either on the lower or upper left end of the block or on the right (pulley) end of the block (see illustration). On 1985 through 1988 V6 models, the engine ID number is located on a flat space on the lower left end of the block (see illustration).

Manual transaxle identification number

The identification number for both the four and five-speed manual transaxles is located on a pad on the top front portion of the housing.

Automatic transaxle identification number

The identification number for the automatic transaxle is located on a pad on top of the housing (see illustration).

Alternator numbers

The alternator ID number is on top of the drive end of the frame.

Starter numbers

The starter ID number is stamped on the outer case towards the rear.

Battery numbers

The battery ID number is on the middle of the cell cover at the left top of the battery.

The Vehicle Identification Number (VIN) number is located on the left side of the dashboard just behind the windshield

The engine ID number (arrow) is located on the crankshaft pulley end of the block on 1984 four-cylinder engines

The engine ID number is located either on the front lower left end (arrow) or the front upper end of the block on all V6 engines

The transaxle optional ID number is located just above the oil pan

Buying parts

Replacement parts are available from many sources, which generally fall into one of two categories - authorized dealer parts departments and independent retail auto parts stores. Our advice concerning these parts is as follows:

Retail auto parts stores: Good auto parts stores will stock frequently needed components which wear out relatively fast, such as clutch components, exhaust systems, brake parts, tune-up parts, etc. These stores often supply new or reconditioned parts on an exchange basis, which can save a considerable amount of money. Discount auto parts stores are often very good places to buy materials and parts needed for general vehicle maintenance such as oil, grease, filters, spark plugs, belts, touch-up paint, bulbs, etc. They also usually sell tools and general accessories, have convenient hours, charge lower prices and can often be found not far from home.

Authorized dealer parts department: This is the best source for parts which are unique to the vehicle and not generally available elsewhere (such as major engine parts, transmission parts, trim pieces, etc.).

Warranty information: If the vehicle is still covered under warranty, be sure that any replacement parts purchased - regardless of the source - do not invalidate the warranty!

To be sure of obtaining the correct parts, have engine and chassis numbers available and, if possible, take the old parts along for positive identification.

Maintenance techniques, tools and working facilities

Maintenance techniques

There are a number of techniques involved in maintenance and repair that will be referred to throughout this manual. Application of these techniques will enable the home mechanic to be more efficient, better organized and capable of performing the various tasks properly, which will ensure that the repair job is thorough and complete.

Fasteners

Fasteners are nuts, bolts, studs and screws used to hold two or more parts together. There are a few things to keep in mind when working with fasteners. Almost all of them use a locking device of some type, either a lockwasher, locknut, locking tab or thread adhesive. All threaded fasteners should be clean and straight, with undamaged threads and undamaged corners on the hex head where the wrench fits. Develop the habit of replacing all damaged nuts and bolts with new ones. Special locknuts with nylon or fiber inserts can only be used once. If they are removed, they lose their locking ability and must be replaced with new ones.

Rusted nuts and bolts should be treated with a penetrating fluid to ease removal and prevent breakage. Some mechanics use turpentine in a spout-type oil can, which works quite well. After applying the rust penetrant, let it work for a few minutes before trying to loosen the nut or bolt. Badly rusted fasteners may have to be chiseled or sawed off or removed with a special nut breaker, available at tool stores.

If a bolt or stud breaks off in an assembly, it can be drilled and removed with a special tool commonly available for this purpose.

Most automotive machine shops can perform this task, as well as other repair procedures, such as the repair of threaded holes that have been stripped out.

Flat washers and lockwashers, when removed from an assembly, should always be replaced exactly as removed. Replace any damaged washers with new ones. Never use a lockwasher on any soft metal surface (such as aluminum), thin sheet metal or plastic.

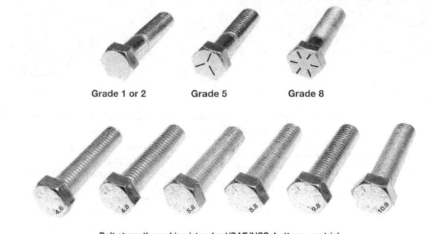

Grade 1 or 2   Grade 5   Grade 8

Bolt strength marking (standard/SAE/USS; bottom - metric)

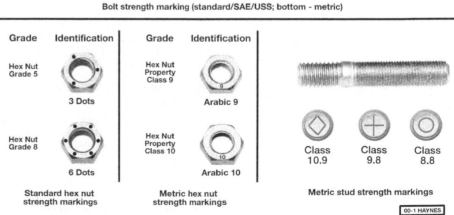

Grade	Identification	Grade	Identification
Hex Nut Grade 5	3 Dots	Hex Nut Property Class 9	Arabic 9
Hex Nut Grade 8	6 Dots	Hex Nut Property Class 10	Arabic 10

Standard hex nut strength markings

Metric hex nut strength markings

Class 10.9   Class 9.8   Class 8.8

Metric stud strength markings

00-1 HAYNES

Fastener sizes

For a number of reasons, automobile manufacturers are making wider and wider use of metric fasteners. Therefore, it is important to be able to tell the difference between standard (sometimes called U.S. or SAE) and metric hardware, since they cannot be interchanged.

All bolts, whether standard or metric, are sized according to diameter, thread pitch and length. For example, a standard 1/2 - 13 x 1 bolt is 1/2 inch in diameter, has 13 threads per inch and is 1 inch long. An M12 - 1.75 x 25 metric bolt is 12 mm in diameter, has a thread pitch of 1.75 mm (the distance between threads) and is 25 mm long. The two bolts are nearly identical, and easily confused, but they are not interchangeable.

In addition to the differences in diameter, thread pitch and length, metric and standard bolts can also be distinguished by examining the bolt heads. To begin with, the distance across the flats on a standard bolt head is measured in inches, while the same dimension on a metric bolt is sized in millimeters (the same is true for nuts). As a result, a standard wrench should not be used on a metric bolt and a metric wrench should not be used on a standard bolt. Also, most standard bolts have slashes radiating out from the center of the head to denote the grade or strength of the bolt, which is an indication of the amount of torque that can be applied to it. The greater the number of slashes, the greater the strength of the bolt. Grades 0 through 5 are commonly used on automobiles. Metric bolts have a property class (grade) number, rather than a slash, molded into their heads to indicate bolt strength. In this case, the higher the number, the stronger the bolt. Property class numbers 8.8, 9.8 and 10.9 are commonly used on automobiles.

Strength markings can also be used to distinguish standard hex nuts from metric hex nuts. Many standard nuts have dots stamped into one side, while metric nuts are marked with a number. The greater the number of dots, or the higher the number, the greater the strength of the nut.

Metric studs are also marked on their ends according to property class (grade). Larger studs are numbered (the same as metric bolts), while smaller studs carry a geometric code to denote grade.

It should be noted that many fasteners, especially Grades 0 through 2, have no distinguishing marks on them. When such is the case, the only way to determine whether it is standard or metric is to measure the thread pitch or compare it to a known fastener of the same size.

Standard fasteners are often referred to as SAE, as opposed to metric. However, it should be noted that SAE technically refers to a non-metric fine thread fastener only. Coarse thread non-metric fasteners are referred to as USS sizes.

Since fasteners of the same size (both standard and metric) may have different

Metric thread sizes	Ft-lbs	Nm
M-6	6 to 9	9 to 12
M-8	14 to 21	19 to 28
M-10	28 to 40	38 to 54
M-12	50 to 71	68 to 96
M-14	80 to 140	109 to 154

Pipe thread sizes		
1/8	5 to 8	7 to 10
1/4	12 to 18	17 to 24
3/8	22 to 33	30 to 44
1/2	25 to 35	34 to 47

U.S. thread sizes		
1/4 - 20	6 to 9	9 to 12
5/16 - 18	12 to 18	17 to 24
5/16 - 24	14 to 20	19 to 27
3/8 - 16	22 to 32	30 to 43
3/8 - 24	27 to 38	37 to 51
7/16 - 14	40 to 55	55 to 74
7/16 - 20	40 to 60	55 to 81
1/2 - 13	55 to 80	75 to 108

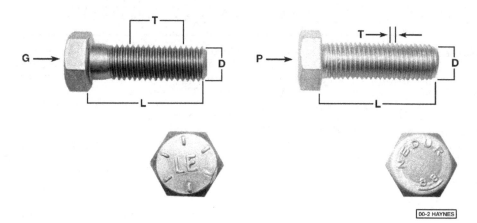

Standard (SAE and USS) bolt dimensions/ grade marks

G   Grade marks (bolt strength)
L   Length (in inches)
T   Thread pitch (number of threads per inch)
D   Nominal diameter (in inches)

Metric bolt dimensions/grade marks

P   Property class (bolt strength)
L   Length (in millimeters)
T   Thread pitch (distance between threads in millimeters)
D   Diameter

strength ratings, be sure to reinstall any bolts, studs or nuts removed from your vehicle in their original locations. Also, when replacing a fastener with a new one, make sure that the new one has a strength rating equal to or greater than the original.

Tightening sequences and procedures

Most threaded fasteners should be tightened to a specific torque value (torque is the twisting force applied to a threaded component such as a nut or bolt). Overtightening the fastener can weaken it and cause it to break, while undertightening can cause it to eventually come loose. Bolts, screws and studs, depending on the material they are

made of and their thread diameters, have specific torque values, many of which are noted in the Specifications at the beginning of each Chapter. Be sure to follow the torque recommendations closely. For fasteners not assigned a specific torque, a general torque value chart is presented here as a guide. These torque values are for dry (unlubricated) fasteners threaded into steel or cast iron (not aluminum). As was previously mentioned, the size and grade of a fastener determine the amount of torque that can safely be applied to it. The figures listed here are approximate for Grade 2 and Grade 3 fasteners. Higher grades can tolerate higher torque values.

Fasteners laid out in a pattern, such as cylinder head bolts, oil pan bolts, differential

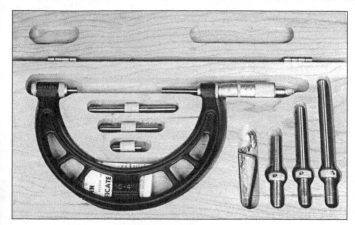

Micrometer set

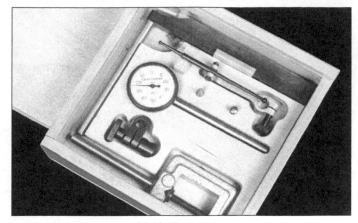

Dial indicator set

cover bolts, etc., must be loosened or tightened in sequence to avoid warping the component. This sequence will normally be shown in the appropriate Chapter. If a specific pattern is not given, the following procedures can be used to prevent warping.

Initially, the bolts or nuts should be assembled finger-tight only. Next, they should be tightened one full turn each, in a crisscross or diagonal pattern. After each one has been tightened one full turn, return to the first one and tighten them all one-half turn, following the same pattern. Finally, tighten each of them one-quarter turn at a time until each fastener has been tightened to the proper torque. To loosen and remove the fasteners, the procedure would be reversed.

Component disassembly

Component disassembly should be done with care and purpose to help ensure that the parts go back together properly. Always keep track of the sequence in which parts are removed. Make note of special characteristics or marks on parts that can be installed more than one way, such as a grooved thrust washer on a shaft. It is a good idea to lay the disassembled parts out on a clean surface in the order that they were removed. It may also be helpful to make sketches or take instant photos of components before removal.

When removing fasteners from a component, keep track of their locations. Sometimes threading a bolt back in a part, or putting the washers and nut back on a stud, can prevent mix-ups later. If nuts and bolts cannot be returned to their original locations, they should be kept in a compartmented box or a series of small boxes. A cupcake or muffin tin is ideal for this purpose, since each cavity can hold the bolts and nuts from a particular area (i.e. oil pan bolts, valve cover bolts, engine mount bolts, etc.). A pan of this type is especially helpful when working on assemblies with very small parts, such as the carburetor, alternator, valve train or interior dash and trim pieces. The cavities can be marked with paint or tape to identify the contents.

Whenever wiring looms, harnesses or connectors are separated, it is a good idea to identify the two halves with numbered pieces of masking tape so they can be easily reconnected.

Gasket sealing surfaces

Throughout any vehicle, gaskets are used to seal the mating surfaces between two parts and keep lubricants, fluids, vacuum or pressure contained in an assembly.

Many times these gaskets are coated with a liquid or paste-type gasket sealing compound before assembly. Age, heat and pressure can sometimes cause the two parts to stick together so tightly that they are very difficult to separate. Often, the assembly can be loosened by striking it with a soft-face hammer near the mating surfaces. A regular hammer can be used if a block of wood is placed between the hammer and the part. Do not hammer on cast parts or parts that could be easily damaged. With any particularly stubborn part, always recheck to make sure that every fastener has been removed.

Avoid using a screwdriver or bar to pry apart an assembly, as they can easily mar the gasket sealing surfaces of the parts, which must remain smooth. If prying is absolutely necessary, use an old broom handle, but keep in mind that extra clean up will be necessary if the wood splinters.

After the parts are separated, the old gasket must be carefully scraped off and the gasket surfaces cleaned. Stubborn gasket material can be soaked with rust penetrant or treated with a special chemical to soften it so it can be easily scraped off. A scraper can be fashioned from a piece of copper tubing by flattening and sharpening one end. Copper is recommended because it is usually softer than the surfaces to be scraped, which reduces the chance of gouging the part. Some gaskets can be removed with a wire brush, but regardless of the method used, the mating surfaces must be left clean and smooth. If for some reason the gasket surface is gouged, then a gasket sealer thick enough to fill scratches will have to be used during reassembly of the components. For most applications, a non-drying (or semi-drying) gasket sealer should be used.

Hose removal tips

Warning: *If the vehicle is equipped with air conditioning, do not disconnect any of the A/C hoses without first having the system depressurized by a dealer service department or a service station.*

Hose removal precautions closely parallel gasket removal precautions. Avoid scratching or gouging the surface that the hose mates against or the connection may leak. This is especially true for radiator hoses. Because of various chemical reactions, the rubber in hoses can bond itself to the metal spigot that the hose fits over. To remove a hose, first loosen the hose clamps that secure it to the spigot. Then, with slip-joint pliers, grab the hose at the clamp and rotate it around the spigot. Work it back and forth until it is completely free, then pull it off. Silicone or other lubricants will ease removal if they can be applied between the hose and the outside of the spigot. Apply the same lubricant to the inside of the hose and the outside of the spigot to simplify installation.

As a last resort (and if the hose is to be replaced with a new one anyway), the rubber can be slit with a knife and the hose peeled from the spigot. If this must be done, be careful that the metal connection is not damaged.

If a hose clamp is broken or damaged, do not reuse it. Wire-type clamps usually weaken with age, so it is a good idea to replace them with screw-type clamps whenever a hose is removed.

Tools

A selection of good tools is a basic requirement for anyone who plans to maintain and repair his or her own vehicle. For the owner who has few tools, the initial investment might seem high, but when compared to the spiraling costs of professional auto maintenance and repair, it is a wise one.

To help the owner decide which tools are needed to perform the tasks detailed in this manual, the following tool lists are offered: *Maintenance and minor repair, Repair/overhaul* and *Special*.

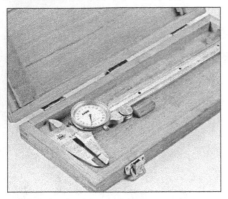

Dial caliper

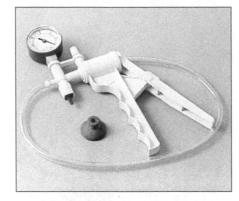

Hand-operated vacuum pump

Timing light

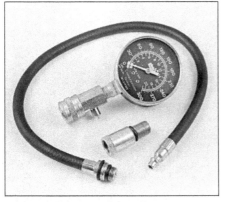

Compression gauge with spark plug hole adapter

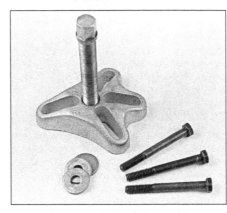

Damper/steering wheel puller

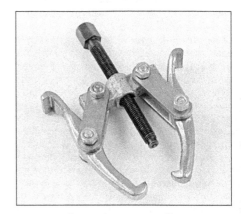

General purpose puller

Hydraulic lifter removal tool

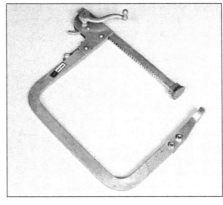

Valve spring compressor

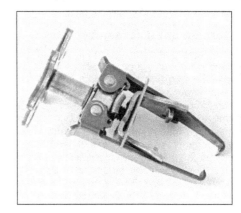

Valve spring compressor

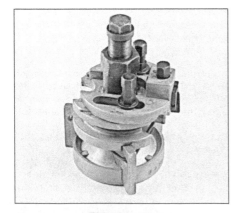

Ridge reamer

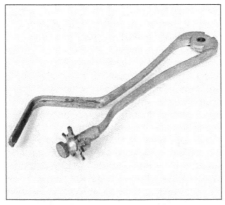

Piston ring groove cleaning tool

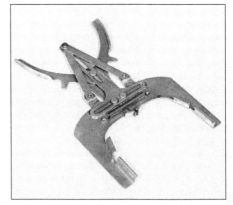

Ring removal/installation tool

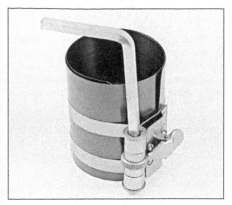

Ring compressor

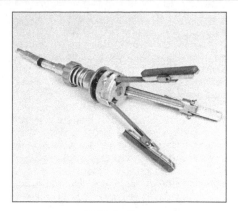

Cylinder hone

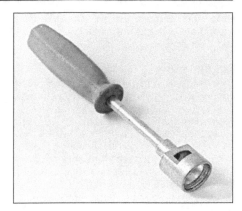

Brake hold-down spring tool

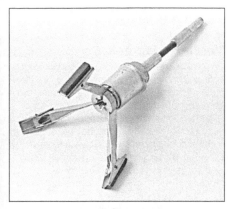

Brake cylinder hone

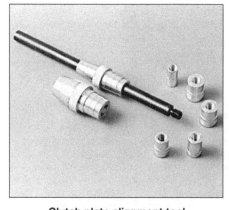

Clutch plate alignment tool

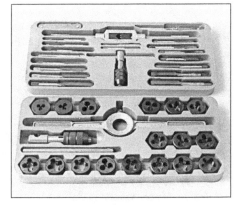

Tap and die set

The newcomer to practical mechanics should start off with the *maintenance and minor repair* tool kit, which is adequate for the simpler jobs performed on a vehicle. Then, as confidence and experience grow, the owner can tackle more difficult tasks, buying additional tools as they are needed. Eventually the basic kit will be expanded into the *repair and overhaul* tool set. Over a period of time, the experienced do-it-yourselfer will assemble a tool set complete enough for most repair and overhaul procedures and will add tools from the special category when it is felt that the expense is justified by the frequency of use.

Maintenance and minor repair tool kit

The tools in this list should be considered the minimum required for performance of routine maintenance, servicing and minor repair work. We recommend the purchase of combination wrenches (box-end and open-end combined in one wrench). While more expensive than open end wrenches, they offer the advantages of both types of wrench.

Combination wrench set (1/4-inch to 1 inch or 6 mm to 19 mm)
Adjustable wrench, 8 inch
Spark plug wrench with rubber insert
Spark plug gap adjusting tool
Feeler gauge set
Brake bleeder wrench
Standard screwdriver (5/16-inch x 6 inch)

Phillips screwdriver (No. 2 x 6 inch)
Combination pliers - 6 inch
Hacksaw and assortment of blades
Tire pressure gauge
Grease gun
Oil can
Fine emery cloth
Wire brush
Battery post and cable cleaning tool
Oil filter wrench
Funnel (medium size)
Safety goggles
Jackstands (2)
Drain pan

Note: *If basic tune-ups are going to be part of routine maintenance, it will be necessary to purchase a good quality stroboscopic timing light and combination tachometer/dwell meter. Although they are included in the list of special tools, it is mentioned here because they are absolutely necessary for tuning most vehicles properly.*

Repair and overhaul tool set

These tools are essential for anyone who plans to perform major repairs and are in addition to those in the maintenance and minor repair tool kit. Included is a comprehensive set of sockets which, though expensive, are invaluable because of their versatility, especially when various extensions and drives are available. We recommend the 1/2-inch drive over the 3/8-inch drive. Although the larger drive is bulky and more expensive, it has the capacity of accepting a very wide range of large sockets. Ideally, however, the mechanic should have a 3/8-inch drive set and a 1/2-inch drive set.

Socket set(s)
Reversible ratchet
Extension - 10 inch
Universal joint
Torque wrench (same size drive as sockets)
Ball peen hammer - 8 ounce
Soft-face hammer (plastic/rubber)
Standard screwdriver (1/4-inch x 6 inch)
Standard screwdriver (stubby - 5/16-inch)
Phillips screwdriver (No. 3 x 8 inch)
Phillips screwdriver (stubby - No. 2)
Pliers - vise grip
Pliers - lineman's
Pliers - needle nose
Pliers - snap-ring (internal and external)
Cold chisel - 1/2-inch
Scribe
Scraper (made from flattened copper tubing)
Centerpunch
Pin punches (1/16, 1/8, 3/16-inch)
Steel rule/straightedge - 12 inch
Allen wrench set (1/8 to 3/8-inch or 4 mm to 10 mm)
A selection of files
Wire brush (large)
Jackstands (second set)
Jack (scissor or hydraulic type)

Note: *Another tool which is often useful is an electric drill with a chuck capacity of 3/8-inch and a set of good quality drill bits.*

Special tools

The tools in this list include those which are not used regularly, are expensive to buy, or which need to be used in accordance with their manufacturer's instructions. Unless these tools will be used frequently, it is not very economical to purchase many of them. A consideration would be to split the cost and use between yourself and a friend or friends. In addition, most of these tools can be obtained from a tool rental shop on a temporary basis.

This list primarily contains only those tools and instruments widely available to the public, and not those special tools produced by the vehicle manufacturer for distribution to dealer service departments. Occasionally, references to the manufacturer's special tools are included in the text of this manual. Generally, an alternative method of doing the job without the special tool is offered. However, sometimes there is no alternative to their use. Where this is the case, and the tool cannot be purchased or borrowed, the work should be turned over to the dealer service department or an automotive repair shop.

> *Valve spring compressor*
> *Piston ring groove cleaning tool*
> *Piston ring compressor*
> *Piston ring installation tool*
> *Cylinder compression gauge*
> *Cylinder ridge reamer*
> *Cylinder surfacing hone*
> *Cylinder bore gauge*
> *Micrometers and/or dial calipers*
> *Hydraulic lifter removal tool*
> *Balljoint separator*
> *Universal-type puller*
> *Impact screwdriver*
> *Dial indicator set*
> *Stroboscopic timing light (inductive pick-up)*
> *Hand operated vacuum/pressure pump*
> *Tachometer/dwell meter*
> *Universal electrical multimeter*
> *Cable hoist*
> *Brake spring removal and installation tools*
> *Floor jack*

Buying tools

For the do-it-yourselfer who is just starting to get involved in vehicle maintenance and repair, there are a number of options available when purchasing tools. If maintenance and minor repair is the extent of the work to be done, the purchase of individual tools is satisfactory. If, on the other hand, extensive work is planned, it would be a good idea to purchase a modest tool set from one of the large retail chain stores. A set can usually be bought at a substantial savings over the individual tool prices, and they often come with a tool box. As additional tools are needed, add-on sets, individual tools and a larger tool box can be purchased to expand the tool selection. Building a tool set gradually allows the cost of the tools to be spread over a longer period of time and gives the mechanic the freedom to choose only those tools that will actually be used.

Tool stores will often be the only source of some of the special tools that are needed, but regardless of where tools are bought, try to avoid cheap ones, especially when buying screwdrivers and sockets, because they won't last very long. The expense involved in replacing cheap tools will eventually be greater than the initial cost of quality tools.

Care and maintenance of tools

Good tools are expensive, so it makes sense to treat them with respect. Keep them clean and in usable condition and store them properly when not in use. Always wipe off any dirt, grease or metal chips before putting them away. Never leave tools lying around in the work area. Upon completion of a job, always check closely under the hood for tools that may have been left there so they won't get lost during a test drive.

Some tools, such as screwdrivers, pliers, wrenches and sockets, can be hung on a panel mounted on the garage or workshop wall, while others should be kept in a tool box or tray. Measuring instruments, gauges, meters, etc. must be carefully stored where they cannot be damaged by weather or impact from other tools.

When tools are used with care and stored properly, they will last a very long time. Even with the best of care, though, tools will wear out if used frequently. When a tool is damaged or worn out, replace it. Subsequent jobs will be safer and more enjoyable if you do.

How to repair damaged threads

Sometimes, the internal threads of a nut or bolt hole can become stripped, usually from overtightening. Stripping threads is an all-too-common occurrence, especially when working with aluminum parts, because aluminum is so soft that it easily strips out.

Usually, external or internal threads are only partially stripped. After they've been cleaned up with a tap or die, they'll still work. Sometimes, however, threads are badly damaged. When this happens, you've got three choices:

1) *Drill and tap the hole to the next suitable oversize and install a larger diameter bolt, screw or stud.*
2) *Drill and tap the hole to accept a threaded plug, then drill and tap the plug to the original screw size. You can also buy a plug already threaded to the original size. Then you simply drill a hole to the specified size, then run the threaded plug into the hole with a bolt and jam nut. Once the plug is fully seated, remove the jam nut and bolt.*
3) *The third method uses a patented thread repair kit like Heli-Coil or Slim-sert. These easy-to-use kits are designed to repair damaged threads in straight-through holes and blind holes. Both are available as kits which can handle a variety of sizes and thread patterns. Drill the hole, then tap it with the special included tap. Install the Heli-Coil and the hole is back to its original diameter and thread pitch.*

Regardless of which method you use, be sure to proceed calmly and carefully. A little impatience or carelessness during one of these relatively simple procedures can ruin your whole day's work and cost you a bundle if you wreck an expensive part.

Working facilities

Not to be overlooked when discussing tools is the workshop. If anything more than routine maintenance is to be carried out, some sort of suitable work area is essential.

It is understood, and appreciated, that many home mechanics do not have a good workshop or garage available, and end up removing an engine or doing major repairs outside. It is recommended, however, that the overhaul or repair be completed under the cover of a roof.

A clean, flat workbench or table of comfortable working height is an absolute necessity. The workbench should be equipped with a vise that has a jaw opening of at least four inches.

As mentioned previously, some clean, dry storage space is also required for tools, as well as the lubricants, fluids, cleaning solvents, etc. which soon become necessary.

Sometimes waste oil and fluids, drained from the engine or cooling system during normal maintenance or repairs, present a disposal problem. To avoid pouring them on the ground or into a sewage system, pour the used fluids into large containers, seal them with caps and take them to an authorized disposal site or recycling center. Plastic jugs, such as old antifreeze containers, are ideal for this purpose.

Always keep a supply of old newspapers and clean rags available. Old towels are excellent for mopping up spills. Many mechanics use rolls of paper towels for most work because they are readily available and disposable. To help keep the area under the vehicle clean, a large cardboard box can be cut open and flattened to protect the garage or shop floor.

Whenever working over a painted surface, such as when leaning over a fender to service something under the hood, always cover it with an old blanket or bedspread to protect the finish. Vinyl covered pads, made especially for this purpose, are available at auto parts stores.

Booster battery (jump) starting

Certain precautions must be observed when using a booster battery to jump start a vehicle.

a) *Before connecting the booster battery, make sure that the ignition switch is in the Off position.*
b) *Turn off the lights, heater and other electrical loads.*
c) *The eyes should be shielded. Safety goggles are a good idea.*
d) *Make sure that the booster battery is the same voltage as the dead one in the vehicle.*
e) *The two vehicles must not touch each other.*
f) *Make sure that the transaxle is in Neutral (manual transaxle) or Park (automatic transaxle).*
g) *If the booster battery is not a maintenance-free type, remove the vent caps and lay a cloth over the vent holes.*

Connect the red jumper cable to the positive (+) terminals of each battery. Connect one end of the black jumper cable to the negative (-) terminal of the booster battery. The other end of this cable should be connected to a good ground on the vehicle to be started, such as a bolt or bracket on the engine block. Use caution to ensure that the cable will not come into contact with the fan, drivebelts or other moving parts of the engine.

Start the engine using the booster battery, then with the engine running at idle speed, disconnect the jumper cables in the reverse order of connection.

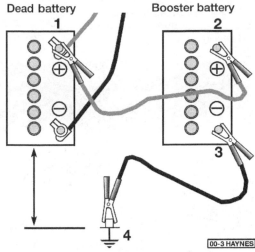

Make the booster battery cable connections in the numerical order shown (note that the negative cable of the booster battery is NOT attached to the negative terminal of the dead battery)

Jacking and towing

Jacking

The jack supplied with the vehicle should only be used for raising the vehicle when changing a tire or placing jackstands under the frame. **Caution:** *Never work under the vehicle or start the engine while this jack is being used as the only means of support.*

The vehicle should be on level ground with the wheels blocked and the transaxle in Park (automatic) or Reverse (manual). If the wheel is being replaced, loosen the wheel nuts one-half turn but leave them in place until the wheel is raised off the ground.

Block the front and rear of the wheel opposite the one being removed before operating the jack. Place the jack under the side of the vehicle in the indicated position and raise it until the jack head groove fits into the rocker flange notch **(see illustrations)**. Refer to Chapter 10 for the wheel/tire changing procedure.

Lower the vehicle, remove the jack and tighten the wheel lug nuts (if loosened or removed) in a criss-cross sequence by turning the wrench clockwise.

There are two notches for the vehicle jack on each side of the vehicle . . .

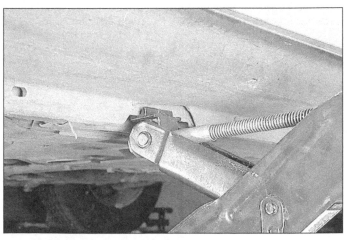

. . . make sure the jack head securely engages with the rocker panel flange

Towing

Towing equipment specifically designed for this purpose should be used and should be attached to the main structural members of he vehicle and not to the bumper or brackets.

Safety is a major consideration when towing and all applicable state and local laws must be obeyed. A safety chain system must be used for all towing.

While towing, the parking brake should be released and the transaxle should be in Neutral. The steering must be unlocked (ignition switch in the Off position). Remember that power brakes will not work with the engine off.

1984 and 1985 vehicles

If equipped with an automatic transaxle, a vehicle may be towed on all four wheels at speeds less than 35 mph for distances up to 50 miles. These speeds and distances do not apply to vehicles with manual transaxles.

1986 through 1988 vehicles

If towing is necessary, contact a professional tow truck service. Do not tow a vehicle on all four wheels.

Front crossmember

Front lower control arm

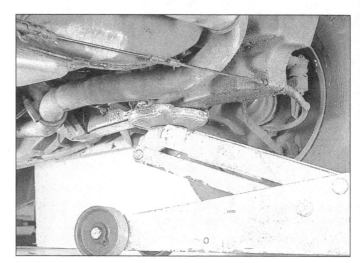

When using a floor jack, these are the only lifting points that are permitted - using any other points may result in damage to the vehicle (rear crossmember on left, rear subframe on right)

Automotive chemicals and lubricants

A number of automotive chemicals and lubricants are available for use during vehicle maintenance and repair. They include a wide variety of products ranging from cleaning solvents and degreasers to lubricants and protective sprays for rubber, plastic and vinyl.

Cleaners

Carburetor cleaner and choke cleaner is a strong solvent for gum, varnish and carbon. Most carburetor cleaners leave a dry-type lubricant film which will not harden or gum up. Because of this film it is not recommended for use on electrical components.

Brake system cleaner is used to remove grease and brake fluid from the brake system, where clean surfaces are absolutely necessary. It leaves no residue and often eliminates brake squeal caused by contaminants.

Electrical cleaner removes oxidation, corrosion and carbon deposits from electrical contacts, restoring full current flow. It can also be used to clean spark plugs, carburetor jets, voltage regulators and other parts where an oil-free surface is desired.

Demoisturants remove water and moisture from electrical components such as alternators, voltage regulators, electrical connectors and fuse blocks. They are non-conductive, non-corrosive and non-flammable.

Degreasers are heavy-duty solvents used to remove grease from the outside of the engine and from chassis components. They can be sprayed or brushed on and, depending on the type, are rinsed off either with water or solvent.

Lubricants

Motor oil is the lubricant formulated for use in engines. It normally contains a wide variety of additives to prevent corrosion and reduce foaming and wear. Motor oil comes in various weights (viscosity ratings) from 0 to 50. The recommended weight of the oil depends on the season, temperature and the demands on the engine. Light oil is used in cold climates and under light load conditions. Heavy oil is used in hot climates and where high loads are encountered. Multi-viscosity oils are designed to have characteristics of both light and heavy oils and are available in a number of weights from 5W-20 to 20W-50.

Gear oil is designed to be used in differentials, manual transmissions and other areas where high-temperature lubrication is required.

Chassis and wheel bearing grease is a heavy grease used where increased loads and friction are encountered, such as for wheel bearings, balljoints, tie-rod ends and universal joints.

High-temperature wheel bearing grease is designed to withstand the extreme temperatures encountered by wheel bearings

in disc brake equipped vehicles. It usually contains molybdenum disulfide (moly), which is a dry-type lubricant.

White grease is a heavy grease for metal-to-metal applications where water is a problem. White grease stays soft under both low and high temperatures (usually from -100 to +190-degrees F), and will not wash off or dilute in the presence of water.

Assembly lube is a special extreme pressure lubricant, usually containing moly, used to lubricate high-load parts (such as main and rod bearings and cam lobes) for initial start-up of a new engine. The assembly lube lubricates the parts without being squeezed out or washed away until the engine oiling system begins to function.

Silicone lubricants are used to protect rubber, plastic, vinyl and nylon parts.

Graphite lubricants are used where oils cannot be used due to contamination problems, such as in locks. The dry graphite will lubricate metal parts while remaining uncontaminated by dirt, water, oil or acids. It is electrically conductive and will not foul electrical contacts in locks such as the ignition switch.

Moly penetrants loosen and lubricate frozen, rusted and corroded fasteners and prevent future rusting or freezing.

Heat-sink grease is a special electrically non-conductive grease that is used for mounting electronic ignition modules where it is essential that heat is transferred away from the module.

Sealants

RTV sealant is one of the most widely used gasket compounds. Made from silicone, RTV is air curing, it seals, bonds, waterproofs, fills surface irregularities, remains flexible, doesn't shrink, is relatively easy to remove, and is used as a supplementary sealer with almost all low and medium temperature gaskets.

Anaerobic sealant is much like RTV in that it can be used either to seal gaskets or to form gaskets by itself. It remains flexible, is solvent resistant and fills surface imperfections. The difference between an anaerobic sealant and an RTV-type sealant is in the curing. RTV cures when exposed to air, while an anaerobic sealant cures only in the absence of air. This means that an anaerobic sealant cures only after the assembly of parts, sealing them together.

Thread and pipe sealant is used for sealing hydraulic and pneumatic fittings and vacuum lines. It is usually made from a Teflon compound, and comes in a spray, a paint-on liquid and as a wrap-around tape.

Chemicals

Anti-seize compound prevents seizing, galling, cold welding, rust and corrosion in

fasteners. High-temperature ant-seize, usually made with copper and graphite lubricants, is used for exhaust system and exhaust manifold bolts.

Anaerobic locking compounds are used to keep fasteners from vibrating or working loose and cure only after installation, in the absence of air. Medium strength locking compound is used for small nuts, bolts and screws that may be removed later. High-strength locking compound is for large nuts, bolts and studs which aren't removed on a regular basis.

Oil additives range from viscosity index improvers to chemical treatments that claim to reduce internal engine friction. It should be noted that most oil manufacturers caution against using additives with their oils.

Gas additives perform several functions, depending on their chemical makeup. They usually contain solvents that help dissolve gum and varnish that build up on carburetor, fuel injection and intake parts. They also serve to break down carbon deposits that form on the inside surfaces of the combustion chambers. Some additives contain upper cylinder lubricants for valves and piston rings, and others contain chemicals to remove condensation from the gas tank.

Miscellaneous

Brake fluid is specially formulated hydraulic fluid that can withstand the heat and pressure encountered in brake systems. Care must be taken so this fluid does not come in contact with painted surfaces or plastics. An opened container should always be resealed to prevent contamination by water or dirt.

Weatherstrip adhesive is used to bond weatherstripping around doors, windows and trunk lids. It is sometimes used to attach trim pieces.

Undercoating is a petroleum-based, tar-like substance that is designed to protect metal surfaces on the underside of the vehicle from corrosion. It also acts as a sound-deadening agent by insulating the bottom of the vehicle.

Waxes and polishes are used to help protect painted and plated surfaces from the weather. Different types of paint may require the use of different types of wax and polish. Some polishes utilize a chemical or abrasive cleaner to help remove the top layer of oxidized (dull) paint on older vehicles. In recent years many non-wax polishes that contain a wide variety of chemicals such as polymers and silicones have been introduced. These non-wax polishes are usually easier to apply and last longer than conventional waxes and polishes.

Conversion factors

Length (distance)

Inches (in)	X	25.4	= Millimetres (mm)	X 0.0394	= Inches (in)
Feet (ft)	X	0.305	= Metres (m)	X 3.281	= Feet (ft)
Miles	X	1.609	= Kilometres (km)	X 0.621	= Miles

Volume (capacity)

Cubic inches (cu in; in^3)	X	16.387	= Cubic centimetres (cc; cm^3)	X 0.061	= Cubic inches (cu in; in^3)
Imperial pints (Imp pt)	X	0.568	= Litres (l)	X 1.76	= Imperial pints (Imp pt)
Imperial quarts (Imp qt)	X	1.137	= Litres (l)	X 0.88	= Imperial quarts (Imp qt)
Imperial quarts (Imp qt)	X	1.201	= US quarts (US qt)	X 0.833	= Imperial quarts (Imp qt)
US quarts (US qt)	X	0.946	= Litres (l)	X 1.057	= US quarts (US qt)
Imperial gallons (Imp gal)	X	4.546	= Litres (l)	X 0.22	= Imperial gallons (Imp gal)
Imperial gallons (Imp gal)	X	1.201	= US gallons (US gal)	X 0.833	= Imperial gallons (Imp gal)
US gallons (US gal)	X	3.785	= Litres (l)	X 0.264	= US gallons (US gal)

Mass (weight)

Ounces (oz)	X	28.35	= Grams (g)	X 0.035	= Ounces (oz)
Pounds (lb)	X	0.454	= Kilograms (kg)	X 2.205	= Pounds (lb)

Force

Ounces-force (ozf; oz)	X	0.278	= Newtons (N)	X 3.6	= Ounces-force (ozf; oz)
Pounds-force (lbf; lb)	X	4.448	= Newtons (N)	X 0.225	= Pounds-force (lbf; lb)
Newtons (N)	X	0.1	= Kilograms-force (kgf; kg)	X 9.81	= Newtons (N)

Pressure

Pounds-force per square inch (psi; lbf/in^2; lb/in^2)	X	0.070	= Kilograms-force per square centimetre (kgf/cm^2; kg/cm^2)	X 14.223	= Pounds-force per square inch (psi; lbf/in^2; lb/in^2)
Pounds-force per square inch (psi; lbf/in^2; lb/in^2)	X	0.068	= Atmospheres (atm)	X 14.696	= Pounds-force per square inch (psi; lbf/in^2; lb/in^2)
Pounds-force per square inch (psi; lbf/in^2; lb/in^2)	X	0.069	= Bars	X 14.5	= Pounds-force per square inch (psi; lbf/in^2; lb/in^2)
Pounds-force per square inch (psi; lbf/in^2; lb/in^2)	X	6.895	= Kilopascals (kPa)	X 0.145	= Pounds-force per square inch (psi; lbf/in^2; lb/in^2)
Kilopascals (kPa)	X	0.01	= Kilograms-force per square centimetre (kgf/cm^2; kg/cm^2)	X 98.1	= Kilopascals (kPa)

Torque (moment of force)

Pounds-force inches (lbf in; lb in)	X	1.152	= Kilograms-force centimetre (kgf cm; kg cm)	X 0.868	= Pounds-force inches (lbf in; lb in)
Pounds-force inches (lbf in; lb in)	X	0.113	= Newton metres (Nm)	X 8.85	= Pounds-force inches (lbf in; lb in)
Pounds-force inches (lbf in; lb in)	X	0.083	= Pounds-force feet (lbf ft; lb ft)	X 12	= Pounds-force inches (lbf in; lb in)
Pounds-force feet (lbf ft; lb ft)	X	0.138	= Kilograms-force metres (kgf m; kg m)	X 7.233	= Pounds-force feet (lbf ft; lb ft)
Pounds-force feet (lbf ft; lb ft)	X	1.356	= Newton metres (Nm)	X 0.738	= Pounds-force feet (lbf ft; lb ft)
Newton metres (Nm)	X	0.102	= Kilograms-force metres (kgf m; kg m)	X 9.804	= Newton metres (Nm)

Vacuum

Inches mercury (in. Hg)	X	3.377	= Kilopascals (kPa)	X 0.2961	= Inches mercury
Inches mercury (in. Hg)	X	25.4	= Millimeters mercury (mm Hg)	X 0.0394	= Inches mercury

Power

Horsepower (hp)	X	745.7	= Watts (W)	X 0.0013	= Horsepower (hp)

Velocity (speed)

Miles per hour (miles/hr; mph)	X	1.609	= Kilometres per hour (km/hr; kph)	X 0.621	= Miles per hour (miles/hr; mph)

Fuel consumption*

Miles per gallon, Imperial (mpg)	X	0.354	= Kilometres per litre (km/l)	X 2.825	= Miles per gallon, Imperial (mpg)
Miles per gallon, US (mpg)	X	0.425	= Kilometres per litre (km/l)	X 2.352	= Miles per gallon, US (mpg)

Temperature

Degrees Fahrenheit = (°C x 1.8) + 32

Degrees Celsius (Degrees Centigrade; °C) = (°F - 32) x 0.56

*It is common practice to convert from miles per gallon (mpg) to litres/100 kilometres (l/100km), where mpg (Imperial) x l/100 km = 282 and mpg (US) x l/100 km = 235

Safety first!

Regardless of how enthusiastic you may be about getting on with the job at hand, take the time to ensure that your safety is not jeopardized. A moment's lack of attention can result in an accident, as can failure to observe certain simple safety precautions. The possibility of an accident will always exist, and the following points should not be considered a comprehensive list of all dangers. Rather, they are intended to make you aware of the risks and to encourage a safety conscious approach to all work you carry out on your vehicle.

Essential DOs and DON'Ts

DON'T rely on a jack when working under the vehicle. Always use approved jackstands to support the weight of the vehicle and place them under the recommended lift or support points.

DON'T attempt to loosen extremely tight fasteners (i.e. wheel lug nuts) while the vehicle is on a jack - it may fall.

DON'T start the engine without first making sure that the transmission is in Neutral (or Park where applicable) and the parking brake is set.

DON'T remove the radiator cap from a hot cooling system - let it cool or cover it with a cloth and release the pressure gradually.

DON'T attempt to drain the engine oil until you are sure it has cooled to the point that it will not burn you.

DON'T touch any part of the engine or exhaust system until it has cooled sufficiently to avoid burns.

DON'T siphon toxic liquids such as gasoline, antifreeze and brake fluid by mouth, or allow them to remain on your skin.

DON'T inhale brake lining dust - it is potentially hazardous (see *Asbestos* below).

DON'T allow spilled oil or grease to remain on the floor - wipe it up before someone slips on it.

DON'T use loose fitting wrenches or other tools which may slip and cause injury.

DON'T push on wrenches when loosening or tightening nuts or bolts. Always try to pull the wrench toward you. If the situation calls for pushing the wrench away, push with an open hand to avoid scraped knuckles if the wrench should slip.

DON'T attempt to lift a heavy component alone - get someone to help you.

DON'T rush or take unsafe shortcuts to finish a job.

DON'T allow children or animals in or around the vehicle while you are working on it.

DO wear eye protection when using power tools such as a drill, sander, bench grinder, etc. and when working under a vehicle.

DO keep loose clothing and long hair well out of the way of moving parts.

DO make sure that any hoist used has a safe working load rating adequate for the job.

DO get someone to check on you periodically when working alone on a vehicle.

DO carry out work in a logical sequence and make sure that everything is correctly assembled and tightened.

DO keep chemicals and fluids tightly capped and out of the reach of children and pets.

DO remember that your vehicle's safety affects that of yourself and others. If in doubt on any point, get professional advice.

Asbestos

Certain friction, insulating, sealing, and other products - such as brake linings, brake bands, clutch linings, torque converters, gaskets, etc. - may contain asbestos. Extreme care must be taken to avoid inhalation of dust from such products, since it is hazardous to health. If in doubt, assume that they do contain asbestos.

Fire

Remember at all times that gasoline is highly flammable. Never smoke or have any kind of open flame around when working on a vehicle. But the risk does not end there. A spark caused by an electrical short circuit, by two metal surfaces contacting each other, or even by static electricity built up in your body under certain conditions, can ignite gasoline vapors, which in a confined space are highly explosive. Do not, under any circumstances, use gasoline for cleaning parts. Use an approved safety solvent.

Always disconnect the battery ground (-) cable at the battery before working on any part of the fuel system or electrical system. Never risk spilling fuel on a hot engine or exhaust component. It is strongly recommended that a fire extinguisher suitable for use on fuel and electrical fires be kept handy in the garage or workshop at all times. Never try to extinguish a fuel or electrical fire with water.

Fumes

Certain fumes are highly toxic and can quickly cause unconsciousness and even death if inhaled to any extent. Gasoline vapor falls into this category, as do the vapors from some cleaning solvents. Any draining or pouring of such volatile fluids should be done in a well ventilated area.

When using cleaning fluids and solvents, read the instructions on the container carefully. Never use materials from unmarked containers.

Never run the engine in an enclosed space, such as a garage. Exhaust fumes contain carbon monoxide, which is extremely poisonous. If you need to run the engine, always do so in the open air, or at least have the rear of the vehicle outside the work area.

If you are fortunate enough to have the use of an inspection pit, never drain or pour gasoline and never run the engine while the vehicle is over the pit. The fumes, being heavier than air, will concentrate in the pit with possibly lethal results.

The battery

Never create a spark or allow a bare light bulb near a battery. They normally give off a certain amount of hydrogen gas, which is highly explosive.

Always disconnect the battery ground (-) cable at the battery before working on the fuel or electrical systems.

If possible, loosen the filler caps or cover when charging the battery from an external source (this does not apply to sealed or maintenance-free batteries). Do not charge at an excessive rate or the battery may burst.

Take care when adding water to a non maintenance-free battery and when carrying a battery. The electrolyte, even when diluted, is very corrosive and should not be allowed to contact clothing or skin.

Always wear eye protection when cleaning the battery to prevent the caustic deposits from entering your eyes.

Household current

When using an electric power tool, inspection light, etc., which operates on household current, always make sure that the tool is correctly connected to its plug and that, where necessary, it is properly grounded. Do not use such items in damp conditions and, again, do not create a spark or apply excessive heat in the vicinity of fuel or fuel vapor.

Secondary ignition system voltage

A severe electric shock can result from touching certain parts of the ignition system (such as the spark plug wires) when the engine is running or being cranked, particularly if components are damp or the insulation is defective. In the case of an electronic ignition system, the secondary system voltage is much higher and could prove fatal.

Troubleshooting

Contents

This section provides an easy reference guide to the more common problems which may occur during the operation of your vehicle. These problems and their possible causes are grouped under headings denoting various components or systems, such as Engine, Cooling system, etc. They also refer you to the Chapter and/or Section which deals with the problem.

Remember that successful troubleshooting is not a mysterious black art practiced only by professional mechanics. It is simply the result of the right knowledge combined with an intelligent, systematic approach to the problem. Always work by a process of elimination, starting with the simplest solution and working through to the most complex - and never overlook the obvious. Anyone can run the gas tank dry or leave the lights on overnight, so don't assume that you are exempt from such oversights.

Finally, always establish a clear idea of why a problem has occurred and take steps to ensure that it doesn't happen again. If the electrical system fails because of a poor connection, check all other connections in the system to make sure that they don't fail as well. If a particular fuse continues to blow, find out why - don't just replace one fuse after another. Remember, failure of a small component can often be indicative of potential failure or incorrect functioning of a more important component or system.

Engine

1 Engine will not rotate when attempting to start

1   Battery terminal connections loose or corroded (Chapter 1).
2   Battery discharged or faulty (Chapter 1).
3   Automatic transaxle not completely engaged in Park (Chapter 7) or clutch not completely depressed (Chapter 8).
4   Broken, loose or disconnected wiring in the starting circuit (Chapters 5 and 12).
5   Starter motor pinion jammed in flywheel ring gear (Chapter 5).
6   Starter solenoid faulty (Chapter 5).
7   Starter motor faulty (Chapter 5).
8   Ignition switch faulty (Chapter 12).
9   Starter pinion or flywheel teeth worn or broken (Chapter 5).

2 Engine rotates but will not start

1   Fuel tank empty.
2   Battery discharged (engine rotates slowly) (Chapter 5).
3   Battery terminal connections loose or corroded (Chapter 1).
4   Leaking fuel injector(s), faulty cold start valve, fuel pump, pressure regulator, etc. (Chapter 4).

5   Fuel not reaching TBI or MPFI system (Chapter 4).
6   Ignition components damp or damaged (Chapter 5).
7   Worn, faulty or incorrectly gapped spark plugs (Chapter 1).
8   Broken, loose or disconnected wiring in the starting circuit (Chapter 5).
9   Loose distributor is changing ignition timing (Chapter 1).
10 Broken, loose or disconnected wires at the ignition coil or faulty coil (Chapter 1).

3 Engine hard to start when cold

1   Battery discharged or low (Chapter 1).
2   Fuel system malfunctioning (Chapter 4).
3   Injector(s) leaking (Chapter 4).
4   Distributor rotor carbon tracked (Chapter 1).

4 Engine hard to start when hot

1   Air filter clogged (Chapter 1).
2   Fuel not reaching the fuel injection system (Chapter 4).
3   Corroded battery connections, especially ground (Chapter 1).
4   Injector(s) leaking (Chapter 4).

5 Starter motor noisy or excessively rough in engagement

1   Pinion or flywheel gear teeth worn or broken (Chapter 5).
2   Starter motor mounting bolts loose or missing (Chapter 5).

6 Engine starts but stops immediately

1   Loose or faulty electrical connections at distributor, coil or alternator (Chapter 5).
2   Insufficient fuel reaching the fuel injector(s) (Chapters 1 and 4).
3   Vacuum leak at the gasket between the intake manifold/plenum and throttle body (Chapters 1 and 4).

7 Oil puddle under engine

1   Oil pan gasket and/or oil pan drain bolt seal leaking (Chapters 1 and 2).
2   Oil pressure sending unit leaking (Chapter 12).
3   Rocker cover gaskets leaking (Chapter 2).
4   Engine oil seals leaking (Chapter 2).
5   Timing cover sealant or sealing flange leaking (Chapter 2)

8 Engine lopes while idling or idles erratically

1   Vacuum leakage (Chapter 4).
2   Leaking EGR valve or plugged PCV valve (Chapters 1 and 6).
3   Air filter clogged (Chapter 1).
4   Fuel pump not delivering sufficient fuel to the fuel injection system (Chapter 4).
5   Leaking head gasket (Chapter 2).
6   Timing chain and/or gears worn (Chapter 2).
7   Camshaft lobes worn (Chapter 2).

9 Engine misses at idle speed

1   Spark plugs worn or not gapped properly (Chapter 1).
2   Faulty spark plug wires (Chapter 1).
3   Vacuum leaks (Chapter 1).
4   Incorrect ignition timing (Chapter 1).
5   Uneven or low compression (Chapter 1).

10 Engine misses throughout driving speed range

1   Fuel filter clogged and/or impurities in the fuel system (Chapter 1).
2   Low fuel output at the injector (Chapter 4).
3   Faulty or incorrectly gapped spark plugs (Chapter 1).
4   Incorrect ignition timing (Chapter 1).
5   Cracked distributor cap, disconnected distributor wires or damaged distributor components (Chapter 1).
6   Leaking spark plug wires (Chapter 1).
7   Faulty emission system components (Chapter 6).
8   Low or uneven cylinder compression pressures (Chapter 1).
9   Weak or faulty ignition system (Chapter 5).
10 Vacuum leak in fuel injection system, intake manifold or vacuum hoses (Chapter 4).

11 Engine stumbles on acceleration

1   Spark plugs fouled (Chapter 1).
2   Fuel injection system needs adjustment or repair (Chapter 4).
3   Fuel filter clogged (Chapters 1 and 4).
4   Incorrect ignition timing (Chapter 1).
5   Intake manifold air leak (Chapter 4).

12 Engine surges while holding accelerator steady

1   Intake air leak (Chapter 4).
2   Fuel pump faulty (Chapter 4).
3   Loose fuel injector harness connections (Chapters 4 and 6).
4   Defective ECM (Chapter 6).

13 Engine stalls

1   Idle speed incorrect (Chapter 6).
2   Fuel filter clogged and/or water and impurities in the fuel system (Chapter 1).
3   Distributor components damp or damaged (Chapter 5).
4   Faulty emissions system components (Chapter 6).
5   Faulty or incorrectly gapped spark plugs (Chapter 1).
6   Faulty spark plug wires (Chapter 1).
7   Vacuum leak in the fuel injection system, intake manifold or vacuum hoses (Chapter 4).

14 Engine lacks power

1   Incorrect ignition timing (Chapter 1).
2   Excessive play in distributor shaft (Chapter 5).
3   Worn rotor, distributor cap or wires (Chapters 1 and 5).
4   Faulty or incorrectly gapped spark plugs (Chapter 1).
5   Fuel injection system out of adjustment or excessively worn (Chapter 4).
6   Faulty coil (Chapter 5).
7   Brakes binding (Chapter 9).
8   Automatic transaxle fluid level incorrect (Chapter 1).
9   Clutch slipping (Chapter 8).
10   Fuel filter clogged and/or impurities in the fuel system (Chapter 1).
11   Emission control system not functioning properly (Chapter 6).
12   Low or uneven cylinder compression pressures (Chapter 1).

15 Engine backfires

1   Emissions system not functioning properly (Chapter 6).
2   Ignition timing incorrect (Chapter 1).
3   Faulty secondary ignition system (cracked spark plug insulator, faulty plug wires, distributor cap and/or rotor) (Chapters 1 and 5).
4   Fuel injection system in need of adjustment or worn excessively (Chapter 4).
5   Vacuum leak at fuel injector(s), intake manifold or vacuum hoses (Chapter 4).
6   Valves sticking (Chapter 2).

16 Pinging or knocking engine sounds during acceleration or uphill

1   Incorrect grade of fuel.
2   Ignition timing incorrect (Chapter 1).
3   Fuel injection system in need of adjustment (Chapter 4).
4   Improper or damaged spark plugs or

wires (Chapter 1).
5   Worn or damaged distributor components (Chapter 5).
6   Faulty emission system (Chapter 6).
7   Vacuum leak (Chapter 4).

17 Engine runs with oil pressure light on

1   Low oil level (Chapter 1).
2   Idle rpm below specification (Chapter 6).
3   Short in wiring circuit (Chapter 12).
4   Faulty oil pressure sender.
5   Worn engine bearings and/or oil pump (Chapter 2).

18 Engine diesels (continues to run) after switching off

1   Idle speed too high (Chapter 6).
2   Thermo-controlled air cleaner heat valve not operating properly (Chapter 6).
3   Excessive engine operating temperature (Chapter 3).

Engine electrical system

19 Battery will not hold a charge

1   Alternator drivebelt defective or not adjusted properly (Chapter 1).
2   Battery terminals loose or corroded (Chapter 1).
3   Alternator not charging properly (Chapter 5).
4   Loose, broken or faulty wiring in the charging circuit (Chapter 5).
5   Short in vehicle wiring (Chapters 5 and 12).
6   Internally defective battery (Chapters 1 and 5).

20 Alternator light fails to go out

1   Faulty alternator or charging circuit (Chapter 5).
2   Alternator drivebelt defective or out of adjustment (Chapter 1).
3   Alternator voltage regulator inoperative (Chapter 5).

21 Alternator light fails to come on when key is turned on

1   Warning light bulb defective (Chapter 12).
2   Fault in the printed circuit, dash wiring or bulb holder (Chapter 12).

Fuel system

22 Excessive fuel consumption

1   Dirty or clogged air filter element (Chapter 1).
2   Incorrectly set ignition timing (Chapter 1).
3   Emissions system not functioning properly (Chapter 6).
4   Fuel injection internal parts excessively worn or damaged (Chapter 4).
5   Low tire pressure or incorrect tire size (Chapter 1).

23 Fuel leakage and/or fuel odor

1   Leak in a fuel feed or vent line (Chapter 4).
2   Tank overfilled.
3   Evaporative canister filter clogged (Chapters 1 and 6).
4   Fuel injector internal parts excessively worn (Chapter 4).

Cooling system

24 Overheating

1   Insufficient coolant in system (Chapter 1).
2   Water pump drivebelt defective or out of adjustment (Chapter 1).
3   Radiator core blocked or grille restricted (Chapter 3).
4   Thermostat faulty (Chapter 3).
5   Electric coolant fan blades broken or cracked (Chapter 3).
6   Radiator cap not maintaining proper pressure (Chapter 3).
7   Ignition timing incorrect (Chapter 1).

25 Overcooling

1   Faulty thermostat (Chapter 3).
2   Inaccurate temperature gauge (Chapter 12)

26 External coolant leakage

1   Deteriorated/damaged hoses; loose clamps (Chapters 1 and 3).
2   Water pump seal defective (Chapters 1 and 3).
3   Leakage from radiator core or header tank (Chapter 3).
4   Engine drain or water jacket core plugs leaking (Chapter 2).

27 Internal coolant leakage

1 Leaking cylinder head gasket (Chapter 2).
2 Cracked cylinder bore or cylinder head (Chapter 2).

28 Coolant loss

1 Too much coolant in system (Chapter 1).
2 Coolant boiling away because of overheating (Chapter 3).
3 Internal or external leakage (Chapter 3).
4 Faulty radiator cap (Chapter 3).

29 Poor coolant circulation

1 Inoperative water pump (Chapter 3).
2 Restriction in cooling system (Chapters 1 and 3).
3 Water pump drivebelt defective/out of adjustment (Chapter 1).
4 Thermostat sticking (Chapter 3).

Clutch

30 Pedal travels to floor - no pressure or very little resistance

1 Master or slave cylinder faulty (Chapter 8).
2 Hose/pipe burst or leaking (Chapter 8).
3 Connections leaking (Chapter 8).
4 No fluid in reservoir (Chapter 8).
5 If fluid is present in master cylinder dust cover, rear master cylinder seal has failed (Chapter 8).
6 If fluid level in reservoir rises as pedal is depressed, master cylinder center valve seal is faulty (Chapter 8).
7 Broken release bearing or fork (Chapter 8).

31 Fluid in area of master cylinder dust cover and on pedal

Rear seal failure in master cylinder (Chapter 8).

32 Fluid on slave cylinder

Slave cylinder plunger seal faulty (Chapter 8).

33 Pedal feels "spongy" when depressed

Air in system (Chapter 8).

34 Unable to select gears

1 Faulty transaxle (Chapter 7).
2 Faulty clutch disc (Chapter 8).
3 Fork and bearing not assembled properly (Chapter 8).
4 Faulty pressure plate (Chapter 8).
5 Pressure plate-to-flywheel bolts loose (Chapter 8).

35 Clutch slips (engine speed increases with no increase in vehicle speed)

1 Clutch plate worn (Chapter 8).
2 Clutch plate is oil soaked by leaking rear main seal (Chapter 8).
3 Clutch plate not seated. It may take 30 or 40 normal starts for a new one to seat.
4 Warped pressure plate or flywheel (Chapter 8).
5 Weak diaphragm spring (Chapter 8).
6 Clutch plate overheated. Allow to cool.

36 Grabbing (chattering) as clutch is engaged

1 Clutch plate lining contaminated with oil, burned or glazed (Chapter 8).
2 Worn or loose engine or transaxle mounts (Chapters 2 and 7).
3 Worn splines on clutch plate hub (Chapter 8).
4 Warped pressure plate or flywheel (Chapter 8).
5 Burned or smeared resin on flywheel or pressure plate (Chapter 8).

37 Transaxle rattling (clicking)

1 Release fork loose (Chapter 8).
2 Clutch plate damper spring failure (Chapter 8).
3 Low engine idle speed (Chapter 6).

38 Noise in clutch area

1 Fork shaft improperly installed (Chapter 8).
2 Faulty bearing (Chapter 8).

39 Clutch pedal stays on floor

1 Fork shaft binding in housing (Chapter 8).
2 Broken release bearing or fork (Chapter 8).

40 High pedal effort

1 Fork shaft binding in housing (Chapter 8).
2 Pressure plate faulty (Chapter 8).
3 Incorrect size master or slave cylinder installed (Chapter 8).

Manual transaxle

41 Knocking noise at low speeds

1 Worn driveaxle constant velocity (CV) joints (Chapter 8).
2 Worn side gear hub counterbore (Chapter 7A).*

42 Noise is most pronounced when turning

Differential gear noise (Chapter 7A).*

43 Clunk on acceleration or deceleration

1 Loose engine mounts (Chapters 2 and 7A).
2 Worn differential pinion shaft in case.*
3 Side gear hub counterbore in case worn oversize (Chapter 7A).*
4 Worn or damaged driveaxle inboard CV joints (Chapter 8).

44 Clicking noise in turns

Worn or damaged outboard CV joint (Chapter 8).

45 Vibration

1 Rough wheel bearing (Chapters 1 and 10).
2 Damaged driveaxle (Chapter 8).
3 Out-of-round tires (Chapter 1).
4 Tire out-of-balance (Chapters 1 and 10).
5 Worn CV joint (Chapter 8).
6 Incorrect driveaxle angle (Chapter 8).

46 Noisy in Neutral with engine running

1 Damaged input gear bearing (Chapter 7A).*
2 Damaged clutch release bearing (Chapter 8).

47   Noisy in one particular gear

1   Damaged or worn constant mesh gears (Chapter 7A).*
2   Damaged or worn synchronizers (Chapter 7A).*
3   Bent reverse fork (Chapter 7A).*
4   Damaged fourth speed gear or output gear (Chapter 7A).*
5   Worn or damaged reverse idler gear or idler bushing (Chapter 7A).*

48   Noisy in all gears

1   Insufficient lubricant (Chapter 1).
2   Damaged or worn bearings (Chapter 7A).*
3   Worn or damaged input gear shaft and/or output gear shaft (Chapter 7A).*

49   Slips out of gear

1   Worn or improperly adjusted linkage (Chapter 7A).
2   Transaxle loose on engine (Chapter 7A).
3   Shift linkage does not work freely, binds (Chapter 7A).
4   Input gear bearing retainer broken or loose (Chapter 7A).*
5   Dirt between clutch cover and engine housing (Chapter 7A).
6   Worn shift fork (Chapter 7A).*

50   Leaks lubricant

1   Inner driveaxle seals (Chapter 8).
2   Excessive amount of lubricant in transaxle (Chapters 1 and 7A).
3   Loose or broken input gear shaft bearing retainer (Chapter 7A).*
4   Input gear bearing retainer O-ring and/or lip seal damaged (Chapter 7A).*

51   Locked in second gear

Lock pin or interlock pin missing (Chapter 7A).*

* **Note:** *Although the corrective action necessary to remedy the symptoms described is beyond the scope of the home mechanic, the above information should be helpful in isolating the cause of the condition so that the owner can communicate clearly with a professional mechanic.*

Automatic transaxle

Note: *Due to the complexity of the automatic transaxle, it is difficult for the home mechanic to properly diagnose and service this component. For problems other than the following, the vehicle should be taken to a dealer service department or transmission shop.*

52   Fluid leakage

1   Automatic transaxle fluid is a deep red color. Fluid leaks should not be confused with engine oil, which can easily be blown by air flow to the transaxle.
2   To pinpoint a leak, first remove all built-up dirt and grime from the transaxle housing with degreasing agents and/or steam cleaning. Then drive the vehicle at low speeds so air flow will not blow the leak far from its source. Raise the vehicle and determine where the leak is coming from. Common areas of leakage are:
   a)   *Pan (Chapters 1 and 7)*
   b)   *Filler pipe (Chapter 7)*
   c)   *Transaxle oil lines (Chapter 7)*
   d)   *Speedometer sensor (Chapter 7)*

53   Transaxle fluid brown or has burned smell

Transaxle fluid burned (Chapter 1).

54   General shift mechanism problems

1   Chapter 7B deals with checking and adjusting the shift linkage on automatic transaxles. Common problems which may be attributed to poorly adjusted linkage are:
   a)   *Engine starting in gears other than Park or Neutral.*
   b)   *Indicator on shifter pointing to a gear other than the one actually being used.*
   c)   *Vehicle moves when in Park.*
2   Refer to Chapter 7B for the shift linkage adjustment procedure.

55   Transaxle will not downshift with accelerator pedal pressed to the floor

Throttle valve cable out of adjustment (Chapter 7B).

56   Engine will start in gears other than Park or Neutral

Neutral start switch malfunctioning (Chapter 7B).

57   Transaxle slips, shifts roughly, is noisy or has no drive in forward or reverse gears

There are many probable causes for the above problems, but the home mechanic should be concerned with only one possibility

- fluid level. Before taking the vehicle to a repair shop, check the level and condition of the fluid as described in Chapter 1. Correct the fluid level as necessary or change the fluid and filter if needed. If the problem persists, have a professional diagnose the probable cause.

Driveaxles

58   Clicking noise in turns

Worn or damaged outboard joint (Chapter 8).

59   Shudder or vibration during acceleration

1   Excessive CV joint angle (Chapter 8).
2   Excessive toe-in (Chapter 10).
3   Incorrect spring heights (Chapter 10).
4   Worn or damaged inboard or outboard CV joints (Chapter 8).
5   Sticking inboard CV joint assembly (double-offset design) (Chapter 8).
6   Sticking spider assembly (tri-pot design) (Chapter 8).

60   Vibration at highway speeds

1   Out-of-balance front wheels and/or tires (Chapters 1 and 10).
2   Out-of-round front tires (Chapters 1 and 10).
3   Worn CV joint(s) (Chapter 8).

Brakes

Note: *Before assuming that a brake problem exists, make sure that the tires are in good condition and properly inflated (Chapter 1), the front end alignment is correct (Chapter 10), the vehicle is not loaded with weight in an unequal manner and the front wheel bearings are properly adjusted (Chapter 9).*

61   Vehicle pulls to one side during braking

1   Incorrect tire pressures (Chapter 1).
2   Front end out of line (have the front end aligned).
3   Unmatched tires on same axle.
4   Restricted brake lines or hoses (Chapter 9).
5   Malfunctioning caliper assembly (Chapter 9).
6   Loose suspension parts (Chapter 10).
7   Loose calipers (Chapter 9).

62 Noise (high-pitched squeal when the brakes are applied)

Front and/or rear disc brake pads worn out. The noise comes from the wear sensor rubbing against the disc (does not apply to all vehicles). Replace pads with new ones immediately (Chapter 9).

63 Brake roughness or chatter (pedal pulsates)

1   Excessive lateral runout (Chapter 9).
2   Parallelism not within specifications (Chapter 9).
3   Wheel bearings not adjusted properly (Chapters 1, 9 and 10).
4   Uneven pad wear caused by caliper not sliding due to improper clearance or dirt (Chapter 9).
5   Wheel bearings not adjusted properly or in need of replacement (Chapters 1 and 9).
6   Defective rotor (Chapter 9).

64 Excessive pedal effort required to stop vehicle

1   Malfunctioning power brake booster (Chapter 9).
2   Partial system failure (Chapter 9).
3   Excessively worn pads (Chapter 9).
4   Piston in caliper stuck or sluggish (Chapter 9).
5   Brake pads contaminated with oil or grease (Chapter 9).
6   New pads installed and not yet seated. It will take a while for the new material to seat against the rotor.

65 Excessive brake pedal travel

1   Partial brake system failure (Chapter 9).
2   Insufficient fluid in master cylinder (Chapters 1 and 9).
3   Air trapped in system (Chapters 1 and 9).

66 Dragging brakes

1   Incorrect adjustment of brake light switch and/or cruise control vacuum release valve assembly could keep the brake pedal from returning fully.
2   Master cylinder pistons not returning correctly (Chapter 9).
3   Restricted brake lines or hoses (Chapters 1 and 9).
4   Incorrect parking brake adjustment (Chapter 9).

67 Grabbing or uneven braking action

1   Malfunction of combination valve (Chapter 9).
2   Malfunction of power brake booster unit (Chapter 9).
3   Binding brake pedal mechanism (Chapter 9).

68 Brake pedal feels "spongy" when depressed

1   Air in hydraulic lines (Chapter 9).
2   Master cylinder mounting bolts loose (Chapter 9).
3   Master cylinder defective (Chapter 9).

69 Brake pedal travels to the floor with little resistance

Little or no fluid in the master cylinder reservoir caused by leaking caliper piston(s), loose, damaged or disconnected brake lines (Chapter 9).

70 Parking brake does not hold

Parking brake linkage improperly adjusted (Chapter 9).

Suspension and steering systems

Note: *Before attempting to diagnose the suspension and steering systems, perform the following preliminary checks:*
a) *Tires for wrong pressure and uneven wear.*
b) *Steering universal joints from the column to the rack and pinion for loose connectors and wear.*
c) *Front and rear suspension and the rack and pinion assembly for loose and damaged parts.*
d) *Out-of-round or out-of-balance tires, bent rims and loose and/or rough wheel bearings.*

71 Vehicle pulls to one side

1   Mismatched or uneven tires (Chapter 10).
2   Broken or sagging springs (Chapter 10).
3   Front wheel or rear wheel alignment incorrect (Chapter 10).
4   Rear axle alignment incorrect (Chapter 10).
5   Front brakes dragging (Chapter 9).

72 Abnormal or excessive tire wear

1   Front wheel or rear wheel alignment incorrect (Chapter 10).
2   Sagging or broken springs (Chapter 10).
3   Tire out-of-balance (Chapter 10).
4   Worn strut damper or shock absorber (Chapter 10).
5   Overloaded vehicle.
6   Tires not rotated regularly.

73 Wheel makes a "thumping" noise

1   Blister or bump on tire (Chapter 10).
2   Improper strut damper or shock absorber action (Chapter 10).

74 Shimmy, shake or vibration

1   Tire or wheel out-of-balance or out-of-round (Chapter 10).
2   Loose, worn or out-of-adjustment wheel bearings (Chapters 1, 8 and 10).
3   Worn tie-rod ends (Chapter 10).
4   Worn lower balljoints (Chapter 10).
5   Excessive wheel runout (Chapter 10).
6   Blister or bump on tire (Chapter 10).

75 Hard steering

1   Lack of lubrication at balljoints, tie-rod ends and rack and pinion assembly (Chapter 10).
2   Front wheel alignment incorrect (Chapter 10).
3   Low tire pressure(s) (Chapters 1 and 10).

76 Steering wheel does not return to center position correctly

1   Lack of lubrication at balljoints and tie-rod ends (Chapter 10).
2   Binding in balljoints (Chapter 10).
3   Binding in steering column (Chapter 10).
4   Lack of lubricant in rack and pinion assembly (Chapter 10).
5   Front wheel alignment incorrect (Chapter 10).

77 Abnormal noise at the front end

1   Lack of lubrication at balljoints and tie-rod ends (Chapters 1 and 10).
2   Damaged shock absorber mount (Chapter 10).
3   Worn control arm bushings or tie-rod ends (Chapter 10).
4   Loose stabilizer bar (Chapter 10).
5   Loose wheel nuts (Chapters 1 and 10).
6   Loose suspension bolts (Chapter 10).

78 Wander or poor steering stability

1   Mismatched or uneven tires (Chapter 10).
2   Lack of lubrication at balljoints and tie-rod ends (Chapters 1 and 10).
3   Worn shock absorbers (Chapter 10).
4   Loose stabilizer bar (Chapter 10).
5   Broken or sagging springs (Chapter 10).
6   Front or rear wheel alignment incorrect (Chapter 10).

79 Erratic steering when braking

1   Wheel bearings worn (Chapters 1, 8 and 10).
2   Broken or sagging springs (Chapter 10).
3   Leaking wheel cylinder or caliper (Chapter 10).
4   Warped rotors (Chapter 10).

80 Excessive pitching and/or rolling around corners or during braking

1   Loose stabilizer bar (Chapter 10).
2   Worn strut dampers, shock absorbers or mount (Chapter 10).
3   Broken or sagging springs (Chapter 10).
4   Overloaded vehicle.

81 Suspension bottoms

1   Overloaded vehicle.
2   Worn strut dampers or shock absorbers (Chapter 10).
3   Incorrect, broken or sagging springs (Chapter 10).

82 Cupped tires

1   Front or rear wheel alignment incorrect (Chapter 10).
2   Worn shock absorbers (Chapter 10).
3   Wheel bearings worn (Chapters 1, 8 and 10).
4   Excessive tire or wheel runout (Chapter 10).
5   Worn balljoints (Chapter 10).

83 Excessive tire wear on outside edge

1   Inflation pressures incorrect (Chapter 1).
2   Excessive speed in turns.
3   Front end alignment incorrect (excessive toe-in). Have professionally aligned.
4   Suspension arm bent or twisted (Chapter 10).

84 Excessive tire wear on inside edge

1   Inflation pressures incorrect (Chapter 1).
2   Front end alignment incorrect (toe-out). Have professionally aligned.
3   Loose or damaged steering components (Chapter 10).

85 Tire tread worn in one place

1   Tires out-of-balance.
2   Damaged or buckled wheel. Inspect and replace if necessary.
3   Defective tire (Chapter 1).

86 Excessive play or looseness in steering system

1   Wheel bearing(s) worn (Chapter 10).
2   Tie-rod end loose (Chapter 10).
3   Rack and pinion loose (Chapter 10).
4   Worn or loose steering intermediate shaft (Chapter 10).

87 Rattling or clicking noise in rack and pinion

1   Insufficient or improper lubricant in rack and pinion assembly (Chapter 10).
2   Rack and pinion attachment loose (Chapter 10).

Chapter 1
Tune-up and routine maintenance

Contents

Specifications

Recommended lubricants, fluids and capacities

Note: *Listed here are manufacturer recommendations at the time this manual was written. Manufacturers occasionally upgrade their fluid and lubricant specifications, so check with your local auto parts store for current recommendations.*

Engine oil type and viscosity	Consult your owner's manual or local dealer for recommendations on the particular service grade and viscosity oil recommended for your area, special driving conditions and climatic parameters
Engine oil capacity (including filter)	
2.5L 4-cylinder engine	4.0 qts
2.8L V6 engine	4.5 qts
Coolant type	Ethylene glycol antifreeze and water
Cooling system capacity	
1984	
With air conditioning	14.2 qts
Without air conditioning	13.7 qts
1985 thru 1987	
Without air conditioning	13.8 qts
With air conditioning	
2.5L 4-cylinder engine with manual transmission	14.1 qts
2.5L 4-cylinder engine with automatic transmission	13.8 qts
2.8L V6 engine	13.8 qts
Fuel tank	
1984	10.5 gal
1985 and 1986	10.3 gal
Manual transaxle (4-speed) oil type	SAE 5W-30 SF engine oil

Recommended lubricants, fluids and capacities

Manual transaxle (5-speed) oil type... SAE 5W-30 or Dexron II ATF (check with your dealer service department or mechanic for the specific fluid used in your transmission)

Automatic transaxle fluid type .... Dexron II automatic transmission fluid

Manual transaxle oil capacity
    1984 thru 1986 4-speed.. 3.0 qts
    1985 and 1986 5-speed.. 2.7 qts

Automatic transaxle fluid capacity
    1984 thru 1987 after pan removal .... 4 qts
    1984 thru 1987 after overhaul .... 6 qts
    1985 thru 1987 after overhaul and converter drain .... 9 qts

Brake fluid type.. Delco Supreme 11 or DOT 3

FRONT

**Firing order
1-3-4-2**

**2.5L
4-CYLINDER
ENGINE**

General

Radiator cap opening pressure .... 15 psi

Thermostat rating .... 195 degrees F

Engine compression pressure .... Lowest reading cylinder should not be less than 70% of the highest and no cylinder reading should be less than 100 psi

Ignition system

Cylinder numbers (right-to-left - transaxle is on left)
    2.5L 4-cylinder engine.. 1-2-3-4
    2.8L V6 engine
        rear bank (trunk).. 1-3-5
        front bank (bulkhead)... 2-4-6

Distributor rotation.. Clockwise

Firing order
    2.5L 4-cylinder engine.. 1-3-4-2
    2.8L V6 engine 1-2-3-4-5-6

Spark plug type*
    1984 and 1985 2.5L 4-cylinder engine...................................... AC Type R43TSX or R44TSX
    1986 and 1987 2.5L 4-cylinder engine...................................... AC Type R43CTS
    2.8L V6 engine .... AC Type R42CTS

Spark plug gap*
    2.5L 4-cylinder engine.. 0.060 in
    2.8L V6 engine .... 0.045 in

Ignition timing .... Refer to the Vehicle Emission Control Information (VECI) label in the engine compartment for the ignition timing procedure and specifications

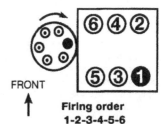

FRONT

**Firing order
1-2-3-4-5-6**

2.8L V6 ENGINE

**Cylinder location and
distributor rotation**

*The blackened terminal shown on the
distributor cap indicates the Number
One spark plug wire position*

* Refer to the Vehicle Emission Control Information (VECI) label in the engine compartment and follow the information on the label if it differs from the specifications here.

Wheel bearings

Front wheel bearing/hub end play.. 0.001 to 0.005 in

Brakes

Minimum refinish thickness
    Front .... 0.440 in
    Rear .... 0.500 in

Replacement (discard) thickness
    Front .... 0.390 in
    Rear .... 0.450 in

Recommended spare tire inflation pressure.................................... 60 psi

Accessory drivebelt tension (with Burroughs gauge)
    2.5L four-cylinder engine
        Alternator belt (without air conditioning)
            New .... 145 lbs
            Used .... 70 lbs
        Alternator belt (with air conditioning)
            New .... 165 lbs
            Used .... 90 lbs
        Air conditioning belt
            New .... 165 lbs
            Used .... 90 lbs

Accessory drivebelt tension (with Burroughs gauge) (continued)

2.8L V6 engine
  Alternator belt
    New .... 145 lbs
    Used .... 70 lbs
  Air conditioning belt
    New .... 145 lbs
    Used .... 70 lbs
  AIR pump belt
    New .... 100 lbs
    Used .... 50 lbs

Torque specifications   **Ft-lbs** (unless otherwise indicated)

Rocker arm cover bolts
  2.5L 4-cylinder engine..................................... 72 in-lbs
  2.8L V6 engine .... 72 to 108 in-lbs
Spark plugs.. 84 to 180 in-lbs
Front wheel bearing hub nut preload
  Step 1 .... 144 in-lbs
  Step 2 .... Back off nut until just loose
  Step 3 .... Snug up nut by hand
  Step 4 .... Loosen until either hole in spindle lines up with a slot in the nut
Automatic transaxle oil pan bolt .... 96 in-lbs
Wheel lug nuts
  Aluminum wheels 100
  Steel wheels 80
TBI mounting nuts .... 120 to 180 in-lbs

The engine compartment of a Fiero with a 2.5L four-cylinder engine

1	EECS vapor canister (under louver)	6	Battery (under louver)
2	Thermostat housing filler cap	7	Battery positive cable
3	THERMAC vacuum motor	8	PCV valve hose
4	Oil level dipstick	9	Throttle cable
5	Oil filler cap	10	Air cleaner

11	Coil
12	Fuel filter
13	Fuel return hose
14	Fuel hose

The engine compartment of a Fiero with a 2.8L V6 engine (viewed from the left side)

1	Throttle linkage	4	Oil filler cap	7	Distributor
2	Throttle body	5	Coil wire	8	Throttle Position Sensor (TPS)
3	Spark plug wires	6	Coil	9	Idle Air Control (IAC) valve

1   Introduction

This Chapter is designed to help the home mechanic maintain his or her vehicle for peak performance, economy, safety and long life.

On the following pages is a maintenance schedule, followed by Sections dealing specifically with each item on the schedule. Visual checks, adjustments, component replacement and other helpful items are included.

Servicing your vehicle in accordance with the mileage/time maintenance schedule and the following Sections entails a planned maintenance program that should result in a long and reliable service life. This is a comprehensive plan, so maintaining some items but not others at the specified service intervals will not produce the same results.

As you service your vehicle, you will discover that many of the procedures can - and should - be grouped together because of the nature of the particular procedure you're performing or because of the close proximity of two otherwise unrelated components to one another.

For example, if the vehicle is raised for chassis lubrication, you should inspect the exhaust, suspension, steering and fuel systems while you're under the vehicle. When you're rotating the tires, it makes good sense to check the brakes and wheel bearings since the wheels are already removed.

Finally, let's suppose you have to borrow or rent a torque wrench. Even if you only need to tighten the rocker arm cover bolts, you might as well check the torque of as many critical fasteners as time allows.

The first step of this maintenance program is to prepare yourself before the actual work begins. Read through all Sections pertinent to the procedures you're planning to do, then make a list of and gather together all the parts and tools you will need to do the job. If it looks like you might run into problems during a particular segment of some procedure, seek advice from your local parts man or dealer service department.

2   Maintenance schedule

The maintenance intervals in this manual are provided with the assumption that you, not the dealer, will be doing the work. These are the minimum maintenance intervals recommended by the factory for vehicles that are driven daily, for at least 15 miles, for at least 30 minutes at a time, in light traffic, on the freeway, in nice mild weather. For vehicles operating in dusty areas, towing a trailer, idling or operating at low speeds for extended periods, operating for short durations (less than four miles) in below freezing temperatures, shorter intervals are recommended. You may wish to perform some of these procedures even more often. Because frequent maintenance enhances the efficiency, performance and resale value of your vehicle, we encourage you to do so.

When your vehicle is new it should be serviced by a factory authorized dealer service department to protect the factory warranty. In many cases, the initial maintenance check is done at no cost to the owner.

The engine compartment of a Fiero with a 2.8L V6 engine (viewed from the right side)

1 Exhaust Gas Recirculation (EGR) valve
2 Engine oil dipstick
3 Exhaust Gas Recirculation (EGR) vacuum control solenoid

4 Thermostat housing filler cap
5 Manifold Absolute Pressure (MAP) sensor
6 Positive Crankcase Ventilation (PCV) valve

General operating checks

Check the operation of the license plate light, the side marker lights, the headlights (including the high beams), the parking lights, the taillights, the brake lights, the turn signals, the backup lights, the instrument panel lights and the hazard warning flashers. Be alert for broken, scratched, dirty or damaged glass, mirrors, lights or reflectors that could reduce your visibility or cause personal injury. If any damage or deterioration is discovered, replace, clean or repair it promptly. Once you're inside the vehicle, note whether all indicator and interior lights are working. Check the operation of all warning lights, buzzers, tone generators and chimes. Check the seat adjuster latches by pushing the seats forward and backward. Make sure that the adhesive backed friction joints of the mir-

rors and sun visors are holding firmly. Check the operation of the doors and the front and rear compartment lids. They should close, latch, lock and seal tightly. Inspect the engine air scoop opening for obstructions such as leaves or paper. While driving make sure that the automatic shift indicator points to the gear chosen (if you have an automatic transaxle). Note the operation and condition of the wiper blades and the flow and aim of the washer spray. Periodically check the air flow from the ducts at the inside base of the windshield. Do this with the heater control set for Defrost and the fan set for High. Sound the horn occasionally to make sure that it works. Check all the button locations. Be alert to abnormal sounds, increased brake pedal travel or repeated pulling to one side when braking. If the brake warning light

comes on, something may be wrong with part of the brake system. Be alert to any changes in the sound of the exhaust system or the smell of fumes. Either of these symptoms could indicate that the system is leaking or overheating. Be alert to any vibration of the steering wheel or seat at normal highway speeds. It could mean a wheel balance is needed. Also, a pull right or left on a straight, level road may indicate that the tire pressure needs to be adjusted or the wheels aligned. Note any changes in steering action. If the steering wheel suddenly becomes harder to turn or has too much free play, or if unusual sounds are noted when turning or parking, the steering system needs to be inspected. Take note of the headlight pattern occasionally. If the beam aim doesn't look right, the headlights should be adjusted.

Routine maintenance schedule

Every 250 miles or weekly, whichever comes first

Check the engine oil level (Section 4)
Check the engine coolant level and condition (Section 4)
Check the windshield washer fluid level (Section 4)
Check the tire pressures (Section 5)

Every 7,500 miles or 12 months, whichever comes first

All of the above, plus:
Check the brake fluid level (Section 6)
Check the clutch fluid level (Section 6)
Check the automatic transaxle fluid level (Section 7)*
Check the manual transaxle fluid level (Section 7)*
Check the throttle linkage (Section 8)*
Check the engine drivebelts (Section 9)*
Check the air filter (Section 10)*
Check the cooling system (Section 11)
Check the battery (Section 12)
Check the TBI mounting torque (Section 13)
Check the underhood hoses (Section 14)
Flush the underbody (Section 15)
Change the engine oil and filter (Section 16)
Check the steering and suspension systems (Section 17)*
Lubricate the chassis (Section 18)
Check the exhaust system (Section 19)*
Rotate the tires (Section 20)*
Check the brake system (Section 21)
Check the windshield wiper blades (Section 22)
Check the spare tire and jack (Section 23)
Check the seatbelts (Section 24)
Check the seatback latch (Section 25)
Check the steering column lock (Section 26)
Check the starter safety switch (Section 27)
Check the parking brake (Section 28)

Every 30,000 miles or 24 months, whichever comes first

All of the above, plus:
Replace the fuel filter (Section 29)
Adjust and replace if necessary the drivebelts (Section 30)
Service the cooling system (Section 31)
Inspect the PCV valve (Section 32)**
Replace the air and PCV filters (Section 33)**
Check the engine compression (Section 34)
Replace the spark plugs (Section 35)
Check the spark plug wires, distributor cap and rotor
   (Section 36)
Service the EGR system (Section 37)**
Check the ignition timing (Section 38)
Change the automatic transmission fluid and filter
   (Section 39)**
Inspect the fuel system (Section 40)
Clean, pack and adjust the front wheel bearings
   (Section 41)**

** Under severe operating conditions, perform these services
   every 3,000 miles or 3 months, whichever comes first.*
*** Under severe operating conditions, perform these ser-
   vices every 15,000 miles or 12 months, whichever
   comes first.*

Severe operating conditions

*Severe operating conditions, under which your vehicle will
   require maintenance at more frequent intervals, is
   defined as operating under one or more of the following
   conditions:*
Operating the vehicle in dusty areas
Towing a trailer
Idling for extended periods
Continuous low speed operation
Operating the vehicle when the outside temperature
   remains below freezing for extended periods
Operating the vehicle when most trips are less than 4 miles
*Note that some procedures in the maintenance schedule
   are marked with a (*) or a (**). These are the procedures
   which will have to be performed more frequently if the
   vehicle is subjected to severe operating conditions.*

3   Tune-up general information

The term tune-up applies to any general operation that restores the engine to its proper running condition. A tune-up is not a specific procedure. Rather, it is a series of individual but interrelated operations such as gapping the spark plugs and setting the ignition timing.

If the routine maintenance schedule (Section 2) is followed closely from the beginning of your vehicle's service life and if fluid levels and high wear items are checked frequently, and replaced if necessary, the engine will remain in good running condition, minimizing the need for unscheduled tune-ups.

More likely than not, however, there will be times when the engine runs poorly because it has not received the regular maintenance. This situation is even more likely if you have bought a used vehicle because it may have received neither regular nor frequent maintenance checks from its previous owner. In any case, an unscheduled engine tune-up will probably be necessary.

The following series of procedures are those most often needed to bring a generally poor running engine back into a proper state of tune.

Minor tune-up

Clean, inspect and test the battery
Check all engine-related fluids
Check and adjust the drivebelts
Replace the spark plugs
Inspect the distributor cap and rotor
Inspect the spark plug and coil wires
Check and adjust the idle speed
Check and adjust the timing
Replace the fuel filter
Check the PCV valve
Tighten the TBI mounting bolts
Check the air and PCV filters
Clean and lubricate the throttle linkage
Check all underhood hoses

Major tune-up

All of the above procedures plus:
Check the EGR system
Check the charging system
Check the fuel system
Check the engine compression
Check the cooling system
Replace the distributor cap and rotor
Replace the spark plug wires
Replace the air and PCV filters

4   Fluid level checks

Refer to illustrations 4.4, 4.9 and 4.13

1   Fluids are an essential part of the lubrication, cooling, braking and several other systems. Because these fluids gradually become depleted and/or contaminated during normal operation of the vehicle, they must be periodically replenished. See Recommended lubricants, fluids and capacities at the beginning of this Chapter before adding

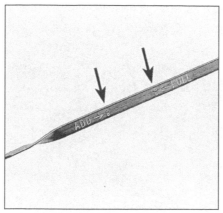

4.4 The oil level should be between the Add and Full marks (arrows) - the difference between the two marks represents one quart

fluids to any of the following components. **Note:** *The vehicle must be parked on level ground before fluid levels can be checked.*

Engine oil

2   The engine oil level is checked with a dipstick located on the front of the engine. The dipstick extends through a metal tube into the bottom of the oil pan.

3   The best time to check the engine oil level is when the engine is cool. If the oil is checked immediately after driving the vehicle, some of it will remain in the upper engine components and the resulting dipstick reading will be inaccurate.

4   Pull the dipstick from the tube and wipe all the oil from the end with a clean rag or paper towel. Insert the clean dipstick all the way back into the oil pan. Pull out the dipstick again. Look at the oil level- it should be above the Add line **(see illustration)**. If it isn't, add oil until the level is at the Full line.

5   It takes one quart of oil to raise the level from the Add to the Full mark. Do not allow the level to drop below the Add mark or engine damage may occur. Do not overfill the engine by adding oil above the Full mark as overfilling can result in oil fouled spark plugs, leaks or seal failures.

6   Oil is added to the engine through a twist-off filler cap located on the rocker arm cover. An oil can spout or funnel will reduce the chance of spilled oil.

7   Checking the oil level is an important part of any preventive maintenance program. A continually dropping oil level is symptomatic of oil leakage through damaged seals, from loose connections or past worn rings or valve guides. If the oil looks milky or has water droplets in it, a cylinder head gasket may be blown. The engine should be checked immediately. The condition of the oil should also be checked regularly. Each time you check the oil level, slide your thumb and index finger up the dipstick before wiping the oil off. If you see small dirt or metal particles clinging to the dipstick, the oil should be drained and fresh oil added.

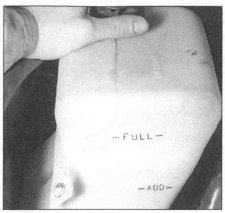

4.9 The coolant level in the coolant recovery tank (unbolted and lifted for a better view) should be between the Add and Full marks

Engine coolant

8   All vehicles covered in this manual are equipped with a pressurized coolant recovery system. A white engine coolant recovery tank is located under the front compartment lid, immediately behind the radiator. It is connected by a hose to the thermostat housing, located in the rear compartment on the left side of the engine. As the engine heats up during operation, coolant flows through the connecting tube and into the recovery tank. As the engine cools, the coolant is drawn back into the radiator.

9   The coolant level should be checked at regular intervals. Lift the front compartment lid and look at the coolant recovery tank. When the engine is cold, the coolant level should be at or slightly above the Cold (or Add) mark on the tank. If you are checking the coolant level after the engine has warmed up to normal operating temperature, the level should be at or above the Full Hot (or Full) mark on the tank **(see illustration)**.

10   **Warning:** *Never remove the radiator cap or the coolant recovery reservoir cap when the system is hot. Escaping steam and scalding liquid could cause serious personal injury. If it is necessary to open the radiator cap, wait until the system has cooled completely, then wrap a thick cloth around the cap and turn it to the first stop. If any steam escapes, wait until the system has cooled further, then remove the cap. The coolant recovery cap may be removed carefully after it is apparent that no further boiling is occurring in the recovery tank. If only a small amount of coolant is required to bring the system up to the proper level, water can be used. However, repeated additions of water will dilute the recommended antifreeze and water solution. In order to maintain the proper ratio, it is advisable to top up the coolant level with the correct mixture of antifreeze and water.*

11   Any time you check the coolant level, note its condition. It should be relatively clear. If it is brown or rust colored, the system should be drained, flushed and refilled. **Note:** *Even if the coolant appears to be normal, the*

anti-corrosion additives wear out with use, so it must still be replaced at the specified intervals.

12   Certain conditions, such as air trapped in the system, may affect the coolant level in the radiator. You should check the coolant level in the radiator - with the engine cold - every time you change the engine oil. Because of the design of the cooling system, the following procedure must be observed if it becomes necessary to add coolant directly to the radiator. **Warning:** *To help avoid the danger of being burned, never remove the radiator cap or thermostat housing cap while the engine and radiator are hot. Scalding fluid and steam can be blown out under pressure if the cap is taken off too soon.*

13   When the engine is cool, remove the thermostat housing cap. Turn the cap slowly to the left until it reaches a stop. Do not press down while turning the cap. Wait until any remaining pressure (indicated by a hissing sound) is relieved, then press down on the cap and continue turning it. Remove the cap, then remove the thermostat by pulling it straight up **(see illustration)**.

14   Add coolant through the thermostat housing until the coolant level reaches the spill point of the radiator neck. Install the radiator cap and tighten.

15   Install the thermostat housing cap but leave the thermostat out at this time. Tighten the thermostat housing cap to the first notch (you will hear a click and you will not be able to turn the cap counterclockwise without pushing it down). Add coolant to the reservoir until it reaches the Full line. Reinstall the recovery tank cap.

16   Run the engine for several minutes to reach normal operating temperature (the radiator hose will become hot).

17   Remove the radiator cap and, with the engine idling, add more coolant to the radiator until the level reaches the bottom of the filler neck. Reinstall the cap, making sure that the arrows on the cap line up with the overflow tube. **Warning:** *Under some conditions, the ethylene glycol in engine coolant is flammable. To avoid being burned when adding coolant, do not spill it on the exhaust system or hot engine parts.*

Windshield washer fluid

18   Fluid for the windshield washer system is contained in a plastic reservoir located on the right side of the front compartment, immediately in front of the windshield. If you live in a climate where temperatures drop below freezing, the reservoir should be kept no more than 2/3 full to allow for expansion should the fluid freeze. The use of an additive such as windshield washer fluid, available at auto parts stores, will help lower the freezing point of the fluid and will result in better cleaning of the windshield surface. **Caution:** *Do not use antifreeze- it will cause damage to the vehicle's paint.*

19   To help prevent icing in cold weather, warm the windshield with the defroster before using the washer.

4.13 When adding coolant, the thermostat must be removed from its housing (note which end is up to ensure correct reinstallation)

Battery electrolyte

20   The battery is located under the removable right side engine compartment louvers. This type of battery is permanently sealed, except for two small vent holes, and has no filler caps. A hydrometer in the top of the battery provides information for testing purposes only. If the vehicle is not going to be driven for 30 days or longer, disconnect the cable from the negative terminal of the battery to prevent discharge.

---

5   Tire and tire pressure checks

---

1   Periodic inspection of the tires may spare you from the inconvenience of being stranded with a flat tire. It can also provide you with vital information regarding possible problems in the steering and suspension systems before major damage occurs.

2   Correct air pressure adds miles to the lifespan of the tires, improves mileage and enhances overall ride quality.

3   Note any abnormal tread wear. Tread pattern irregularities - cupping, flat spots, more wear on one side than the other - are

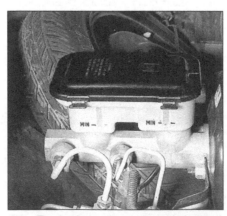

6.1a The brake master cylinder is located on the left side of the front compartment - keep the fluid level above the Min marks in the sight windows

indications of front end alignment and/or balance problems.

4   Watch closely for cuts, punctures and embedded nails or tacks. Sometimes a tire will hold its air pressure for a short time - or leak down very slowly - even after a nail has embedded itself into the tread. In most cases, a repair shop or gas station can repair the punctured tire.

5   Carefully inspect the inboard sidewall of each tire for evidence of brake fluid leakage. If you find any, inspect the brakes immediately.

6   Tire pressure cannot be accurately estimated by looking at a tire, particularly if it is a radial. A tire pressure gauge is essential. If you do not already have an accurate gauge, get one and keep it in the glovebox. The pressure gauges built into the nozzles of air hoses at gas stations are often inaccurate.

7   Always check tire pressure when the tires are cold. Cold, in this case, means the vehicle has not been driven over a mile in the three hours preceding a tire pressure check. A pressure rise of four to eight pounds is not uncommon once the tires are warm.

8   Unscrew the valve cap protruding from the wheel or hubcap and press the gauge firmly onto the valve. Observe the reading on the gauge and compare this figure to the recommended tire pressure shown on the tire placard on the left door.

9   Check all four tires and, if necessary, add enough air to bring them up to the recommended pressure levels. Don't forget to keep the spare tire inflated to the specified pressure. Be sure to reinstall the valve caps, which keep dirt and moisture out of the valve stem mechanism.

---

6   Brake and clutch master cylinder fluid level checks

Refer to illustrations 6.1a and 6.1b

1   The brake master cylinder is mounted on the front of the power booster unit located on the left side of the front compartment **(see illustration)**. The clutch master cylinder is

6.1b The clutch master cylinder is located just behind and to the left of the brake booster - keep the fluid level between the Add and Full marks

7.1 The speedometer fitting (the speed sensor harness has already been unplugged) is removed by removing the bolt (arrow) and retaining plate and pulling straight up on the fitting

just behind and above the brake master cylinder power booster **(see illustration)**.

2   If the fluid level of the brake master cylinder falls below the Min mark in the sight windows on the side of the reservoir, add enough fluid to raise the level to the top of the sight window. If the level of the clutch master cylinder falls below the Add mark, add enough fluid to bring the level between the Add and the Full marks. Don't overfill either reservoir.

3   Before adding fluid to either reservoir, wipe away any dirt or debris from the lid to prevent it from falling into the reservoir. Pour the fluid carefully. Do not spill it onto the surrounding painted surfaces. Use only the specified fluid (see *Recommended lubricants and fluids* or your owner's manual). Mixing different types of brake fluid can damage the system.

4   The fluid in both reservoirs should also be checked for any contamination. If dirt particles or water droplets are seen in the fluid, drain and refill the system.

5   The brake master cylinder has a snap type lid that must be pushed down until all four tangs snap into place around the rim of the reservoir. The clutch master cylinder has a screw-type lid. After filling either reservoir to the proper level, make sure that the lid is properly fastened to prevent fluid leakage.

6   The brake fluid in the master cylinder will drop slightly as the pads at each wheel wear down during normal operation. The fluid in the clutch master cylinder may drop slightly as the clutch friction material wears down. If either the brake or clutch master cylinder requires repeated replenishing to keep it at the proper level, there is a leak somewhere in the system and it must be repaired immediately.

7   If, upon checking the brake master cylinder fluid level, you discover one or both reservoirs empty or nearly empty, the brake system should be thoroughly inspected, then repaired and bled (Chapter 9).

7 Transaxle fluid level check

Refer to illustrations 7.1, 7.3 and 7.9

Manual transaxle

1   The manual transaxle does not have a dipstick. The fluid level is checked by removing the speedometer fitting on the driver's side of the transaxle case, above the driveaxle **(see illustration)**. **Note:** *Check the fluid level only when the engine is off, the vehicle is level and the transaxle is cool enough to proceed with the check without burning your fingers.*

2   Unplug the speedometer sensor harness from the speedometer fitting. Remove the retaining bolt and hold down plate. Wipe off any dirt or sludge around the speedometer fitting to prevent it from falling into the transaxle when the fitting is removed. Carefully extract the speedometer fitting by pulling it straight up.

3   The fluid level should be between the L and H marks **(see illustration)**.

4   Reinstall the speedometer fitting, hold down plate and retaining bolt. Be sure the fitting is fully seated. Tighten the bolt securely. Plug the speedometer sensor wiring harness into the top of the speedometer fitting. Drive the vehicle a short distance, then check for leaks.

Automatic transaxle

5   The fluid level should be checked with the transaxle at operating temperature. It takes approximately 15 miles of highway driving to bring the transaxle up to this temperature range.

6   An accurate fluid level reading cannot be obtained if the vehicle has recently been operated in hot weather (above 90~F), at sustained high speed, in heavy city traffic in hot weather or pulling a trailer. If any of these conditions apply, wait 30 minutes before checking the fluid level.

7   Park the vehicle on level ground. Apply the parking brake. Run the engine at idle and move the gear selector through all gear positions.

8   Move the selector to Park, remove the dipstick from the filler tube and wipe the fluid from the dipstick with a clean rag. Reinsert the dipstick into the filler tube.

9   Remove the dipstick once more and read the fluid level. It should be in the crosshatch area between the Add and Full hot marks **(see illustration)**.

10   Add just enough fluid to fill the transaxle to the proper level. It takes about one pint to raise the level from the Add mark to the Full mark when the transaxle is at normal operating temperature. Add fluid a little at a time and keep checking the level until it is correct. **Caution:** *Do not overfill. Overfilling causes foaming and loss of fluid through the vent and may damage the automatic transaxle.*

11   The condition of the fluid should also be noted when checking the level. If the fluid is a dark reddish-brown color or has a burned odor, it should be changed. If you are in doubt about the condition of the fluid, purchase some new fluid and compare the two for color and smell.

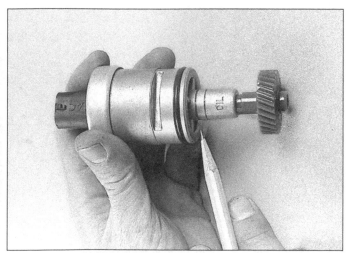

7.3 The transaxle fluid level should be between the two marks on the speedometer fitting

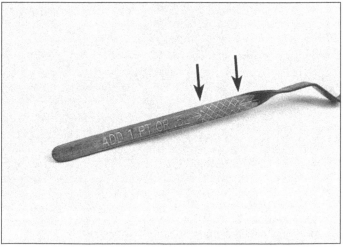

7.9 The fluid level should fall within the crosshatch area (arrows) on the automatic transaxle dipstick

8   Throttle linkage inspection

1   Check for the correct opening and clos-ing positions by operating the accelerator pedal. Make sure that the EFI unit reaches wide open throttle position. If it doesn't, inspect the throttle linkage components for damaged or bent brackets, levers, or other components. Check for poor carpet fit under the accelerator pedal or any interference or binding that might cause the cable or linkage to hang up.
2   If any binding is present in the linkage, make sure that the cable is properly routed and is free of kinks or damage. Check for free movement of the EFI lever at the EFI unit, the cable at the EFI lever stud, the accelerator lever at the bearing support and the pedal at the lever.

9   Drivebelt check

1   The drivebelts, or V-belts as they are often called, are located on the timing chain (right) end of the engine. Because of their composition and the high stresses to which they are subjected, drivebelts stretch and deteriorate as they get older. Therefore, they must be inspected periodically.
2   The number of belts used on a particular vehicle depends upon the accessories installed. They transmit power from the crankshaft to components such as the alter-nator, water pump and air conditioning com-pressor. Depending on the pulley arrange-ment, a single belt may be used to drive more than one of these components, while two belts may drive the same component on other models.
3   Using your fingers, and a flashlight if necessary, move along the belts checking for fraying, cracks and separation of the belt plies. Also watch for glazing, which gives the belt a shiny appearance. Both sides of each

10.8 The air filter on the 2.8L V6 is located inside the cylinder in front of the charcoal canister

belt should be inspected. Twist each belt to check its underside.

10   Air filter and Thermac check

Refer to illustrations 10.8, 10.10a and 10.10b
1   At the specified intervals, check the air intake, air cleaner and filter thoroughly. Be sure that all clamps and the air cleaner cover retaining nuts are tight. Loose connections or out-of-place parts could allow dirty air into the engine.

Air filter

2   The air filter is located inside the air cleaner housing atop the throttle body on four-cylinder engines and in the remote air cleaner housing on the left side of the engine compartment on V6 powered models.
3   To replace the filter on four-cylinder models, unscrew the flanged hex nuts from the studs protruding through the top of the air cleaner assembly and lift off the top plate. **Caution:** *Whenever the top plate is removed,*

be careful not to drop anything down into the throttle body.
4   Lift the air filter element out of the hous-ing.
5   Wipe out the inside of the air cleaner housing with a clean rag.
6   Place the new filter in the air cleaner housing. Make sure it seats properly in the bottom of the housing.
7   Lay the air cleaner top plate back in place and tighten the nuts finger tight.
8   Replacement of the V6 air filter is accomplished by locating the air filter hous-ing on the left side of the rear compartment, removing the nut securing the top cover, lift-ing off the cover and removing the air filter element **(see illustration)**.

Thermac (thermostatically controlled air cleaner) check

9   The four-cylinder engine is equipped with a thermostatic air cleaner (Thermac). The purpose of a heated intake air system is to provide a uniform inlet air temperature.
10   The Thermac assembly, located inside the long snorkel of the air cleaner, is actuated by heated air and intake manifold vacuum. Air can enter the air cleaner from outside the engine compartment or from a heat stove built around the exhaust manifold. When the temperature is below 86 degrees F the damper door will be up, shutting off outside air and allowing only heated air from the exhaust manifold to enter the air cleaner **(see illustration)**. When the temperature is above 131degrees F only outside air enters the air cleaner **(see illustration)**. Between 86 degrees F and 131degrees F, the damper door allows both heated and outside air to enter the air cleaner.
11   Make sure that the vacuum hoses between the intake manifold and the temper-ature sensor and the temperature sensor and the vacuum motor are connected. Make sure that the flexible air hose between the Ther-mac and the exhaust manifold is connected

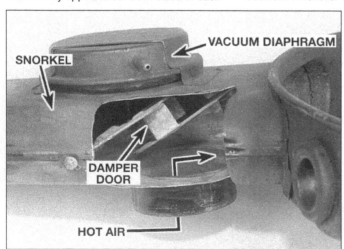

10.10a When the temperature at the air sensor is below 86 degrees F, the door in the air cleaner snorkel prevents outside air from entering the throttle body

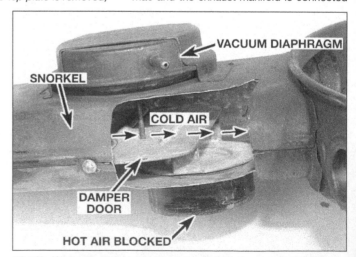

10.10b When the temperature exceeds 131 degrees F, the door in the snorkel opens to admit cooler outside air (between 86 degrees F and 131 degrees F, the door is neither fully closed nor open, permitting both heated and outside air to enter)

11.4a The underside of the front of the Fiero reveals the radiator hoses and metal pipes carrying coolant to the water pump and from the cylinder head back to the radiator, the heater hoses and the air conditioning lines

1 *Lower radiator hose*	4 *Heater hose*
2 *Upper radiator hose*	5 *Air conditioning line*
3 *Return coolant tube (coming from cylinder head)*	6 *Supply coolant tube (going to water pump)*

to the heat stove tube on the underside of the air cleaner snorkel.

12   Remove the outside air duck from the snorkel. Look through the end of the snorkel at the vacuum motor damper door.

13   The damper door should be open to the outside air. A functional check of the vacuum motor damper door operation should begin with a cold engine.

14   Start the engine and watch the damper door through the mouth of the snorkel. When the engine is first started, the damper door should move to a closed position, shutting off the supply of outside air.

15   As the engine warms the damper should open slowly to admit outside air. **Note:** *Depending on the ambient temperature, it may take 10 to 15 minutes before the damper door opens. To speed up this check, you can reconnect the snorkel air duct, drive the vehicle and then check to see if the damper is completely open.*

11   Cooling system check

Refer to illustrations 11.4a and 11.4b

Warning 1: *Do not allow antifreeze to come in contact with your skin or painted surfaces of the vehicle. Flush contaminated areas immediately with plenty of water. Don't store new coolant or leave old coolant lying around where it's accessible to children or pets – they're attracted by its sweet smell. Ingestion of even a small amount of coolant can be*

fatal! Wipe up garage floor and drip pan spills immediately. Keep antifreeze containers covered and repair cooling system leaks as soon as they're noticed.

Warning 2: *DO NOT remove the radiator cap or the coolant recovery cap while the cooling system is hot as escaping steam could cause serious injury.*

1   Many major engine failures can be attributed to a faulty cooling system. If your vehicle is equipped with an automatic transmission, the cooling system also plays an important role in prolonging transmission life.

2   The engine should be cold for a cooling system check, so perform the following procedures before the vehicle is driven for the day or after it has been shut off for at least three hours.

3   Remove the radiator cap and clean it thoroughly, inside and out, with water. Also clean the filler neck on the radiator. All traces of corrosion should be removed.

4   Carefully check the radiator hoses and the smaller diameter heater hoses. Because of its front mounted radiator and mid-engine design, metal tubes connect the front radiator hoses to the engine coolant hoses **(see illustration)**. Inspect each coolant hose along its entire length, replacing any hose which is cracked, swollen or otherwise deteriorated. Cracks are more apparent if the hose is squeezed **(see illustration)**. Inspect all metal tubing for dents, kinks, and deterioration that might block the flow of coolant.

5   Make sure that all hose connections are

Check for a chafed area that could fail prematurely.

Check for a soft area indicating the hose has deteriorated inside.

Overtightening the clamp on a hardened hose will damage the hose and cause a leak.

Check each hose for swelling and oil-soaked ends. Cracks and breaks can be located by squeezing the hose.

11.4b Simple checks can detect radiator hose problems

tight. A leak in the cooling system will usually show up as white or rust colored deposits.

6   Clean the front of the radiator core and the air conditioning condenser with compressed air, if available, or a soft brush. Remove all bugs, leaves, road grit or gravel that may have imbedded itself into the cooling fins of the radiator. Be extremely careful not to damage the delicate cooling fins or cut your fingers on the sharp edges.

7   Pressure test the radiator cap and the cooling system with a pressure tester, if available, or have them tested by a repair shop or gas station.

8   Check the coolant for proper level and condition. It should be clear and free of impurities. If the coolant looks murky or rust colored, the system should be drained, flushed and refilled.

12   Battery check and maintenance

Refer to illustration 12.10
Note: *There are certain precautions to be taken when working on or near the battery: a) Never expose a battery to open flame or sparks which could ignite the hydrogen gas given off by the battery b) Wear protective clothing and eye protection to reduce the possibility of the corrosive sulfuric acid solution inside the battery harming you (if the fluid is splashed or spilled, flush the contacted area immediately with plenty of water); c) Remove all metal jewelry which could contact the positive terminal and another grounded metal source, thus causing a short circuit; d) Always keep batteries and battery acid out of the reach of children.*

1   Before performing any battery maintenance, you should make sure you have proper eye and hand protection, baking soda and petroleum jelly on hand.

2   A sealed maintenance-free battery is standard equipment. Although this type of battery doesn't require the addition of water, it should still be given routine preventive maintenance in accordance with the following procedures. **Warning:** *Don't smoke or allow open flames anywhere near the vents on sealed batteries. Small quantities of hydrogen gas are present.*

3   Inspect the case and cover of the battery for physical damage such as cracks which could cause a loss of electrolyte. If you find evidence of cracking, replace the battery.

4   Check the tightness of the battery cables to ensure good electrical connections. Inspect the entire length of each cable for cracks and frayed conductors. The positive cable is sheathed in a protective plastic shroud to protect it from the engine block and the water pump. If this protective sheathing is broken, inspect the positive cable closely for places where the insulation may have worn through.

5   If terminal corrosion - it looks like white or green fluffy deposits - is evident, remove

12.10 The built-in hydrometer (arrow) should be green - if it is any other color, the battery must be charged or replaced

the cables from the terminals, clean both the terminals and the ends of the cables with a battery brush and reinstall the cables. Corrosion can be minimized by applying a layer of petroleum jelly or grease to the terminals and cable clamps after they are assembled.

6   Inspect the battery carrier. It should be in good condition and the hold-down clamp bolt should be tight.

7   If the battery is removed from its tray, make sure that no parts remain in the bottom of the tray when the battery is reinstalled. When reinstalling the hold-down clamp bolt, do not overtighten it.

8   Corrosion of any metal components near the battery may be removed with a solution of water and baking soda. Thoroughly wash all cleaned areas with plain water to rinse away residual battery acid. **Warning:** *Do not splash any solution into your eyes or onto your skin or clothes. Protective gloves should be worn.*

9   Any metal parts of the vehicle damaged by corrosion should be covered with a zinc-based primer, then painted.

10   Once the battery, terminals, cables, carrier and hold down clamp are clean and in good condition, check the built-in hydrometer **(see illustration)**. You should see a green dot in the window, indicating that the battery is in good condition.

11   If the window is dark and no green dot is visible, charge the battery in accordance with the instructions included with your charger. Charge the battery until the green dot is visible in the hydrometer window once more. Charging time will vary, depending upon such factors as the temperature, the charger capacity and the initial state-of-charge (see Chapter 5 for a complete description of the charging procedure).

13   Throttle Body Injector mounting torque check

Refer to illustration 13.4
1   The tightness of the throttle body

13.4 There are two mounting bolts inside the throttle body and a stud and nut outside on the flanged base - check all three to be sure they are tightened to the specified torque

mounting bolts must be checked every 7500 miles to insure that there are no vacuum leaks.

2   If you suspect that a vacuum leak has developed between the throttle body and the intake manifold, obtain a length of hose about the diameter of fuel hose. Start the engine and place one end of the hose next to your ear as you probe around the base of the throttle body with the other end. You will hear a hissing sound if a leak exists. **Warning:** *When probing with a vacuum hose stethoscope, be careful not to allow your body or the hose to come into contact with moving engine components such as drivebelts.*

3   Remove the air cleaner assembly. **Caution:** *Do not allow foreign objects to fall into the throttle body while the air cleaner is off.*

4   There are two mounting bolts inside the throttle body and one mounting stud and nut on the right side of the throttle body mounting flange **(see illustration)**. Check all three to insure that they are tightened to the specified torque. If the bolts have loosened, tighten them evenly. Do not overtighten or you may strip the threads in the cast aluminum throttle body.

5   If the vacuum leak persists after the mounting bolts have been tightened, the throttle body will have to be disassembled and inspected more closely (Chapter 4).

6   Reinstall the air cleaner assembly.

14   Underhood hose check and replacement

Warning: *Replacement of air conditioning hoses must be left to a dealer or air conditioning specialist who has the proper equipment to depressurize the system safely. Never remove air conditioning components or hoses until the system has been depressurized.*

Vacuum hoses

1   High temperatures in the engine com-

partment can cause the deterioration of the rubber and plastic hoses used for engine, accessory and emission systems operation.

2   A periodic inspection should be made for pinches, cuts, cracks, loose clamps, disconnects, leaks and hardened hose material.

3   Some, but not all, vacuum hoses use clamps to secure the hoses to fittings. Check all hose clamps for tension to prevent leakage. Make sure that hoses which don't use clamps have not expanded or hardened, allowing them to leak or slip off.

4   Various fuel and emission systems require hoses with different wall thicknesses, collapse resistance and temperature resistance. It is critical, therefore, to replace old hoses with new ones of identical specification.

5   Often the only effective way to inspect a hose is to remove it completely from the vehicle. If it's necessary to remove more than one hose, be sure to label the hoses and their respective connections to insure proper reattachment.

6   When checking vacuum hoses, don't forget to inspect plastic T-fittings. Check the fittings for cracks and tight hose fit.

7   A small piece of vacuum hose (1/4-inch inside diameter) can be used as a stethoscope to detect vacuum leaks. Hold one end of the hose to your ear and probe around vacuum hoses and fittings, listening for the "hissing" sound characteristic of a vacuum leak. **Warning:** *When probing with the vacuum hose stethoscope, be careful not to allow your body or the hose to come into contact with moving engine components like drivebelts.*

Fuel lines

Warning: *Gasoline is extremely flammable, so take extra precautions when you work on any part of the fuel system. Don't smoke or allow open flames or bare light bulbs near the work area, and don't work in a garage where a natural gas-type appliance (such as a water heater or a clothes dryer) with a pilot light is present. Since gasoline is carcinogenic, wear latex gloves when there's a possibility of being exposed to fuel, and, if you spill any fuel on your skin, rinse it off immediately with soap and water. Mop up any spills immediately and do not store fuel-soaked rags where they could ignite. When you perform any kind of work on the fuel system, wear safety glasses and have a Class B type fire extinguisher on hand.*

Warning: *Before removing the fuel filter, the fuel system pressure must be relieved (Chapter 4).*

8   Check the rubber hose sections of the fuel lines for deterioration and chafing. Watch for cracks, especially where the hoses bend downward along the surface of the left rear wheel well. Check all threaded connectors for a tight fit.

9   Deteriorated rubber fuel hose must be replaced with hose of the same specification.

Metal lines

10   Unlike the rubber sections, the metal sections of fuel line are not likely to deteriorate but they can develop vibration induced cracks. Inspect them carefully for kinks and crimps.

11   Make sure that all the threaded connectors on both ends of the metal lines are tight.

12   If a section of metal fuel line is cracked, replace it. Do not substitute metal tubing of "similar" length and diameter. Replace it with an identical section of the correct specification. Tubing of the wrong material, wall thickness, etc. may not be designed to withstand the pressure and vibration to which it will be subjected.

13   Check the metal brake lines connected to the brake master cylinder and proportioning valve in the front compartment for cracks in the lines or loose fittings. Any sign of brake fluid leakage calls for an immediate and thorough inspection of the brake system.

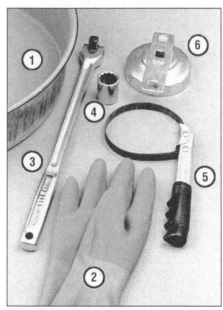

16.3 The tools needed for an oil change

1   **Drain pan** - *It should be fairly shallow in depth, but wide to prevent spills*
2   **Rubber gloves** - *When removing the drain plug and filter, you will get oil on your hands (the gloves will prevent burns)*
3   **Breaker bar** - *Sometimes the oil drain plug is tight, and a long breaker bar is needed to loosen it*
4   **Socket** - *To be used with the breaker bar or a ratchet (must be the correct size to fit the drain plug - six-point preferred)*
5   **Filter wrench** - *This is a metal band-type wrench, which requires clearance around the filter to be effective*
6   **Filter wrench** - *This type fits on the bottom of the filter and can be turned with a ratchet or breaker bar (different-size wrenches are available for different types of filters)*

15   Underbody flushing

Note: *If you live in regions with severe winter seasons where the roads are salted, perform the following procedure at least every spring.*

Flush the underbody with plain water to remove any corrosive materials used for ice and snow removal and dust control. Take care to thoroughly clean any areas where mud and debris can collect. Sediment packed into closed areas of the vehicle should be loosened before flushing.

16   Engine oil and filter change

Refer to illustrations 16.3 and 16.14

1   Frequent oil changes may be the best form of preventive maintenance available to the home mechanic.

2   Although the manufacturer recommends oil filter changes every other oil change for normal driving conditions, we feel that the minimal cost of an oil filter, the relative ease with which it is installed and the importance of keeping the oil free of contaminants make a compelling case for changing the filter every time the oil is changed.

3   The tools necessary for an oil and filter change are a wrench to fit the drain plug at the bottom of the oil pan, an oil filter wrench to remove the old filter, a container (with at least a six quart capacity) into which the old oil can be drained and a funnel to prevent spills while pouring fresh oil into the engine **(see illustration).**

4   Access to the underside of the vehicle is greatly improved if the vehicle can be lifted on a hoist, driven onto ramps or supported by jackstands. **Warning:** *Do not work under a vehicle which is supported only by a bumper, hydraulic or scissors-type jack.*

5   If this is your first oil change, get under the vehicle and familiarize yourself with the locations of the oil drain plug and the oil filter. The engine and exhaust components will be warm during the actual work, so figure out any potential problems before the engine and accessories are hot.

6   Warm the engine to normal operating temperature. Warm oil and sludge will flow out more easily.

7   Once the oil is warmed up, raise and support the vehicle. **Warning:** *To avoid personal injury, never get beneath the vehicle when it is supported only by a jack. The jack provided with your vehicle is designed for use only when changing the wheels. Always use jackstands to support the vehicle if it is necessary to get underneath.*

8   Move all necessary tools, rags and newspapers under the vehicle. Position the drain pan under the drain plug. Keep in mind that the oil will initially flow from the pan with some force, so position the pan accordingly.

9   Being careful not to touch any of the hot exhaust components, remove the drain plug in the bottom of the oil pan. Depending on

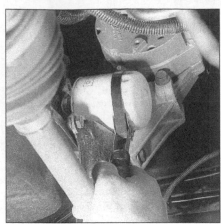

16.14 Use an oil filter wrench to remove the old filter - try to keep your hands away from hot engine parts and be prepared to catch some hot oil when the filter comes off

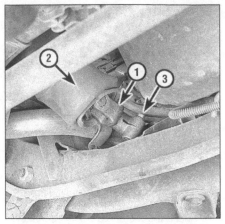

17.4 Inspect the steering intermediate shaft, universal joint couplings and the plastic boot seal covering the assembly

1	Intermediate shaft U-joint
2	Boot seal
3	Pinch bolt

the temperature of the oil, you may want to wear gloves while unscrewing the plug the final few turns.

10   Allow the old oil to drain into the pan. It may be necessary to move the pan farther under the engine as the oil flow slows to a trickle. Check the old oil for metal shavings and chips.

11   After all the oil has drained, wipe off the drain plug with a clean rag. Small metal particles may cling to the plug which would immediately contaminate the new oil.

12   Clean the area around the drain plug opening and reinstall the plug.

13   Move the drain pan into position under the oil filter.

14   Loosen the oil filter with a filter wrench **(see illustration)**. Chain or metal band filter wrenches may distort the filter canister, but this is of no concern as the filter will be discarded anyway. **Note:** *If your vehicle is powered by a V6, you may find the oil filter easier to change if the heat shield is removed first.*

Note: *On later models equipped with the 2.5L four cylinder engine, an element type filter in the oil pan replaces the spin-on type filter on the side of the engine block. Some of these engines have a separate drain plug to make draining the oil easier. Unscrew and remove the large oil filter access plug from the center of the oil pan, then reach up inside the hole with a pair of pliers, grasp the filter securely and pull it down using a twisting motion and remove it from the oil pan. Check the filter to make sure the rubber O-ring has come out with it. If it hasn't, reach up inside the pan opening and remove it.*

15   Sometimes the oil filter is on so tight it cannot be loosened. As a last resort, you can punch a long drift punch or screwdriver directly through the walls of the canister and use it as a T-bar to turn the filter. Be prepared for oil to spurt out of the canister as it is punctured.

16   Unscrew the old filter carefully - it's full of oil. Empty the contents into the drain pan.

17   Compare the old filter with the new one to make sure they are the same type.

18   Use a clean rag to remove all oil, dirt and sludge from the recess and sealing face of the filter bracket. Check the old filter to make sure the rubber gasket is not stuck to the engine mounting surface. If the gasket is stuck to the engine (use a flashlight if necessary), remove it.

19   Apply a light coat of oil to the sealing gasket of the new oil filter.

20   Screw on the new filter by hand until it's snug. The manufacturer recommends against the use of a filter wrench for installation because the seal will be damaged if the filter is overtightened (follow the directions on the filter or package). If you are changing the oil on a V6 and you removed the heat shield, replace it at this time. **Note:** *On later models with the element type filter, coat the new O-ring with engine oil and slide it into position all the way up the filter opening. Also, coat the inside of the grommet on the top of the new filter with engine oil. Clean the area around the drain plug opening, reinstall the plug and tighten it by hand until the gasket just touches the oil pan. Tighten the filter an additional 1/4-turn using a wrench.*

21   Remove all tools, rags, etc. from under the vehicle, being careful not to spill the oil in the drain pan, then lower the vehicle.

22   Move to the engine compartment and locate the oil filler cap on the rocker arm cover.

23   Remove the oil filler cap from the valve cover and add the new oil to the engine. Use a funnel, if necessary, to prevent spills.

24   Wait a few minutes to allow the oil to drain into the pan, then check the level on the dipstick. If the oil level is at or near the Full mark, start the engine and allow the new oil to circulate.

25   Run the engine for only about a minute and then shut it off. Look under the vehicle and check for leaks at the oil pan drain plug and around the oil filter. If either is leaking,

tighten it a little more.

26   Once the new oil has circulated and the filter is completely full, recheck the level on the dipstick. If necessary, add enough oil to bring the level to the Full mark on the dipstick.

27   During the first few trips after an oil change, make it a point to check frequently for leaks and proper oil level.

28   The old oil drained from the engine cannot be reused in its present state and should be disposed of. Check with your local refuse disposal company, environmental agency or recycling center to see if they will accept the oil for recycling. After the oil has cooled, it can be drained into a suitable container (capped plastic jugs, topped bottles, milk cartons, etc.) for transport to one of these disposal sites.

17   Suspension and steering check

Refer to illustrations 17.4, 17.7 and 17.9

1   When changing the oil - or lubricating the chassis - inspect the steering, suspension and driveaxle components for damaged, loose or missing parts, signs of wear and lack of lubrication.

2   If there is excessive play in the steering wheel, excessive sway (body lean) around corners or binding at some point as the steering wheel is turned, the steering and/or suspension components are probably at fault.

3   Raise the front of the vehicle and support it securely on jackstands placed under the frame rails.

4   Inspect the steering universal joints at both ends of the intermediate shaft between the steering column and the rack and pinion assembly for loose connectors and wear **(see illustration)**.

5   Have an assistant turn the steering wheel from side-to-side and check the steering linkage for free movement, chafing and binding.

6   If the steering linkage doesn't react positively to the movement of the steering wheel, try to determine which component is causing the free play.

7   Check the rack and pinion steering assembly boot seals for damage. Inspect the pinion seal for leakage. Inspect all steering and suspension components **(see illustration)** for loose bolts, broken or disconnected parts and deteriorated rubber bushings. Check the balljoints for wear. Replace broken components (Chapter 10).

8   Check the wheel bearings.

9   Clean and inspect the driveaxle boots for damage, tears and leakage **(see illustration)**. Replace them if they're torn or deteriorated (Chapter 8).

18   Chassis lubrication

Refer to illustrations 18.5, 18.7 and 18.9

1   A grease gun and a cartridge filled with the proper grease (see *Recommended lubricants and fluids*) are necessary to lubricate the chassis components.

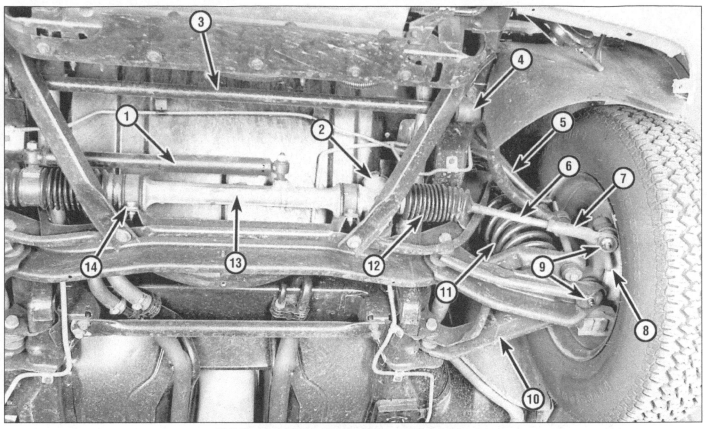

17.7 Front suspension and steering component layout

1	Steering damper	6	Inner tie rod	11	Coil spring
2	Steering pinion seal	7	Outer tie rod	12	Boot seal
3	Stabilizer bar	8	Steering knuckle	13	Rack and pinion steering assembly
4	Bushing	9	Grease fitting	14	Mounting grommet and clamp
5	Upper control arm	10	Lower control arm		

2   Ideally, you should lubricate the chassis when the vehicle is on a hoist. If you don't have access to one, raise the vehicle with a jack and place it securely on jackstands.

3   Balljoints should only be lubricated at temperatures of 10 degrees F and higher. During cold weather, the vehicle should be allowed to warm up in a heated garage before the balljoints are lubricated.

4   Before proceeding, pump a little grease out of the nozzle of the grease gun to remove any dirt from the end of the gun. Wipe the nozzle clean with a rag.

5   There are six grease fittings - three at each wheel - which must be lubricated on the front end **(see illustration)**. Locate the upper and lower balljoints on the suspension arms. There is a grease fitting on the upper arm and another on the bottom of the lower arm. A third grease fitting can be found on the outer tie-rod.

6   On the back of the lower arms there is a steering stop that should also be lubricated.

7   On the rear suspension there are four

17.9 The outboard driveaxle boot (arrow) should also be cleaned and inspected for damage, tears and leakage

18.5 There are three grease fittings on each front suspension assembly (arrows)

18.7 There are two grease fittings on each rear suspension assembly (arrows)

18.9 Lube each grease fitting by shoving the grease gun nozzle onto the nipple and pumping grease into the rubber reservoir until it's firm to the touch

grease fittings - two per side - that must be lubricated **(see illustration)**. One is on the toe link rod where it attaches to the knuckle. The other, which lubricates the balljoint stud, is on the lower control arm.

8   With the grease gun and plenty of rags, crawl under the vehicle and begin lubricating the components.

9   Wipe the grease fitting nipple clean and push the nozzle firmly over the fitting nipple **(see illustration)**. Pump the grease gun to force grease into the component.

10   If grease seeps out around the grease gun nozzle, the nipple is clogged or the nozzle is not seated on the fitting nipple. Press the gun nozzle onto the fitting and try again. If the nipple won't accept grease, replace it (Chapter 10).

11   Wipe the excess grease from each fitting when you have finished lubricating it.

12   While you are under the vehicle, clean and lubricate the parking brake cable by smearing a little chassis grease onto the cable and equalizer nut groove with a small brush.

13   Lower the vehicle for the remaining lubrication procedures.

14   Lubricate the door, front compartment lid and engine compartment lid hinge pins and latch mechanisms, the headlight mechanism and the manual transaxle shift linkage (if applicable) with a light oil.

15   Apply a coat of chassis grease to the door hinge detents and the engine compartment latch mechanism.

16   Lubricate the key lock cylinders in the doors and engine compartment with graphite dry lubricant, which is available at auto parts stores.

19   Exhaust system check

1   With the engine cold (at least three hours after the vehicle has been driven), check the complete exhaust system from the exhaust manifold to the tailpipe.

2   Check the exhaust pipe and connections for signs of leakage or corrosion. Make sure that all brackets and hangers are in good condition and tight.

3   Inspect the underside of the body for holes, corrosion, open seams, etc. which may allow exhaust gases to enter the passenger compartment. Seal all body openings.

4   Rattles and other noises can often be traced to the exhaust system, especially the mounts and hangers. Try to move the pipes, muffler and catalytic converter. If the components can come in contact with the body or suspension parts, secure the exhaust system with new mounts (Chapter 4).

5   Check the muffler heat shield and catalytic converter heat shield and splash shield retaining bolts for tightness.

6   Check the running condition of the engine by inspecting inside the end of the tailpipe. The exhaust deposits here are an indication of engine state-of-tune. If the pipe is black and sooty or coated with white deposits, the engine is in need of a tune-up.

20   Tire rotation

Refer to illustration 20.2

1   The tires should be rotated at the specified intervals and whenever uneven wear is noticed. Since the vehicle will be raised and the tires removed, this is a good time to check the brakes (Section 21) and the wheel bearings (Section 41).

2   Refer to the accompanying illustration for the recommended tire rotation pattern.

3   Refer to the information in *Jacking and towing* at the front of this manual for the proper procedures to follow when raising the vehicle and changing a tire. If the brakes are to be checked, do not apply the parking brake. Make sure the tires are blocked to prevent the vehicle from rolling if you are only raising one end at a time.

4   Preferably, the entire vehicle should be raised at the same time. Always use four jackstands and make sure the vehicle is firmly supported.

5   After rotation, check and adjust the tire pressures as necessary and check the lug nut tightness.

6   For further information on the wheels and tires, refer to Chapter 10.

21   Brake check

Refer to illustration 21.6

1   The brakes should be inspected every time the wheels are removed or whenever a defect is suspected. Any of the following symptoms could indicate a potential brake

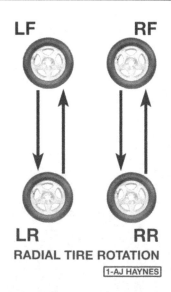

LF   **RF**

LR   **RR**

RADIAL TIRE ROTATION

1-AJ HAYNES

20.2 The preferred rotation pattern for vehicles equipped with radial tires

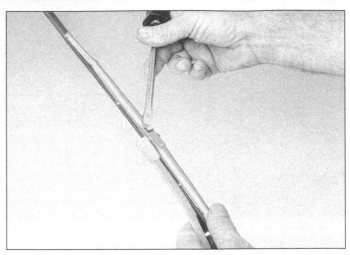

21.6 Inspect brake pad thickness by looking through the small window (arrows) in the caliper (pads should be at least 1/16-inch thick - the backing material is not included in this measurement)

22.8 Using a small screwdriver, pry up on the release pin and detach the blade assembly from the wiper arm pin

system defect: The vehicle pulls to one side when the brake pedal is depressed; the brakes make squealing or dragging noises when applied; brake travel is excessive; the pedal pulsates; brake fluid leaks, usually on the inside of the tire or wheel.

2   The disc brake pads have built-in wear indicators which should make a high-pitched squealing when the brake pads are worn to where new pads are needed. **Caution:** *Expensive rotor damage can result if the pads are not replaced soon after the wear indicators start squealing.*

3   It's not necessary to wait until the wear indicators let you know that the pads are worn out. Pad thickness is easy to check any time the wheels are removed.

4   Raise the vehicle and place it securely on jackstands.

5   Remove the wheels. The disc brake calipers containing the pads are now visible. There is an outer and inner pad in each caliper.

6   Check the pad thickness by looking through the inspection hole in the caliper body **(see illustration)**. If the lining material is 1/16-inch or less in thickness, the pads should be replaced. **Note:** *The lining material of the pad is riveted or bonded to a metal backing, which is not included in this measurement.*

7   If it proves impossible to determine the thickness of the remaining lining material by the above method, or if you are concerned about the condition of the pads, remove them for further inspection (Chapter 9).

8   Before installing the wheels, check all brake lines for leakage, particularly in the vicinity of the threaded connections at the calipers. Look for cracks or splits in the brake lines. Replace the lines and/or fittings as necessary (Chapter 9).

9   Check the condition of the rotors. Look for scoring, gouging and burned spots. If these conditions exist, the hub/rotor assembly should be removed and serviced (Chapter 9).

22 Wiper blade inspection and replacement

Refer to illustration 22.8

1   The windshield wiper and blade assembly should be inspected periodically for damage, loose components and cracked or worn wiper blade elements.

2   Road film can build up on the wiper blades and affect their efficiency, so they should be washed regularly with a mild detergent solution.

3   The action of the wiping mechanism can loosen the bolts, nuts and fasteners, so they should be checked and tightened, as necessary, at the same time the wiper blades are checked.

4   If the wiper blade elements are cracked, worn or warped, they should be replaced with new ones.

5   Lift the arm assembly away from the glass for clearance and remove the wiper blade. Three methods are used to retain wiper blades to wiper arms.

6   The first type blade uses an internal spring. To remove it, press down on the blade, release the spring and remove the blade from the arm.

7   The second type of blade uses a press-type release lever. When the release lever is depressed, the blade assembly can be slid off the wiper arm pin.

8   The third type of blade uses an exterior spring **(see illustration)**. To remove the blade, insert a screwdriver under the spring and then push downward on the screwdriver to raise the spring. The blade can then be removed from the arm.

23 Spare tire and jack check

1   Be alert to rattles in the front compartment of the vehicle. Make sure that the spare tire and the jacking equipment are securely

stowed at all times.

2   Lubricate the jack mechanism with white grease after each use.

24 Seat belt check

1   Inspect the belt system, including the webbing, buckles, latch plates, retractors and anchors. Look for torn or frayed webbing. If any component of the system looks questionable, replace it.

2   The best way to check the buckles is to snap them together and try to jerk them apart forcefully. If either buckle fails to hold, replace it.

25 Seatback latch check

1   Be sure the seatbacks latch by pushing back and forth on them (see your owner's manual for latch operating information).

2   Make sure that the recliner is holding by pushing and pulling on the top of the seatback while it is reclined.

26 Steering column lock check

While parked, try to turn the key to Lock in each gear range. The key should turn to Lock only when the gear indicator is in Park on automatic transaxle vehicles. On manual transaxle vehicles, try to turn the key to Lock without depressing the key release lever. The key should turn to Lock only with the key release lever depressed. On all vehicles, the key should come out only in Lock.

27 Starter safety switch check

Firmly apply both the parking brake and the regular brakes. If the engine starts, be ready to turn off the ignition promptly. Take

29.1 Disconnect the threaded fittings (arrows) at both ends of the fuel filter with a flare nut wrench and a backup wrench to hold the filter

these precautions because the vehicle could move without warning and possibly cause personal injury or property damage. On automatic transaxle vehicles, try to start the engine in each gear. The starter should turn only in Park or Neutral. On manual transaxle vehicles, place the shift lever in Neutral, push the clutch halfway and try to start. The starter should turn only when the clutch is fully depressed.

28 Parking brake check

Caution: *Before checking the holding ability of the parking brake and the automatic transaxle Park mechanism, park on a steep hill with enough room for movement in the downhill direction. To reduce the risk of personal injury or property damage, be prepared to apply the brakes promptly if the vehicle begins to move.*

To check the parking brake, with the engine running and the transaxle in Neutral, slowly remove foot pressure from the brake

pedal until the vehicle is held by only the parking brake. To check the automatic transaxle Park mechanism holding ability, release all brakes after shifting the transaxle to Park.

29 Fuel filter replacement

Refer to illustrations 29.1 and 29.11

Warning: *Gasoline is extremely flammable, so take extra precautions when you work on any part of the fuel system. Don't smoke or allow open flames or bare light bulbs near the work area, and don't work in a garage where a natural gas-type appliance (such as a water heater or a clothes dryer) with a pilot light is present. Since gasoline is carcinogenic, wear latex gloves when there's a possibility of being exposed to fuel, and, if you spill any fuel on your skin, rinse it off immediately with soap and water. Mop up any spills immediately and do not store fuel-soaked rags where they could ignite. When you perform any kind of work on the fuel system, wear safety glasses and have a Class B type fire extinguisher on hand.*

Warning: *Before removing the fuel filter the fuel system pressure must be relieved (Chapter 4).*

Four-cylinder engine

1   Loosen the threaded connectors on both ends of the fuel filter with a flare nut wrench **(see illustration)**. Disconnect both ends.

2   Loosen the clamp bolt underneath the filter.

3   Remove the old filter and discard it.

4   Install the new filter. Make sure that the arrow indicating the direction of fuel flow points toward the fuel injection system.

5   Reconnect the fuel line fittings to both ends of the filter. Snug them finger tight.

6   Tighten the clamp bolt underneath the filter.

7   Tighten the fuel line fittings securely.

V6 engine

8   Relieve the pressure in the fuel system be-fore disconnecting any fuel lines (Chapter 4).

9   Raise the rear of the vehicle, support it securely on jackstands and block the front wheels.

10   Place a container under the fuel filter to catch any spilled fuel.

11   Loosen the fittings on either side of the fuel filter **(see illustration)**, allow the fuel in the filter to drain, then remove the lines from the filter. A backup wrench must be used on the fuel filter canister to prevent it from turning.

12   Remove the fuel filter clamp bolt and slide the filter out of the bracket.

13   Position the new filter in the bracket. The arrow on the canister must be pointed in the direction of fuel flow. Install, but do not tighten, the bracket clamp bolt.

14   Start the fuel line fittings on either side of the filter, but do not tighten them at this time.

15   Tighten the fuel filter clamp bolt.

16   Tighten the fuel line fittings, using a backup wrench to keep the filter from turning.

17   Cycle the ignition switch On and Off several times at two second intervals to pressurize the fuel system, then check for fuel leaks.

30 Drivebelt adjustment and replacement

V-drivebelts

Refer to illustration 30.1

1   Check the tension of each belt by pushing on it at a distance halfway between the pulleys. Push firmly with your thumb and see how much the belt deflects. Measure this deflection with a ruler **(see illustration)**. The belt should deflect 1/4-inch if the distance from pulley center to pulley center is between 7 and 11 inches. The belt should deflect 1/2-inch if the distance from pulley

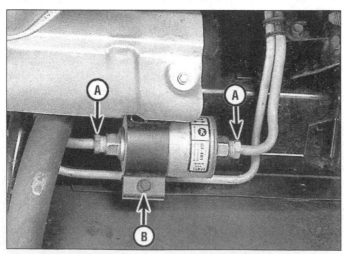

29.11 Loosen and remove the fuel line fittings (A), then remove the filter clamp bolt (B)

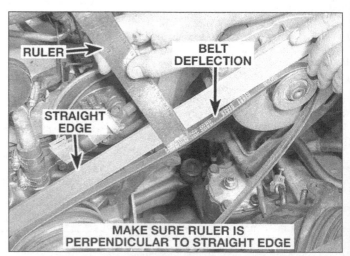

30.1 Drivebelt deflection can be checked with a straightedge and a ruler

31.7a The radiator drain valve (arrow) is located at the lower right corner of the radiator

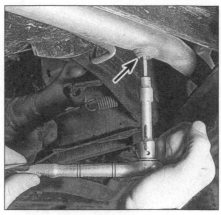

31.7b There is a drain plug in each coolant tube just forward of the rear wheels

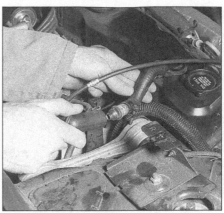

32.2a The PCV valve on the four-cylinder engine is located inside a rubber elbow on the right side of the throttle body and is attached to the end of a short hose that fits inside a rubber grommet atop the rocker arm cover

center to pulley center is between 12 and 16 inches.

2   Another, more precise method of measuring and adjusting belt deflection is accomplished with special belt tension tools. These tools measure the amount of force required to deflect a belt a specified distance.

3   To adjust belt tension, loosen the pivot bolt(s) securing the beltdriven component to the engine block and the retaining nut or bolt securing the component to the slotted adjustment bracket. Use a levering device, such as a large screwdriver or pry bar, to pivot the component until the specified tension is achieved. **Caution:** *If a pry bar is used, be careful not to damage either the component or the engine.*

4   Tighten the retaining nut or bolt securing the component to its slotted adjustment bracket snug enough to hold it. Check the belt tension again. Adjust as necessary. Tighten the component securely.

5   If inspection reveals that a belt must be replaced, loosen the pivot bolt of the appropriate accessory (alternator, air conditioning compressor, etc.) in accordance with the procedure outlined in Step 3 above, slip the old belt off the crankshaft pulley and accessory pulley, then lift it out.

6   Before installing the new belt, make sure it's the same width and length as the old one.

7   Once the new belt is in place, tighten and adjust it just like you would a used belt.

Serpentine drivebelts

8   The serpentine drivebelt is automatically adjusted by a spring loaded tensioner pulley. The belt should be inspected regularly for missing ribs and frayed plies. To replace the belt, insert a 1/2-inch drive breaker bar onto the bolt head in the center of the tensioner pulley. Rotate the pulley counterclockwise releasing the belt tension. Be sure to note the belt routing arrangement before removing the belt from the pulleys. Draw a schematic if necessary. Install the new belt starting with the bottom pulleys, then release the tensioner. Make sure each belt is properly centered onto each pulley.

31 Cooling system servicing (draining, flushing and refilling)

Refer to illustrations 31.7a and 31.7b

Warning 1: *Do not allow antifreeze to come in contact with your skin or painted surfaces of the vehicle. Flush contaminated areas immediately with plenty of water. Don't store new coolant or leave old coolant lying around where it's accessible to children or pets – they're attracted by its sweet smell. Ingestion of even a small amount of coolant can be fatal! Wipe up garage floor and drip pan spills immediately. Keep antifreeze containers covered and repair cooling system leaks as soon as they're noticed.*

Warning 2: *DO NOT remove the radiator cap or the coolant recovery cap while the cooling system is hot as escaping steam could cause serious injury.*

1   Periodically, the cooling system should be drained, flushed and refilled to replenish the antifreeze mixture and prevent formation of rust and corrosion.

2   At the same time the cooling system is serviced, all hoses and the radiator cap should be inspected and replaced if defective (Chapter 3).

3   With the engine cold, remove the radiator cap and the thermostat housing cap by slowly rotating them counterclockwise to their respective detents (do not press down while rotating). Wait until any residual pressure (indicated by a hissing sound) is relieved. After all pressure has been relieved, press down on each cap while continuing to rotate it counterclockwise.

4   Remove the thermostat.

5   Reinstall the thermostat housing cap.

6   Move a large container under the radiator to catch the coolant as it is drained. Place smaller containers under the coolant tube plugs and the engine block drain plug.

7   Open the radiator drain valve **(see illustration)**, the coolant tube plugs **(see illustration)** and the engine drain plug to drain the coolant. Be careful that none of the solution is splashed on your skin or into your eyes.

8   After coolant has ceased to drain from these openings, close them.

9   Refill the system with water, run the engine, then drain and refill the system repeatedly until the drained liquid is clear.

10   Allow the system to drain completely and then close the radiator drain valve tightly. Install the block and coolant pipe drain plugs and tighten them securely.

11   Remove the coolant recovery tank retaining bolts. Remove the tank and empty it of old coolant. Flush the tank out with clean water.

12   Fill the radiator with antifreeze and water in the required proportions to the base of its fill neck and add sufficient coolant to the recovery tank to raise the level to the Full mark. Reinstall the recovery tank cap.

13   Run the engine, with the radiator cap removed, until normal operating temperature is reached.

14   With the engine idling, add coolant until the level reaches the bottom of the filler neck and reinstall the cap, making certain that the arrows line up with the filler tube.

15   Remove the thermostat housing cap, install the thermostat and screw on the cap.

32 Positive crankcase ventilation (PCV) valve check and replacement

Refer to illustrations 32.2a, 32.2b and 32.11

1   If the engine is idling roughly, a clogged PCV valve, filter or hose might be the reason.

Four-cylinder engine

2   The PCV valve is located in the rubber elbow at the throttle body end of the rubber hose connecting the rocker arm cover to the throttle body **(see illustration)**. The PCV valve filter is located between the rocker arm cover and the air cleaner snorkel **(see illustration)**.

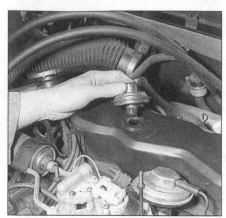

32.2b The PCV filter on the four-cylinder engine is plugged into a rubber grommet in the rocker arm cover

32.11 The PCV valve and filter on the V6 engine are located in the rear valve cover and the front valve cover, respectively

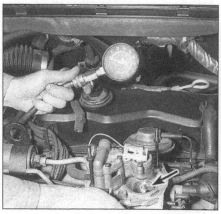

34.6 Crank the engine over at least four times while holding the throttle plate wide open to get an accurate compression reading

3   With the engine idling at normal operating temperature, pull the PCV valve hose plastic fitting from its rubber grommet in the rocker arm cover.

4   Place your finger over the end of the PCV hose fitting to check for intake manifold vacuum. You should feel suction at the fitting and the engine speed should drop. If there is no vacuum and engine speed doesn't drop, check the hose. It may be plugged. If the hose is clear, inspect the PCV filter. If the filter isn't clogged, the valve itself must be checked.

5   Turn off the engine and remove the rubber elbow from the throttle body. Pull the PCV valve from the rubber elbow. Unscrew the hose clamp and remove the hose from the other end of the valve.

6   Shake the valve and listen for the rattle of the ball inside the valve. If the valve does not rattle, replace it. **Note:** *When purchasing a replacement PCV valve, make sure it is for your particular vehicle, model year and engine size.*

7   Push the new valve into the end of the elbow until it is seated. **Caution:** *Make sure you install it facing the same direction that it did when you removed it. A PCV valve won't work if installed facing the wrong way.*

8   Slip the hose over the other end of the PCV valve and tighten the hose clamp securely.

9   Inspect the rubber grommet at the rocker arm cover for damage and replace it with a new one if warranted.

10   Push the rubber elbow over its fitting on the side of the throttle body and slip the plastic fitting at the other end of the PCV hose into the rubber grommet.

V6 engine

11   The procedure for checking the PCV valve on a V6 is identical, but the valve itself is located between a rubber elbow and a rubber grommet located on the top of the rear valve cover. The filter is located at the other end of the system, where it attaches to the top of the front valve cover **(see illustration)**.

33   Air filter replacement

1   On the four-cylinder engine, remove the flanged hex nuts on the air cleaner top plate. Remove the top plate. Remove the old air filter. Wipe out the inside of the air cleaner housing. Install the new air filter. Install the top plate. Tighten the flanged hex nuts securely.

2   The air filter for the V6 engine is located inside the cylindrical air cleaner housing at the left front corner of the engine compartment. Remove the top nut, lift out the old filter, wipe out the housing and install the new filter.

34   Compression check

Refer to illustration 34.6

1   A compression check will tell you the mechanical condition of the engine. It can tell you if the compression is down due to leakage caused by worn piston rings, defective valves and seats or a blown head gasket.

2   Warm the engine to normal operating temperature, shut it off and allow it to sit for ten minutes to allow the catalytic converter temperature to drop. **Warning:** *Disconnect the ignition switch feed wire at the distributor (the electrical leads to the coil) to reduce the risk of electrical shock.*

3   Clean the area around the spark plugs before removing them to prevent dirt from falling into the cylinders.

4   Remove the spark plugs.

5   Remove the air cleaner housing. **Note:** *When checking cylinder compression, the battery should be at or near full charge.*

6   With the compression gauge in the number one spark plug hole and the throttle plate wide open **(see illustration)**, crank the engine over at least four times and note the reading on the gauge.

7   Compression should build quickly. Low compression on the first stroke which does not increase during successive strokes indicates leaking valves, a blown head gasket or a cracked head. Record the highest gauge reading tand compare it to the specified com-

pression.

8   Repeat the above procedure for the remaining cylinders. The lowest compression reading should not be less than 70% of the highest reading. No reading should be less than 100 psi.

9   Pour a couple of teaspoons of engine oil (a squirt can is handy for this purpose) through the spark plug hole in each cylinder and repeat the test.

10   If the compression increases markedly after the oil is added, the piston rings are worn. If it does not increase significantly, the leakage is occurring at the valves or through the head gasket. Leakage past the valves may be caused by burned valve seats or faces, or by warped, cracked or bent valves.

11   If two adjacent cylinders have equally low compression, there is a strong possibility that the head gasket between them is blown. The appearance of coolant in the combustion chambers or the crankcase would verify this condition.

12   If the compression is higher than normal, the combustion chambers are probably coated with carbon deposits. If that is the case, the cylinder head(s) should be removed and decarbonized (Chapter 2).

13   If compression is way down or varies greatly between cylinders, it would be a good idea to have a leakdown test performed by an automotive repair shop to pinpoint the exact location and severity of the leakage.

35   Spark plug replacement

Refer to illustrations 35.4a and 35.4b

1   Spark plug replacement requires a spark plug socket which fits onto a ratchet wrench. This socket is lined with a rubber grommet to protect the porcelain insulator of the spark plug and to hold the plug while you direct it to the spark plug hole. You also will need a wire-type feeler gauge to check and adjust the spark plug gap.

2   Purchase the new plugs, adjust them to the proper gap and then replace each plug

one at a time. When buying new spark plugs, it's essential that you obtain the correct plugs for your specific engine. This information can be found on the Vehicle Emissions Control Information (VECI) label located under the hood or in the owner's manual. If differences exist between these sources, purchase the spark plug type specified on the VECI label, because that information is provided specifically for your engine.

3   Inspect each of the new plugs for defects. If there are any signs of cracks in the porcelain insulator of a plug, don't use it.

4   Check the electrode gaps of the new plugs. The gap is checked by inserting the proper thickness gauge between the electrodes at the tip of the plug **(see illustrations)**. The gap between the electrodes should be the same as that given in the Specifications or on the VECI label. If the gap is incorrect, use the notched adjuster on the feeler gauge body to bend the curved side electrode slightly.

5   If the side electrode is not exactly over the center electrode, use the notched adjuster to align them.

6   To prevent the possibility of mixing up spark plug wires, work on one spark plug at a time. Remove the wire from one spark plug. Grasp the boot at the end of the wire, not the wire itself. Sometimes it is necessary to use a twisting motion while pulling the plug wire boot off the spark plug.

7   If compressed air is available, use it to blow any dirt or foreign material away from the spark plug area. A common bicycle pump will also work.

8   Remove the spark plug.

9   Compare the spark plug with those shown on the inside back cover of this manual to get an indication of the overall running condition of the engine.

10   It's often difficult to insert spark plugs into their holes without cross-threading them. To avoid this possibility, fit a short piece of snug-fitting rubber tubing over the end of the spark plug. The flexible tubing acts as a universal joint to help align the plug with the plug hole. Should the plug begin to cross-thread, the hose will slip on the spark plug, preventing thread damage. Use a torque wrench, if available, to tighten the plug to the specified torque.

11   Attach the plug wire to the new spark plug, again using a twisting motion on the boot until it is firmly seated on the end of the spark plug.

12   Follow the above procedure for the remaining spark plugs, replacing them one at a time to prevent mixing up the spark plug wires.

---

36 Spark plug wire, distributor cap and rotor check and replacement

Refer to illustrations 36.11 and 36.12

1   The spark plug wires should be checked whenever new spark plugs are installed in the

35.4a Spark plug manufacturers recommend using a wire type gauge when checking the gap - if the wire does not slide between the electrodes with a slight drag, adjustment is required

35.4b To change the gap, bend the side electrode only, as indicated by the arrows, and be very careful not to crack or chip the porcelain insulator surrounding the center electrode

36.11 Examine the inside of the distributor cap closely for signs of carbon tracks and for wear at the terminals

engine.

2   Begin this procedure by making a visual check of the spark plug wires while the engine is running. In a darkened garage (make sure there is ventilation) start the engine and observe each plug wire. Be careful not to come into contact with any moving engine parts. If there is a break in the wire, you will see arcing or a small spark at the damaged area. If arcing is noticed, make a note to obtain new wires, then allow the engine to cool and check the distributor cap and rotor.

3   The spark plug wires should be inspected one at a time to prevent mixing up the order, which is essential for proper engine operation. Each original plug wire should be numbered to help identify its location. If the number is illegible, a piece of tape can be marked with the correct number and wrapped around the plug wire.

4   Disconnect the plug wire from the spark plug. A removal tool can be used for this purpose or you can grasp the rubber boot, twist the boot half a turn and pull the boot free. Do not pull on the wire itself.

5   Check inside the boot for corrosion, which will look like a white crusty powder.

6   Push the wire and boot back onto the end of the spark plug. It should be a tight fit on

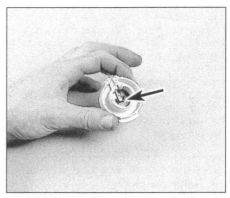

36.12 Make sure the rotor contact (arrow) isn't pitted or corroded excessively

the plug end. If it isn't, remove the wire and use pliers to carefully crimp the metal connector inside the wire boot until the fit is snug.

7   Using a clean rag, wipe the entire length of the wire to remove built-up dirt and grease. Once the wire is clean, check for burns, cracks and other damage. Do not bend the wire sharply, because the conductor might break.

8   Disconnect the wire from the distributor. Again, pull only on the rubber boot. Check for corrosion and a tight fit. Replace the wire in the distributor.

9   Inspect the remaining spark plug wires, making sure that each one is securely fastened at the distributor and spark plug when the check is complete.

10   If new spark plug wires are required, purchase a set for your specific engine model. Pre-cut wire sets with the boots already installed are available. Remove and replace the wires one at a time to avoid mixups in the firing order.

11   Remove the distributor cap by loosening the two phillips retaining screws. Look inside the distributor cap for cracks, carbon tracks and worn, burned or loose contacts **(see illustration)**.

12   Pull the rotor off the distributor shaft and examine it for cracks and carbon tracks **(see illustration)**.

37.3 With the engine cold, reach up inside the EGR housing (arrow) from underneath with your fingers and push up on the diaphragm - moderate pressure should move it upward

13 Replace the cap and rotor if defects are discovered. **Note:** *It is common practice to install a new cap and rotor whenever new spark plug wires are installed.*
14 When installing a new cap, remove the wires from the old cap one at a time and attach them to the new cap in the exact same location- do not simultaneously remove all the wires from the old cap or firing order mix-ups may occur.

37 Exhaust gas recirculation (EGR) valve checking

Refer to illustrations 37.3 and 37.6
1 The EGR valve is located on the intake manifold between the rocker arm cover and the throttle body on the rear side of the four-cylinder engine and between the intake duct and the exhaust crossover pipe on the left end of the V6 engine.
2 Usually, when a problem develops in this emissions system, it is due to a stuck or corroded valve.
3 With the engine cold to prevent burns, reach underneath the EGR valve and manu-

38.9 Aligned timing marks on the pointer and pulley

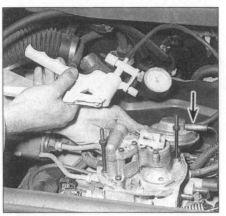

37.6 With the engine running, hook a vacuum pump up to the EGR at the intake vacuum fitting (arrow) and apply vacuum - the engine should sputter and die

ally push on the diaphragm inside the housing **(see illustration)**. You should be able to move the diaphragm with moderate pressure.
4 Disconnect the intake vacuum hose from the port on top of the EGR valve and attach a vacuum pump to the port.
5 Start the engine and allow it to reach normal temperature.
6 With the engine running, apply vacuum to the EGR valve with the vacuum pump **(see illustration)**.
7 The engine should sputter and die.
8 If the EGR valve fails either of the above tests, it must be replaced (Chapter 6).

38 Ignition timing check and adjustment

Refer to illustrations 38.6, 38.9 and 38.14
Note: *It is imperative that the procedures included on the Vehicle Emissions Control Information label be followed when adjusting the ignition timing. The VECI label includes all information concerning preliminary steps to be performed before adjusting the timing, as well as the timing specifications. The following procedure is typical of the timing proce-*

38.14 The distributor hold-down bolt (arrow) should be loosened just enough to rotate the distributor slightly

38.6 When checking the ignition timing, connect a jumper wire between terminals A and B of the diagnostic (ALCL) connector

dure for both four-cylinder and V6 engines, but it not necessarily the only one which might be outlined on the VECI label. If the procedure outlined on the VECI label of your vehicle is different, follow it.
1 Set the parking brake and block the drive wheels.
2 Start the engine and allow it to reach normal operating temperature.
3 Verify that the check engine light on the dashboard is off.
4 Switch the engine off when it's warmed up.
5 Remove the panel in the console between the seats by loosening the two retaining screws.
6 Ground the diagnostic connector in the console with a paper clip between terminals A and B **(see illustration)**.
7 The check engine light should begin flashing.
8 Connect the timing light in accordance with the manufacturer's instructions. Be careful not to tangle the wires in moving engine parts.
9 Clamp the timing light inductive pickup around the no. 1 spark plug wire, start the engine and point the timing light at the crankshaft timing marks located on the edge of the crank pulley **(see illustration)**. The stationary timing plate on the face of the timing chain cover has eight marks at two degree increments. When a stroboscopic timing light is pointed at the crank pulley and the stationary timing plate while the engine is running, the notch on the pulley will appear to be stationary and in close proximity to one of the marks on the plate. Record your reading.
10 Unclamp the timing light inductive pickup from the no. 1 spark plug wire and attach it to the no. 4 plug wire. Repeat the above step and record your second reading.
11 Add the timing figures you got for both cylinders and divide that sum by 2.
12 If the average of the two cylinders is not within the specified timing on your vehicle's VECI label, it must be reset.
13 Put the inductive pickup back on spark

plug wire no. 1.

14   Loosen the distributor hold-down bolt **(see illustration)** and rotate the distributor slightly to bring the timing within the figure specified on the VECI label. If the average of cylinders no. 1 and no. 4 was lower than that specified on the VECI label, rotate the distributor counterclockwise. If the average was too high, rotate it clockwise.

15   When the timing mark on the crankshaft pulley is at the specified number of degrees before TDC, tighten the distributor hold-down bolt.

16   Recheck the timing to make sure that it wasn't disturbed by tightening the distributor hold-down bolt.

17   Check the timing for the no. 1 and no. 4 cylinders again.

18   Add your readings together and divide by 2 again. Compare the average to the specified timing on the VECI label. It should be within specification. If it isn't, repeat the above procedure until it is.

19   Remove the paper clip from the diagnostic connector. Verify that the check engine light is off.

20   Replace the diagnostic connector cover panel in the console.

39   Automatic transaxle fluid and filter change

Refer to illustrations 39.5

1   Before beginning work, purchase the specified type and amount of transmission fluid (see Recommended lubricants, fluids and capacities), a gasket and a filter.

2   In order to remove any sediment buildup, the automatic transaxle fluid should be drained immediately after the vehicle has been driven. **Warning:** *Because fluid temperatures can exceed 350 degrees in a hot transaxle, protective gloves should be worn when performing the following procedure.*

3   Raise the vehicle and support it on jackstands.

4   Place a drain pan under the transaxle oil pan.

5   Remove all bolts from the oil pan except bolts A and B **(see illustration)**. **Note:** *A special bolt will be required to remove a Turbo 125C oil pan assembled with RTV sealant. This special bolt can be made from an oil pan bolt by grinding down a section of the shank diameter to approximately 3/16-inch just below the bolt head .*

6   Remove bolt A and install the special bolt in its place.

7   Loosen bolt B. With a rubber mallet, strike the oil pan corner. **Caution:** *Do not try to pry the oil pan loose from the case as damage to the pan flange or case will occur.*

8   Remove the special bolt and allow the fluid to drain.

9   Remove the remaining bolt and detach the oil pan. Remove the old filter and O-ring seal.

10   Inspect the oil pan and filter for foreign material such as metal particles, clutch facing

39.5 Remove all the automatic transaxle oil pan bolts except bolts A and B (arrows)

material, rubber particles and engine coolant. If any of the above are found, the transaxle should be inspected and, if necessary, overhauled by a dealer or transmission shop.

11   Clean the gasket mating surface of the transaxle case. Remove all traces of old gasket. If the pan was previously assembled with RTV sealant, use a sharp edge plastic scraper to remove any old sealant. Use a clean rag to dry the case.

12   Clean the oil pan and flanges with solvent and blow it dry with compressed air, if available, or a clean rag. Make sure no RTV sealant is left in the pan or on the flanges.

13   Install the new filter and O-ring seal. Coat the seal with petroleum jelly.

14   Install the oil pan, using a new gasket. Tighten the oil pan bolts to the specified torque. 1984 models built with the 125C transmission may use RTV sealer or gaskets on the oil pan and side cover. If the transmission has the new style oil pan or side cover and bolts with conical washers, it must be installed with a gasket. If the old style pan or side cover is used, it can also be installed with a gasket.

15   Lower the vehicle.

16   Fill the transaxle with the proper quantity and type of fluid (see *Recommended lubricants, fluids and capacities*) through the filler tube. Use a funnel to avoid spills. Add a little at a time and check the indicated level on the dipstick continually (Section 4). Allow the fluid time to drain into the transaxle oil pan.

17   With the vehicle on level ground, place the gear selector in Park and apply the parking brake.

18   Start the engine without depressing the accelerator pedal (if possible). Run it at a slow idle. Don't race it.

19   Move the gear selector through all gear positions.

20   Move the gear selector to Park and, with the engine running at idle, check the fluid level. It should be in the cross-hatched area on the dipstick. **Caution:** *Do not overfill the transaxle. Overfilling causes foaming and loss of fluid through the vent and may damage the transaxle.*

40.2 Examine the gasoline filler cap for corrosion and damage and make sure the sealing ring (arrow) is unbroken

21   Look under the vehicle for leaks around the transaxle oil pan mating surface.

22   Push the dipstick firmly back into its tube and drive the vehicle far enough to reach normal operating temperature in the transmission. This should take approximately 15 miles of highway driving or slightly less in the city.

23   Park the vehicle on a level surface and check the underside of the transaxle for leaks.

24   Check the fluid level on the dipstick with the engine idling and the transaxle in Park. The level should now be at the upper end of the crosshatched area on the dipstick. If it isn't, add fluid to bring the level to this point. Again, do not overfill.

40   Fuel system check

Refer to illustration 40.2

Warning 1: *Gasoline is extremely flammable, so take extra precautions when you work on any part of the fuel system. Don't smoke or allow open flames or bare light bulbs near the work area, and don't work in a garage where a natural gas-type appliance (such as a water heater or a clothes dryer) with a pilot light is present. Since gasoline is carcinogenic, wear latex gloves when there's a possibility of being exposed to fuel, and, if you spill any fuel on your skin, rinse it off immediately with soap and water. Mop up any spills immediately and do not store fuel-soaked rags where they could ignite. When you perform any kind of work on the fuel system, wear safety glasses and have a Class B type fire extinguisher on hand.*

Warning 2: *Before removing the fuel filter the fuel system pressure must be relieved (Chapter 4).*

1   If you smell gasoline while driving or after the vehicle has been sitting in the sun, inspect the fuel system immediately.

2   Remove the gas filler cap and inspect it for damage and corrosion. The gasket should have an unbroken sealing imprint **(see illustration)**.

41.5 To remove the hub assembly without dropping the tapered roller bearing in the middle, put your thumb over the bearing and pull the hub towards you, then push the hub back onto the spindle

3   Inspect the fuel feed and return lines for cracks. Make sure that the threaded connectors which secure the fuel lines to the fuel injection system and to the in-line fuel filter are tight.

4   Since some components of the fuel system - the fuel tank and part of the fuel feed and return lines for example - are underneath the vehicle, they can be checked more easily with the vehicle raised on a hoist. If that's not possible, raise the vehicle and secure it on jackstands.

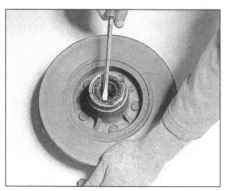

41.7 Pry the dust seal off the inner race to get at the inner wheel bearing

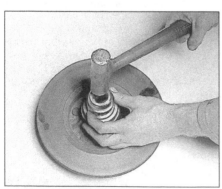

41.12 Install a new race with a large socket of a slightly smaller diameter than the outside diameter of the race

5   With the vehicle raised, inspect the gas tank and filler neck for punctures, cracks and other damage. The connection between the filler neck and the tank is particularly critical. Sometimes a rubber filler neck will leak because of loose clamps or deteriorated rubber. These are problems a home mechanic can usually rectify. **Warning:** *Do not, under any circumstances, try to repair a fuel tank (except rubber components). A welding torch or any open flame can easily cause fuel vapors inside the tank to explode.*

6   Carefully check all rubber hoses and metal lines leading away from the fuel tank. Check for loose connections, deteriorated hoses, crimped lines and other damage. Carefully inspect the lines from the tank to the fuel injection system. Repair or replace damaged sections as necessary (Chapter 4).

41   Wheel bearing adjustment and lubrication

Refer to illustrations 41.5, 41.7, 41.10, 41.12 and 41.19

1   Raise the vehicle on a hoist or jack up the front end and place it on jackstands.
2   Remove the wheel and tire.
3   Remove the brake caliper from the knuckle and suspend it by a wire from the upper arm (refer to Chapter 9).
4   Remove the hub dust cup, cotter pin, spindle nut and washer.

41.10 If either race is scored or galled, knock it out with a brass drift punch

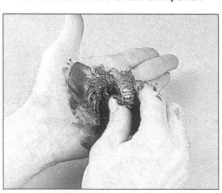

41.19 When packing a bearing, work the grease in between all the rollers and the cage

5   Remove the hub and bearing. **Caution:** *Do not allow the bearing to fall out of the hub when removing the hub from the spindles* **(see illustration).**
6   Remove the outer bearing.
7   Remove the inner bearing by prying out the grease seal **(see illustration).** Discard the old seal.
8   Repeat steps 2 through 7 for the other wheel.
9   Wash all parts thoroughly in solvent.
10   Look at the old races. If either of them is galled or scored, drive it out of the hub with a brass drift inserted behind the race in the notch in the hub **(see illustration).**
11   Lubricate the new race with a light film of grease.
12   Start the race squarely into the hub and carefully seat the race using an appropriate tool **(see illustration).**
13   Check the bearings for a cracked cage or pitting.
14   If it is necessary to replace either the outer or inner bearing, also replace the race for that bearing.
15   Clean off any grease in the hub and spindle.
16   Use high temperature front wheel bearing grease (see *Recommended lubricants, fluids and capacities*) to lubricate wheel bearings.
17   Apply a thin film of grease to the spindle at the outer bearing seat and at the inner bearing seat, shoulder and seal seat.
18   Put a small quantity of grease inboard of each bearing race in the hub. This can be applied with your finger forming a dam to provide extra grease availability to the bearing and to keep thinned grease from flowing out of the bearing.
19   Pack the bearing with grease. It is extremely important to work the grease thoroughly into the area between the rollers **(see illustration).**
20   Place the inner bearing in the hub. Using your finger, put an additional quantity of grease inboard of the bearing.
21   Install a new grease seal using a flat plate until the seal is flush with the hub. Lubricate the seal lip with a thin layer of grease.
22   Carefully install the hub and rotor assembly. Place the outer bearing in the outer bearing race. Install the washer and nut.
23   Tighten the spindle nut to the specified torque while turning the wheel hub assembly forward by hand to fully seat the bearings.
24   Back off the nut to the "just loose" position.
25   Hand tighten the spindle nut. Loosen the spindle nut until either hole in the spindle lines up with a slot in the nut (not more than 1/2 flat).
26   Install a new cotter pin.
27   Install the dust cap on the hub.
28   Install the brake caliper.
29   Install the wheel and tire.
30   Lower the vehicle to the ground.

Chapter 2 Part A
Four-cylinder engine

Contents

Specifications

Torque specifications

	Ft-lbs (unless otherwise indicated)
Oil pan bolt	
1984	75 in-lbs
1985	54 in-lbs
1986 through 1988	90 in-lbs
Oil pan drain plug	25
Oil screen support block	37
Oil pump-to-block bolt	22
Pushrod cover-to-engine block nut	90 in-lbs
Harmonic balancer/crankshaft pulley hub bolt	
1984 and 1985	200
1986 through 1988	162
Transaxle-to-engine block	55
Flywheel-to-crankshaft bolt	
1984 through 1986	44
1987 and 1988	
Manual transaxle	55
Automatic transaxle	69

**Firing order
1-3-4-2**

**2.5L
4-CYLINDER
ENGINE**

Cylinder location and distributor rotation

*The blackened terminal shown on the distributor cap
indicates the Number One spark plug wire position*

Torque specifications (continued)

Ft-lbs (unless otherwise indicated)

Intake manifold-to-cylinder head bolt **(refer to illustration 5.20)**	
1984 and 1985	29
1986	
Bolt 1...	25
Bolt 2...	37
Bolt 3...	28
1987 and 1988	25
Exhaust manifold-to-cylinder head bolt **(refer to illustration 8.13)**	
1984 through 1986 (all bolts)..	44
1987 and 1988	
Bolts 1, 2, 6 and 7...	31
Bolts 3, 4 and 5...	37
EGR valve-to-manifold bolt	
1984 and 1985	10
1986 through 1988	16
Water outlet housing bolt	20
Thermostat housing bolt..	20
Water pump-to-engine block bolt	25
Timing cover bolt	90 in-lbs
Radiator hose clamps...	17 in-lbs
Rocker arm bolt	
1984 and 1985	20
1986 through 1988	24
Rocker arm cover bolt	45 in-lbs
Cylinder head-to-block bolt **(refer to illustration 9.19)**	
1984 and 1985	92
1986 and 1987	
Step 1	18
Step 2	
All but bolt 9	22
Bolt 9	29
Step 3	
All but bolt 9	Turn an additional 120 degrees
Bolt 9	Turn an additional 90 degrees
1988	
Step 1	18
Step 2	
All but bolt 9	26
Bolt 9	18
Step 3 (all bolts)	Turn an additional 90 degrees
Force balancer assembly-to-block bolts	
Short bolts	
Step 1	108 in-lbs
Step 2	Turn an additional 75 degrees
Long bolts	
Step1	108 in-lbs
Step 2	Turn an additional 90 degrees
Strut rod bolts..	42
Cradle bolts	
Front	67
Rear	76
Engine mounts	
1984 and 1985 engine mount-to-cradle nuts...................................	41
1986 through 1988 engine mount-to-cradle nuts	
Automatic...	41
Manual	35
Automatic transaxle mount and support bracket (1986 through 1988)	
Mount-to-bracket nuts	33
Upper bracket-to-transaxle case bolt	40
Bracket-to-transaxle case bolt..	55
Transaxle-to-bracket nut...	18
Later model vehicles equipped with automatic transaxles have one large support bracket	
Automatic transaxle mounts and support brackets (1984, 1985 and some 1986)	
Forward support bracket-to-transaxle bolt......................................	48
Forward mount-to-support bracket nut	41
Forward transaxle insulator-to-crossmember nut............................	41
Rear support bracket-to-transaxle case bolts	47
Rear mount-to-support bracket nut...	41

1   General information

This part of Chapter 2 is devoted to in-vehicle repair procedures and engine removal and installation procedures for the 2.5L four-cylinder engine. Information regarding engine block and cylinder head servicing is in Part C of this Chapter.

The repair procedures are based on the assumption that the engine is still installed in the vehicle. So if you are performing a complete overhaul; that is the engine is already out of the vehicle and mounted on a stand, many of the steps outlined in this part of Chapter 2 will not apply.

The specifications included in this part of Chapter 2 apply only to the procedures contained in this part. Part C of Chapter 2 contains the specifications necessary for engine block and cylinder head rebuilding.

2   Repair operations possible with the engine in the vehicle

Many major repair operations can be accomplished without removing the engine from the vehicle.

It is a good idea to clean the engine compartment and the exterior of the engine with some type of pressure washer before any work is begun. A clean engine will make the job easier and will prevent the possibility of getting dirt into internal areas of the engine.

Remove the engine compartment lid (Chapter 11) and cover the fenders to provide as much working room as possible and to prevent damage to the painted surfaces.

If oil or coolant leaks develop, indicating a need for gasket or seal replacement, the repairs can generally be made with the engine in the vehicle. The oil pan gasket, the cylinder head gasket, intake and exhaust manifold gaskets, timing cover gaskets and the front crankshaft oil seal are accessible with the engine in place.

Exterior engine components such as the water pump, the starter motor, the alternator, the distributor and the fuel injection (TBI or MPFI), as well as the intake and exhaust manifolds, can be removed for repair with the engine in place.

Since the cylinder head can be removed without pulling the engine, valve component servicing can also be accomplished with the engine in the vehicle.

Replacement of repairs to or inspection of the timing gears and the oil pump are all possible with the engine in place.

In extreme cases caused by a lack of necessary equipment, repair or replacement of piston rings, pistons, connecting rods and rod bearings and reconditioning of the cylinder bores is possible with the engine in the vehicle. However, this practice is not recommended because of the cleaning and preparation work that must be done to the components involved.

Detailed removal, inspection, repair and installation procedures for the above mentioned components can be found in the appropriate Part of Chapter 2 or the other Chapters in this manual.

3   Rocker arm cover - removal and installation

Refer to illustrations 3.3, 3.4, 3.7, 3.8 and 3.9

1   Remove the air cleaner assembly (refer to Chapter 4).

2   Disengage the throttle cable from its bracket and put it aside. Carefully note the exact relationship of the cable and hardware components to one another to ensure correct installation.

3   Disconnect the PCV valve filter, rubber elbow and hose from the rocker arm cover and TBI assembly **(see illustration)**.

4   Label each spark plug wire before removal to ensure correct installation. Unsnap the retaining clips **(see illustration)**

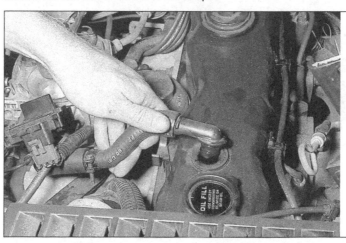

3.3 Pull the PCV valve hose from the rocker arm cover

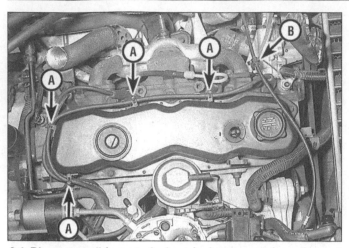

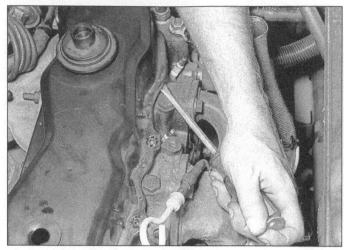

3.4 Disconnect all four spark plug wire harness retainer clips and the exhaust manifold throttle cable retainer from the rocker arm cover

A   Spark plug wire clips   B   Throttle cable retainer

3.7 The rocker arm cover flange is glued to the cylinder head with RTV sealant - if you have to pry it loose, try to avoid using too much force or you will bend the flange

on the rocker arm cover to free the wire harness from the cover. Unplug the wires from the spark plugs and from the distributor terminals and set the wire harness aside.

5   Loosen the throttle body mounting nut and bolts to provide clearance for removal of the EGR valve, then remove the EGR valve (Refer to Chapter 6).

6   Remove the rocker arm cover bolts.

7   Remove the rocker arm cover. If the cover is stuck to the cylinder head, use a soft-face hammer or a block of wood and a hammer to dislodge it. If the cover still won't come loose, pry on it carefully at several points until the sealant is broken loose **(see illustration)**, but do not distort the sealing flange surface. Gently pry on the flange at several locations to avoid bending it. **Note:** *If you bend the flange, set it on a flat surface after removal and tap it with a soft-face hammer until it's flat again.*

8   Clean all dirt, oil and old sealant material from the sealing surfaces of the cover and cylinder head with a scraper and degreaser **(see illustration)**.

9   Apply a continuous 3/16-inch wide bead

of RTV-type sealant to the sealing flange of the cover. Be sure to apply the sealant inboard of the bolt holes **(see illustration)**.

10   Place the rocker arm cover on the cylinder head while the sealant is still wet and install the mounting bolts. Tighten the bolts a little at a time to the specified torque.

11   Install the EGR valve and tighten the bolts to the specified torque.

12   Tighten the throttle body mounting nut and bolts to the specified torque.

13   Install the spark plug wire boots onto the spark plugs and the distributor terminals in the same order in which they were removed. Snap the wire harness retaining clips back onto the rocker arm cover.

14   Install the PCV valve filter, rubber hose and elbow between the rocker arm cover and the throttle body assembly.

15   Push the throttle cable grommet back into its bracket.

16   Install the air cleaner assembly. **Note:** *Make sure that the manifold absolute pressure (MAP) sensor vacuum line and the Thermac motor temperature sensor line are properly installed.*

4   Valve train components - replacement (cylinder head in place)

Refer to illustrations 4.2, 4.4a, 4.4b, 4.8 and 4.10

1   Remove the rocker arm cover (Section 3).

2   If only a pushrod is to be replaced, loosen the rocker arm bolt enough to allow the rocker arm to be rotated to the side so that it will clear the pushrod. Pull the pushrod out of the hole in the cylinder head **(see illustration)**.

3   If a valve spring or valve guide oil seal is to be replaced, remove the rocker bolt and pivot and lift off the rocker arm. Then remove the spark plug from the affected cylinder.

4   If all the pushrods and rocker arms are removed, they must be kept in the proper order for reinstallation. The best way to organize them is to store them in clearly labeled boxes **(see illustration)**. **Note:** *Do not mix up the pushrod guides* **(see illustration)**. *They must be kept in order for proper reinstallation.*

5   There are two ways to hold the valve in place while the valve spring is removed. If

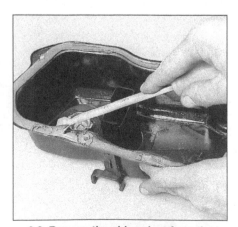

3.8 Remove the old sealant from the rocker arm cover flange with a gasket scraper

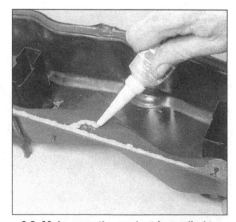

3.9 Make sure the sealant is applied to the INSIDE of the bolt holes or oil will leak out around the bolt threads

4.2 Loosen the rocker arm bolt, rotate the rocker arm, then lift out the pushrod

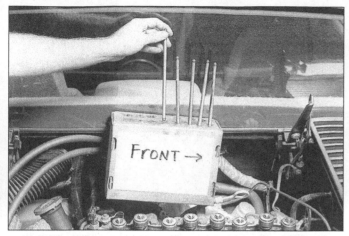

4.4a If you are removing more than one pushrod, store them in a clearly labeled box to prevent mixups during reinstallation

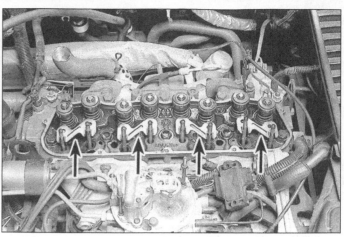

4.4b Make sure that the pushrod guides are kept in order too, if they are removed

you have access to compressed air, install an air hose adapter in the spark plug hole. These adapters are sold by most auto parts stores. Most quality compression gauges have a lower extension hose that will serve the same purpose. One end of the extension screws into the spark plug hole and the other end has a quick-connect air fitting that attaches to the compression gauge. This quick-connect fitting can be attached to a compressor air hose. Bring the piston to top dead center on the firing stroke to ensure that both valves are closed before applying air pressure. Once the cylinder is pressurized, the valves will remain closed while the spring is removed.

6   If you don't have access to compressed air, bring the piston of the appropriate cylinder to approximately 45-degrees before top dead center on the compression (firing) stroke. Feed a long piece of 1/4-inch rope or cord into the cylinder through the spark plug hole until it fills the combustion chamber. Be sure to leave the end of the rope hanging out of the spark plug hole so that it can be removed when the procedure is completed.

7   Turn the crankshaft with a wrench in the normal direction of rotation until a slight resistance is felt. This will be the rope compressing

between the piston top and the valves, effectively holding the valves closed while the valve spring is removed. **Caution:** *Make sure before inserting the rope that the piston is coming up on the compression stroke with both valves closed. If the rope is inserted on the exhaust stroke when the exhaust valve is open to expel the burned gas and the intake valve is starting to open to allow a fresh charge into the engine, the valves can easily be bent when the piston is brought up to compress the rope.*

8   Install a valve spring compressor over the valve spring **(see illustration)**.

9   Depress the spring and retainer and remove the keeper halves, then release pressure on the tool.

10   Remove the valve spring compressor, valve stem O-ring seal, retainer, spring and damper (inner spring). Note the relationship between the components before removing them **(see illustration)**.

11   Assemble the spring, damper, shield and retainer, place them in position and compress them again with the spring compressor.

12   Carefully install the O-ring seal in the lower groove of the valve stem. Make sure the seal is not twisted - it must lie perfectly flat in the groove.

13   Position the keepers in the upper groove. If necessary, apply a small dab of grease to each keeper to hold it in place.

14   Carefully release the spring compressor pressure. Be sure that the keepers are properly seated.

15   Coat the bearing surfaces of the rocker arm and pivot with moly-base grease or engine assembly lube.

16   Install the rocker arm. The engine valve mechanism requires no special valve lash adjustment. Simply tighten the rocker arm bolt to the specified torque.

5   Intake manifold - removal and installation

Refer to illustrations 5.8a, 5.8b, 5.11, 5.12, 5.15, 5.16 and 5.20

Removal

1   Disconnect the cable from the negative battery terminal.

2   Remove the air cleaner assembly. Tag each hose with a piece of numbered or colored tape as it is disconnected to simplify installation.

4.8 A lever-type valve spring compressor is used to compress the spring and remove the keepers to replace seals or springs while the head is installed

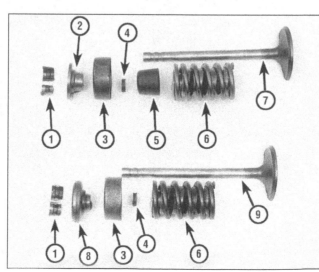

4.10 The components of a typical valve spring assembly

1   Keepers
2   Retainer
3   Oil shield
4   O-ring stem seal
5   Umbrella or positive type seal
6   Spring and damper
7   Intake valve
8   Retainer/rotator
9   Exhaust valve

5.8a Pry the linkage rod from the pivot ball using a flat bladed screwdriver (arrow) at the bottom . . .

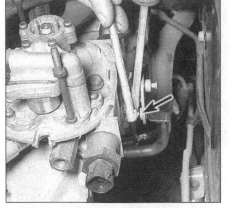

5.8b . . . and at the top (arrow) of the linkage assembly

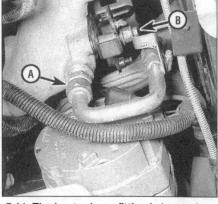

5.11 The heater hose fitting is located on the underside of the intake manifold (arrow A) and is supported by a bracket attached to the throttle linkage pivot (arrow B)

3   Relieve the fuel pressure (Chapter 4).
4   Drain the cooling system (Chapter 1).
5   Remove the PCV valve and hose at the throttle body housing.
6   Label and disconnect the fuel line fittings at the throttle body and remove the fuel filter bracket bolt underneath the filter.
7   Label and disconnect the vacuum lines and electrical leads from the fuel injection assembly.
8   Disconnect the throttle cable from the throttle linkage by prying the C-clip off with a screwdriver. Remove the linkage rod by popping it off with a screwdriver (see illustrations). Carefully note how the linkage rod is installed before removing it.
9   If your vehicle is equipped with an automatic transmission, disconnect and remove the transaxle downshift linkage. Note how the components are installed.
10   If your vehicle is equipped with a cruise control, disconnect the linkage.
11   Disconnect the heater hose fitting (see illustration).
12   Remove the forward strut rod through-bolt at the end adjacent to the cylinder head. Remove the access plate in the top of the

right fender well (see illustration) and remove the rear through bolt. Remove the strut rod.
13   Disconnect the alternator by removing the belt tensioning nut and bolt. Remove the alternator bracket.
14   Disconnect the coil mounting nuts, coil lead and coil wire and remove the coil.
15   Remove the intake manifold mounting bolts and separate the manifold from the cylinder head (see illustration). Do not pry between the manifold and the head as damage to the gasket sealing surfaces may result.
16   Remove the intake manifold gasket with a gasket scraper (see illustration).

Installation

17   If the intake manifold itself is being replaced, transfer all components still attached to the old manifold onto the new one.
18   Clean the cylinder head and manifold gasket surfaces. All gasket material and sealing compound must be removed prior to installation.
19   Place a new intake manifold gasket on the manifold, hold the manifold in position

5.12 To get at the rear strut rod through-bolt, remove the access cover in the top of the right fender well

against the cylinder head and install the mounting bolts finger tight.
20   Tighten the mounting bolts a little at a time in the proper sequence (see illustration) to the specified torque.

5.15 Location of the intake manifold mounting bolts

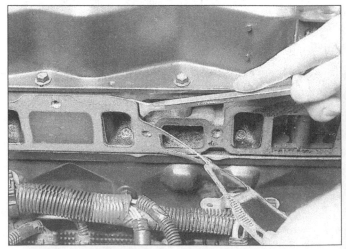

5.16 Remove the old intake manifold gasket with a gasket scraper - don't leave any material on the mating surface

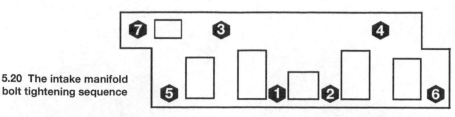

5.20 The intake manifold bolt tightening sequence

24070-2c-8.15 HAYNES

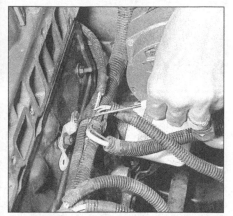

6.3 Disconnect the wire harness clip and push it out of the way to get at the pushrod cover retaining nuts

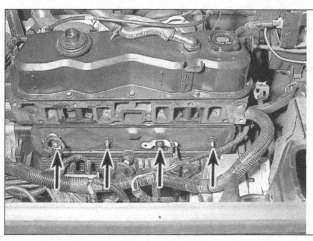

6.4 There are four pushrod cover retaining nuts (arrows)

21   Push the electrical connectors back up through the intake manifold.
22   Install the alternator bracket. Install the alternator belt tension adjustment nut and bolt and adjust the belt (Chapter 1).
23   Install the strut rod and both the through-bolts. Tighten the throughbolts to the specified torque. Replace the fender well through-bolt access plate.
24   Place the TBI linkage rod into place and reattach the throttle cable.
25   Reattach the heater hose fitting to the underside of the intake manifold.
26   Reconnect the fuel feed and return lines. Install the fuel filter bracket mounting bolt.
27   Reattach the PCV valve, elbow and hose between the rocker arm cover and the TBI housing.

28   Reattach the remaining vacuum lines and electrical connectors.
29   Install the coil assembly. Reconnect the coil lead and coil wire.
30   Install the air cleaner (Chapter 4).
31   Fill the radiator with coolant, start the engine and check for leaks.

6   Pushrod cover - removal and installation

Refer to illustrations 6.3, 6.4, 6.8 and 6.9

1   Remove the air cleaner assembly (Chapter 4). Tag each hose with a piece of numbered or colored tape as it is disconnected to ensure correct installation.
2   Remove the intake manifold (Section 5).
3   Remove the wire harness from the pushrod cover **(see illustration)**.
4   Loosen the four pushrod cover nuts **(see illustration)**.
5   Remove the pushrod cover by carefully prying it off with a screwdriver. **Caution:** *Careless prying may damage the sealing surface of the cover. If you bend the cover during removal, place it on a flat surface and straighten it with a soft-face hammer.*
6   Clean the sealing surfaces on the pushrod cover and cylinder block with a gasket scraper and solvent.
7   Apply a continuous 3/16-inch diameter bead of RTV-type sealant to the sealing surface of the pushrod cover.
8   Install new rubber pushrod cover mounting stud washers **(see illustration)**.
9   Install the cover while the sealant is still wet. Make sure that the semi-circular cutout in the edge of the pushrod cover is facing down **(see illustration)**.
10   Tighten the retaining nuts gradually until they're snug, then tighten them to the specified torque.
11   Reattach the wire harness to the pushrod cover.
12   Install the intake manifold (Section 5).
13   Install the air cleaner (Chapter 4).

7   Hydraulic lifters - removal, inspection and installation

Refer to illustrations 7.7, 7.8 and 7.9

Removal

1   A noisy valve lifter can be isolated when the engine is idling. Place a length of hose or tubing near the position of each valve while listening at the other end of the tube. Or remove the rocker arm cover and, with the engine idling, place a finger on each of the valve spring retainers, one at a time. If a valve lifter is defective, it will be evident from the shock felt at the retainer as the valve seats.
2   The most likely cause of a noisy valve lifter is a piece of dirt trapped between the plunger and the lifter body.
3   Remove the rocker arm cover (Section 3).

6.8 Don't forget to install new rubber sealing washers around the pushrod cover mounting studs or oil will leak past the studs

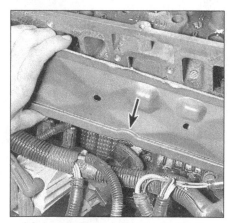

6.9 Install the pushrod cover while the sealant is still tacky - be sure that the semi-circular cutout (arrow) is facing down

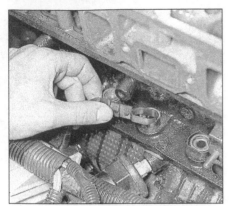

7.7 Remove the lifter guide - if you are removing more than one guide, keep them in order to ensure that they are installed in the same location from which they were removed

7.8 On newer engines which haven't become sticky with sludge and varnish, the lifters can usually be removed without any special tool

7.9 If you are removing more than one lifter, keep them in order with a clearly labeled box

4   Loosen both rocker arm bolts at the cylinder with the noisy lifter and rotate the rocker arms away from the pushrods. Remove the pushrod guide plates and pushrods (Section 4).

5   Remove the intake manifold (Section 5).

6   Remove the pushrod cover (Section 6).

7   Remove the lifter guide retainer by unscrewing the pushrod cover studs. Remove the lifter guide **(see illustration)**.

8   There are several ways to extract a lifter from its bore. A special hydraulic lifter removal tool is available, but isn't absolutely necessary. On newer engines without a lot of varnish buildup, lifters can often be removed with a small magnet or even with your fingers **(see illustration)**. A small scribe can also be used to pry the lifter out of the bore. **Caution:** *Do not use pliers of any type to remove a lifter unless you intend to replace it with a new one because they will damage the precision machined and hardened surface finish of the lifter, rendering it useless.*

9   Store the lifters in a clearly labeled box to insure their reinstallation in the same lifter bores **(see illustration)**.

Inspection

10   It is easier to simply replace a worn lifter with a new one than to repair a defective lifter. The internal components are not available separately - you must buy a lifter as a complete assembly. But sometimes, disassembling and cleaning the internal components of a dirty lifter will restore normal operation. For complete disassembly, inspection and reassembly procedures, refer to Chapter 2C.

Installation

11   The lifters must be installed in their original bores. Coat each lifter foot with moly-base grease or engine assembly lube.

12   Lubricate the bearing surfaces of the lifter bores with engine oil.

13   Install the lifter(s) in the lifter bore(s). **Note:** *Make sure that the oil orifice is facing toward the front of the engine.*

14   Install the lifter guide(s) and retainer(s).

15   Install the pushrods, pushrod guide plates, rocker arms and rocker arm retaining bolts (Section 4). **Caution:** *Make sure that each pair of lifters is on the base circle of the camshaft; that is, with both valves closed, before tightening the rocker arm bolts.*

16   Tighten the rocker arm bolts to the specified torque.

17   Install the pushrod cover (Section 6).

18   Install the intake manifold (Section 5).

19   Install the rocker arm cover (Section 3).

20   Install the air cleaner (Chapter 4).

8   Exhaust manifold - removal and installation

Refer to illustration 8.13

1   Remove the air cleaner assembly (Chapter 4) and the pre-heat ducting between the exhaust manifold and the air cleaner.

2   Remove the EFI throttle cable bracket bolt, the dipstick tube bracket bolt and the ground strap bolt.

3   Raise the vehicle and support it securely on jackstands.

4   Remove the catalytic converter splash shield.

5   Disconnect the exhaust pipe-to-exhaust manifold flange bolts. **Note:** *These bolts are often corroded, so you may have to apply penetrating oil to break them loose.*

6   If necessary, support the exhaust system with a piece of wire. If all the mounting springs are in place and in good shape, the

exhaust system will remain in place without additional support.

7   Remove the oxygen sensor connector wire from its retaining clip and disconnect the electrical connector from the wire harness. **Note:** *It is not necessary to remove the oxygen sensor from the manifold unless you are replacing the manifold. If you do remove the sensor from the old manifold, be sure to install it with an anti-seize compound in the new one.*

8   Label the four spark plug wires, then disconnect them and set them out of the way.

9   Loosen and remove the exhaust manifold retaining bolts and washers. Loosen the end ones first and then the middle ones. You may have to apply penetrating oil to the fastener threads, because they're often corroded.

10   Remove the exhaust manifold and gasket from the engine. **Note:** *The oil dipstick tube may be knocked loose during removal of the exhaust manifold. Be sure to push it back into the block before reinstallation of the exhaust manifold.*

11   Clean the gasket mating surfaces on the cylinder head and the exhaust manifold with a wire brush. All leftover gasket material and carbon deposits must be removed.

12   Place a new exhaust manifold gasket in position on the cylinder head, then place the manifold in position and install the mounting bolts finger tight. **Note:** *You can drop the factory gasket into position with the exhaust manifold already in place and the bottom retaining bolts loosely installed.*

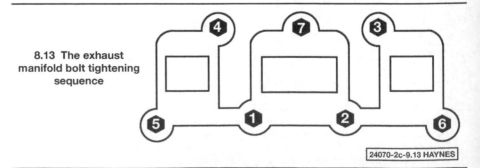

8.13 The exhaust manifold bolt tightening sequence

24070-2c-9.13 HAYNES

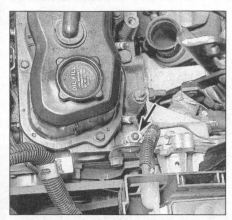

9.7 There is a ground strap (arrow) on the right end of the cylinder head that must come off before removal of the head

9.10b Push the head bolts through the "bolt holes" of your cardboard tracing in the exact locations they occupy in the head

13 Tighten the exhaust manifold bolts in the correct sequence **(see illustration)**.
14 Reattach the four spark plug wires.
15 Reconnect the oxygen sensor lead wire to the wire harness.
16 Install a new exhaust manifold flange gasket between the exhaust manifold and the exhaust pipe.

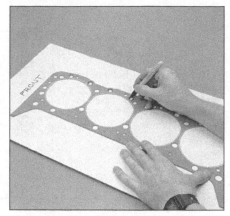

9.10a To store cylinder head bolts in the right order, use a new cylinder head gasket to trace the bolt hole pattern onto a piece of cardboard

17 Tighten the exhaust manifold-to-exhaust pipe flange bolts.
18 Install the catalytic converter splash shield and tighten it securely.
19 Install the EFI throttle cable bracket and tighten it securely.
20 Install the pre-heat ducting between the exhaust manifold and the air cleaner.
21 Install the air cleaner assembly (Chapter 4).
22 Start the engine and check for exhaust leaks between the exhaust manifold and cylinder head and between the exhaust manifold and the exhaust pipe.

9 Cylinder head - removal and installation

Refer to illustrations 9.7, 9.10a, 9.10b, 9.11, 9.15, 9.19 and 9.22

Removal

1 Remove the cable from the negative terminal of the battery.
2 Drain the cooling system (Chapter 1).
3 Remove the air cleaner assembly (Chap-

ter 4). Tag each hose with numbered or colored tape as it is disconnected to facilitate reassembly.
4 Detach the exhaust manifold from the cylinder head (Section 8).
5 Remove the intake manifold (Section 5).
6 Disconnect the electrical connectors and the radiator hose at the left end of the cylinder head.
7 Remove the ground strap at the right end of the cylinder head **(see illustration)**.
8 Remove the rocker arm cover (Section 3).
9 Remove the rocker arms and pushrods (Section 4).
10 Using the new head gasket, trace an outline of the cylinders and the bolt pattern onto a piece of cardboard **(see illustration)**. Loosen the head bolts 1/4-turn at a time each until the can be removed by hand. Remove the cylinder head bolts. Insert the bolts into the cardboard outline in the same order in which they are installed in the head **(see illustration)**.
11 Using a hammer and a block of wood, tap the cylinder head free and remove it. If the cylinder head is stuck to the engine block, pry it free only at the overhang on the thermostat end of the head **(see illustration)**. **Caution:** *If you pry on the cylinder head anywhere else, you may damage the sealing surfaces.*
12 Place the head on a block of wood to prevent damage to the gasket surface and/or valves.
13 Remove the cylinder head gasket and discard it.

Installation

14 If a new cylinder head is being installed, transfer all of the external components from the old cylinder head to the new one.
15 Stuff clean rags into the cylinders to prevent debris from falling into them. Remove all dirt, oil and old gasket material from the gasket surfaces of the cylinder head and block **(see illustration)**. The surfaces must also be free of nicks and heavy scratches. Clean up all retaining bolt threads and cylinder block threaded holes with a tap and die

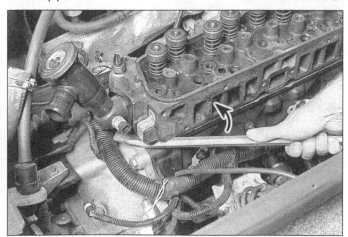

9.11 If you can't knock the head loose, pry it up with a large screwdriver at the overhang just behind and below the thermostat housing on the left end of the head

9.15 Once the head is off, stuff the cylinders with clean shop rags to prevent debris from falling into them and scrape off the old gasket material with a gasket scraper

set. Dirt will affect bolt torque.

16   Lay a new gasket on the cylinder block, making sure that the gasket is positioned correctly.

17   Carefully lower the cylinder head into place, making sure that it is positioned over the dowel pins.

18   Coat the threads of the cylinder head retaining bolts and the under sides of the bolt heads with sealing compound. Install them in the cylinder head finger tight. Do not tighten any of the bolts at this time.

1984 and 1985 engines

19   Tighten the cylinder head bolts in at least three steps, following the sequence shown, until the specified torque is reached **(see illustration)**.

1986 and 1987 engines

20   Gradually tighten the cylinder head bolts in the sequence shown in **illustration 9.19** to the step 1 torque figure shown in the Specifications.

21   Repeat the sequence, bringing them to the second specified torque on all bolts except number 9. Tighten number 9 to its specified torque.

22   Repeat the sequence. Turn all bolts, except number 9, 120 degrees (1/3-turn). Turn number 9 an additional 90 degrees (1/4-turn) **(see illustration)**.

23   Install the pushrods and the rocker arms (Section 4).

24   Install the rocker arm cover (Section 3).

25   Reconnect the radiator hose to the thermostat housing.

26   Install the intake manifold (Section 5).

27   Install the exhaust manifold (Section 8).

28   Install the air cleaner assembly (Chapter 4).

29   Refill the cooling system with fresh coolant (Chapter 1).

30   Connect the cable to the negative terminal of the battery.

31   Start the engine and check for oil, compression and coolant leaks.

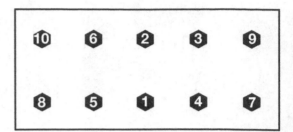

9.19 The cylinder head bolts must be tightened in three stages to the torque listed in this Chapter's Specifications using the numerical sequence shown here

24070-2c-10.25 HAYNES

10   Crankshaft pulley and hub - removal and installation

Refer to illustration 10.5

1   Remove the cable from the negative battery terminal.

2   Loosen the accessory drivebelt tension adjusting bolts and remove the drivebelts (Chapter 1). Tag each belt as it is removed to simplify reinstallation.

3   Jack up the vehicle and place it securely on jackstands.

4   Remove the right rear wheel.

5   Remove the right rear inner fender splash shield by pulling out the plastic pop fasteners **(see illustration)**.

6   If your vehicle is equipped with a manual transaxle, apply the parking brake and put the transaxle in gear to prevent the engine from turning over, then remove the crank pulley bolts. If your vehicle is equipped with an automatic transaxle, remove the starter motor (Chapter 5) and immobilize the starter ring gear with a large screwdriver. **Note:** *On both manual and automatic transaxle equipped vehicles, it may be necessary to use a breaker bar because the bolts are very tight.*

7   To break the crankshaft hub retaining bolt loose, thread a bolt into one of the pulley bolt holes. Place a large breaker bar and socket on the crankshaft hub retaining bolt. Immobilize the hub by wedging a large screwdriver between the bolt and the socket, then remove the bolt.

8   Remove the crank hub. Use a puller if necessary.

9   Carefully pry the oil seal out of the front cover with a large screwdriver. **Caution:** *Do not gouge or distort the cover or it won't seal properly around the new seal.*

10   Install the new seal with the lip toward the engine. Drive the seal into place using a seal installation tool or a large socket. A block of wood and a hammer will also work.

11   Apply a thin layer of multi-purpose grease to the seal contact surface of the hub.

12   Position the pulley hub on the crankshaft nose and slide it on until it bottoms against the crankshaft timing gear. Note that the slot in the hub must be aligned with the Woodruff key in the end of the crankshaft. The crankshaft hub bolt can also be used to press the hub into position.

13   Tighten the hub-to-crankshaft bolt to the specified torque.

14   Install the crank pulley on the hub. Coat the pulley-to-hub bolts with thread locking compound before installation.

9.22 The final step in tightening the cylinder head bolts is known as "angle torquing" - all bolts except bolt 9 should be tightened an additional 120 degrees, while bolt 9 should be turned an additional 90 degrees

10.5 The right rear fender well inner splash shield must be removed to gain access to the crankshaft pulley - pull off the plastic fasteners (arrows) with a pliers

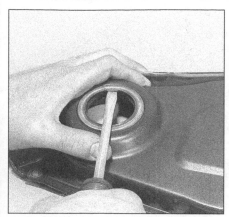

11.7 Once the timing cover is removed, place it on a flat surface and gently pry the old seal out with a large screwdriver

15 Install the drivebelts (Chapter 1).
16 Install the right rear inner fender splash shield.

11 Timing gear cover and front oil seal - removal and installation

Refer to illustrations 11.7 and 11.8
1 Remove the cable from the negative terminal of the battery.
2 Raise the vehicle and secure it on jackstands.
3 Remove the crankshaft pulley and hub (Section 10).
4 Remove the timing cover-to-block bolts.
5 Remove the timing cover by carefully prying it loose. **Note:** *The timing cover is installed with RTV sealant so it isn't easy to pry it off. The sealing flange between the timing cover and the oil pan will probably be bent during removal. Try to minimize the damage to the sealing flange during removal or it may be too damaged to be straightened.*
6 Using a scraper and degreaser, remove all the old sealant from the mating surfaces of

11.8 Place the new seal squarely in the bore, then use a seal driver or a large socket (slightly smaller than the outside diameter of the seal) to drive it into position in the timing cover

the timing gear cover, engine block and oil pan.
7 Remove the front oil seal by carefully prying it out of the timing gear cover with a large screwdriver **(see illustration)**. Do not distort the cover.
8 Install the new seal with the lip facing toward the inside of the cover. Drive the seal into place using a seal installation tool or a large socket and hammer **(see illustration)**. A block of wood will also work.
9 Apply a 3/8-inch wide by 3/16-inch thick bead of RTV-type gasket sealant to the timing gear cover flange at the sealing surface between the timing cover and the oil pan, a 1/4-inch wide by 1/8-inch thick bead of RTV-type gasket sealant between the cover and the block and a dab of sealant at the joints between the oil pan and the engine block.
10 Place the timing gear cover in position and loosely install a couple of bolts to support it.
11 Lubricate the cover seal and insert the hub through the seal. As the hub slides onto

the end of the crankshaft, it will center the timing gear cover.
12 Install the remaining timing gear cover mounting bolts and tighten them to the specified torque.
13 Install the crankshaft hub and pulley.
14 Lower the vehicle.
15 Reattach the cable to the negative terminal of the battery.

12 Oil pump driveshaft - removal and installation

Refer to illustrations 12.3 and 12.7
Note: *1988 models are equipped with a force balancer/oil pump assembly. This combination balance shaft and oil pump system must be removed as an assembly if the balance shafts must be replaced. However, the oil pump assembly can be removed as a separate component. Refer to Chapter 2C for the removal procedure.*
1 Disconnect the cable from the negative terminal of the battery.
2 Raise the vehicle and secure it on jackstands.
3 Remove the oil pump driveshaft retainer plate bolts **(see illustration)**.
4 Remove the oil pump driveshaft and bushing with a magnet.
5 Clean the sealing surfaces on the cylinder block and the retainer plate.
6 Inspect the bushing and driveshaft for wear.
7 Install the bushing and oil pump driveshaft in the block. The shaft driven gear must mesh with the camshaft drive gear and the slot in the lower end of the shaft must mate with the oil pump gear tang **(see illustration)**.
8 Apply a 1/16-inch bead of RTV-type sealant to the retainer plate so that it completely seals around the oil pump driveshaft hole in the block.
9 Lay the retainer plate in position on the block and tighten the mounting bolts securely.

12.3 The oil pump driveshaft retainer plate is on the rear of the cylinder block, just below the pushrod cover and just above the oil filter

12.7 If the slotted oil pump driveshaft is properly mated with the oil pump gear tang, the top of the bushing will be flush with the retainer plate mounting surface

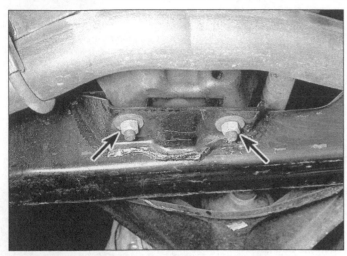

13.4 The front engine mount studs protrude through the right cradle member and are secured with large nuts and washers (arrows)

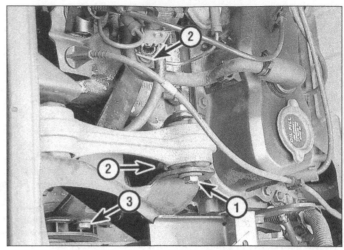

13.7a Details of the upper alternator mounting bracket and bolts on the 2.5L engine

1	*Upper alternator/engine strut rod bracket*	*2*	*Mounting bolts*
		3	*Alternator pivot bolt*

13 Oil pan - removal and installation

Refer to illustrations 13.4, 13.7a, 13.7b, 13.9 and 13.11

Removal

1   Disconnect the cable from the negative battery terminal.
2   Raise the vehicle and place it securely on jackstands.
3   Drain the engine oil (Chapter 1).
4   Remove the front engine mount nuts **(see illustration)**.
5   Remove the catalytic converter splash shield.
6   Remove the strut rod (Section 5).
7   Remove the alternator (Chapter 5). Remove the alternator bracket bolts and the remaining engine support bracket bolt **(see illustrations)**.
8   Remove the air conditioning compres-

sor bracket bolts.
9   Remove the starter (Chapter 5) and the flywheel dust cover **(see illustration)**.
10   With a hoist, raise the engine and support bracket/front mount assembly slightly off the cradle.
11   Remove the forward (toward the passenger compartment) bolts of the engine front support bracket. Remove the mount and support bracket as an assembly **(see illustration)**.
12   Remove the oil pan retaining bolts.
13   Remove the oil pan.

Installation

14   Clean the mating surfaces of the oil pan and cylinder block.
15   Apply a 1/8-inch wide bead of RTV sealant to the oil pan flange.
16   Install the oil pan and tighten the retaining bolts to the specified torque.
17   Install the engine front support bracket

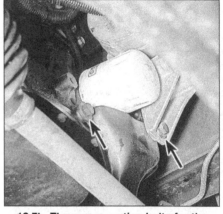

13.7b The rear mounting bolts for the front engine support bracket (arrows) - note that the bolt to the right of the oil filter is also a retaining bolt for the lower alternator bracket

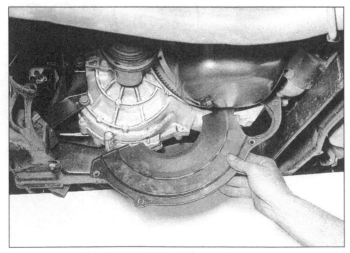

13.9 Removing the flywheel dust cover

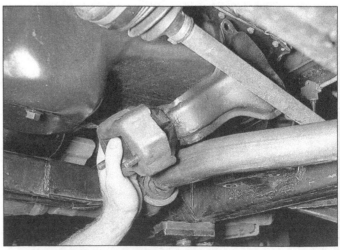

13.11 Push the mount and support bracket upward until the mount studs clear the cradle, then remove both mount and bracket as an assembly

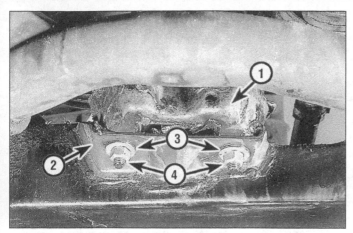

15.4 Typical engine mount-to-cradle mounting details

1	Engine mount	3	Washer
2	Engine cradle	4	Nuts

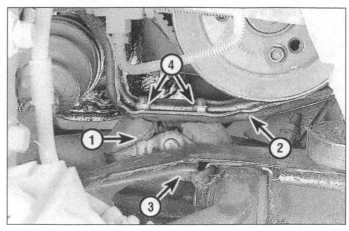

15.5 Typical engine mount-to-engine support bracket mounting details

1	Engine mount	3	Cradle
2	Engine support bracket	4	Mounting nuts

and tighten the mounting bolts to the specified torque.

18 Lower the engine until the front support bracket is resting on the cradle. Make sure that the engine mount studs protrude through the holes in the cradle. Tighten the engine mount nuts to the specified torque.

19 Install the flywheel dust cover and the starter.

20 Install the air conditioning compressor bracket.

21 Install the alternator (Chapter 5).

22 Install the catalytic converter splash shield.

23 Lower the vehicle.

24 Add engine oil.

25 Install the negative battery cable.

26 Start the engine and check for leaks.

14 Oil pump - removal and installation

Note: *1988 models are equipped with a force balancer/oil pump assembly. This combination balance shaft and oil pump system must be removed as an assembly if the balance shafts must be replaced. However, the oil pump assembly can be removed as a separate component. Refer to Chapter 2C for the removal procedure.*

1 Remove the oil pan (Section 13).

2 Remove the two oil pump flange mounting bolts and the filter bracket nut from the main bearing cap bolt.

3 Remove the oil pump and screen assembly.

4 For inspection procedures, refer to Part C of this Chapter.

5 To install the pump, align the shaft so that the gear tang mates with the slot on the lower end of the oil pump driveshaft. The oil pump should slide easily into place over the oil pump driveshaft lower bushing. If it doesn't, pull it off and turn the tang until it is aligned with the pump driveshaft slot.

6 Install the pump mounting bolts and the filter bracket nut. Tighten them to the specified torque.

7 Reinstall the oil pan (Section 13).

15 Engine and transaxle mounts - replacement with engine in vehicle

Refer to illustrations 15.4, 15.5, 15.15a, 15.15b, 15.23a, 15.23b and 15.59

Note: *If the rubber mounts have become hard, split or separated from the metal backing, they must be replaced. This operation can be performed with the engine/transaxle assembly in the vehicle.*

Engine mount

1 Support the engine with a jack or other suitable equipment.

2 Remove the forward torque reaction strut bolt.

3 Raise the vehicle and support it securely on jackstands.

4 Remove the engine mount-to-cradle nuts **(see illustration)**.

5 Remove the upper mount-to-engine

support bracket nuts **(see illustration)**. **Note:** *It is not necessary to detach the support bracket from the engine.*

6 Remove the engine mount.

7 Place the new mount in position.

8 Tighten the mount-to-engine bracket nuts to the specified torque.

9 Tighten the mount-to-chassis nuts to the specified torque.

10 Lower the vehicle.

11 Install the forward torque reaction strut bolt and tighten it to the specified torque.

12 Remove the jack or support fixture.

Automatic transaxle mounts - earlier model vehicles

Note: *Earlier vehicles equipped with an automatic transaxle use two smaller transaxle mounts - one at the front and one at the rear.*

Forward transaxle mount

13 Raise the vehicle and support it securely on jackstands.

14 Support the engine and transaxle with a floor jack or suitable support fixture.

15 Remove the mount-to-cradle nuts and the support bracket **(see illustrations)**.

16 Remove the mount.

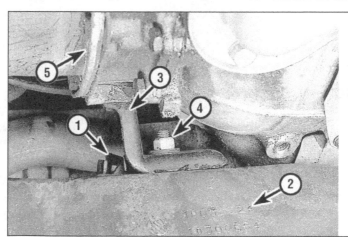

15.15a Forward transaxle mount-to-support bracket mounting details

1	Forward transaxle mount
2	Cradle
3	Support bracket
4	Mounting nut
5	Transaxle

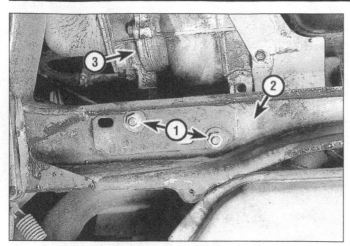

15.15b Forward transaxle mount-to-engine cradle mounting details

1	Mounting nut	3	Transaxle
2	Cradle		

15.23a Rear transaxle mount-to-support bracket mounting details

1	Mounting nut	3	Cradle
2	Support bracket	4	Transaxle

17   Position the new mount.
18   Tighten the mount-to-support bracket and the mount-to-cradle nuts to the specified torque.
19   Lower the vehicle.
20   Remove the support fixture.

Rear transaxle mount

21   Raise the vehicle.
22   Place a cradle stand or floor jack under the rear cradle.
23   Remove the rear cradle-to-chassis bolts **(see illustrations)**.
24   Lower the cradle.
25   Remove the mount-to-bracket nut and the mount-to-cradle nuts.
26   Remove the mount.
27   Place the new mount in position and install the mount-to-cradle nuts.
28   Install the mount-to-bracket nut and tighten it to the specified torque.
29   Raise the cradle, install the cradle bolts and tighten them to the specified torque.

30   Tighten the mount-to-cradle nuts to the specified torque.
31   Remove the cradle support.
32   Lower the vehicle.

Automatic transaxle mounts - later model vehicles

Note: *Some 1986 and all 1987 vehicles equipped with an automatic transaxle have one large support bracket and mount instead of two smaller ones (a front and a rear mount/bracket).*

33   Support the engine and transaxle with a jack or other suitable support fixture.
34   Remove the forward torque reaction strut bolt.
35   Remove the upper mount nuts.
36   Raise the vehicle.
37   Remove the lower mount nuts.
38   Remove the mount.
39   Position the new mount.
40   Install the lower mount nuts and tighten

them to the specified torque.
41   Install the forward torque reaction strut bolt and tighten it to the specified torque.
42   Remove the jack or support fixture.

Manual transaxle mounts (all model years)

Note: *The following procedure allows you to remove both front and rear mounts simultaneously.*

43   Raise the vehicle.
44   Support the engine and transaxle with a jack or other suitable support fixture.
45   Remove the forward torque reaction strut bolt.
46   Remove the front engine mount-to-frame nuts.
47   Remove the rear engine mount-to-frame nuts.
48   Remove the front engine mount-to-engine bracket nut.
49   Remove the rear engine mount-to-

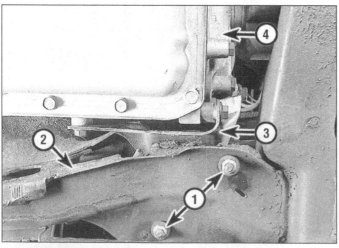

15.23b Rear transaxle mount-to-cradle mounting details

1	Mounting nuts	3	Support bracket
2	Cradle	4	Transaxle

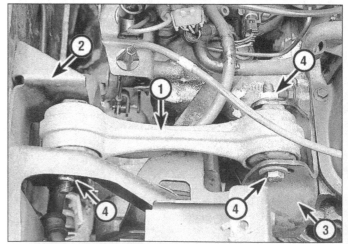

15.59 Engine torque reaction strut and related components

1	Torque reaction strut	3	Engine bracket
2	Chassis bracket	4	Bolt/nut assemblies

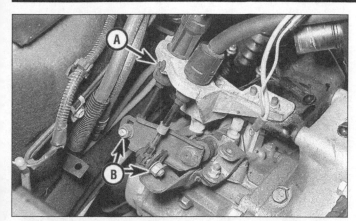

17.8 Remove the nuts from the shifter cable pins and disconnect the cables from the shift linkage, then split the shifter cable bracket by removing the retaining bolt and hang the shifter cables out of the way

A   Shifter cable bracket bolt   B   Shifter cable pin nuts

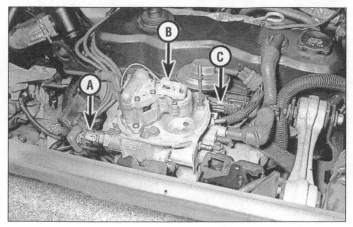

17.12 Disconnect the idle air control solenoid connector (A), the injector connector (B) and the throttle position sensor connector (C) from the throttle body

engine bracket nut.
50   Remove the front and rear mounts.
51   Position the new front and rear mounts.
52   Tighten the rear engine mount-to-engine bracket nut to the specified torque.
53   Tighten the front engine mount-to-engine bracket nut to the specified torque.
54   Tighten the rear engine mount-to-frame nuts to the specified torque.
55   Tighten the front engine mount-to-frame nuts to the specified torque.
56   Remove the jack or support fixture.
57   Lower the vehicle.
58   Install the forward torque reaction strut bolt and tighten it to the specified torque.

Engine torque reaction strut

59   Remove the rear strut bolt **(see illustration)**.
60   Remove the front strut bolt.
61   Remove the strut.
62   Loosely insert the front strut bolt.
63   Push the engine back until the rear strut bolt can be installed.
64   While the load is still being applied, tighten the front and rear strut bolts to the specified torque.

16   Engine removal - methods and precautions

If it has been decided that an engine must be removed for overhaul or major repair work, certain preliminary steps should be taken.

Locating a suitable work area is extremely important. A shop is, of course, the most desirable place to work. Adequate work space, along with storage space for the vehicle, is very important. If a shop or garage is not available, at the very least a flat, level, clean work surface made of concrete or asphalt is required.

Cleaning the engine compartment and engine prior to removal will help keep tools clean and organized.

A vehicle hoist will also be necessary to remove the engine on this vehicle.

If the engine is being removed by a novice, a helper should be available. Advice and aid from someone more experienced would also be helpful. There are many instances when one person cannot simultaneously perform all of the operations required when removing the engine from the vehicle.

Plan the operation ahead of time. Arrange for or obtain all of the tools and equipment you will need prior to beginning the job. Some of the equipment necessary to perform engine removal and installation safely and with relative ease are (in addition to a vehicle hoist) complete sets of wrenches and sockets as described in the front of this manual, wooden blocks and plenty of rags and cleaning solvent for mopping up the inevitable spills.

Plan for the vehicle to be out of use for a considerable amount of time. A machine shop will be required to perform some of the work which the do-it-yourselfer cannot accomplish due to a lack of special equipment. These shops often have a busy schedule, so it would be wise to consult them before removing the engine in order to accurately estimate the amount of time required to rebuild or repair components that may need work.

Always use extreme caution when removing and installing the engine. Serious injury can result from careless actions. Plan ahead. Take your time and a job of this nature, although major, can be accomplished successfully.

17   Engine removal and installation

Refer to illustrations 17.8, 17.12, 17.14a, 17.14b, 17.14c, 17.16, 17.20, 17.22, 17.31, 17.32, 17.33, 17.34, 17.38, 17.39, 17.41 and 17.43
Note: *Do not attempt to remove the engine from this vehicle unless you have access to a hoist. This procedure is extremely difficult to*

perform using a floor jack and jackstands.
Caution: *If your vehicle is equipped with air conditioning, take it to a dealer or air conditioning specialist and have the air conditioning system depressurized before you begin to remove the engine.*

Removal

1   If the engine is not already warm, warm it up and drain the oil (Chapter 1).
2   Disconnect the negative battery cable and the ground strap from the right front corner of the cylinder head, then disconnect the cable from the positive battery terminal.
3   Allow the engine sufficient time to cool down and drain the engine coolant (Chapter 1).
4   Disconnect the rear compartment lid ground strap. Mark the alignment of the lid mounting bolts and remove them. Remove the rear compartment lid. **Caution:** *Do not remove the torsion rod retaining bolts.*
5   Remove the vapor canister (left side) and the battery (right side) cover panels.
6   Remove the air cleaner (Chapter 4).
7   Disconnect the throttle cable (Section 5) and set it aside.
8   Remove the shifter cable bracket bolt, disconnect the shifter cables from the shift linkage **(see illustration)** and secure the cables out of the way with wire.
9   Disconnect the heater hose from the intake manifold and disconnect the heater hose bracket from the throttle linkage assembly.
10   Disconnect the vacuum hoses between the intake system and all non-engine components.
11   Disconnect the fuel lines at the throttle body and the filter and remove the filter bracket retaining bolt.
12   Disconnect the throttle position sensor, the injector connector and the idle air control solenoid connector from the throttle body **(see illustration)**.
13   Before disconnecting the remaining electrical connectors, label them clearly for

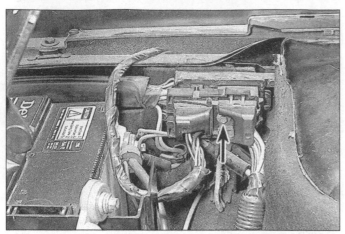

17.14a The main engine harness junction box is located on the right side of the engine compartment just behind the battery - the bolt in the middle of the junction box (arrow) must be removed before the box can be unplugged

17.14b Disconnect the two-part plug from the junction box, then unplug the two sections of the plug from one another

17.14c Disconnect the two fusible link connectors (arrows) right below the junction box

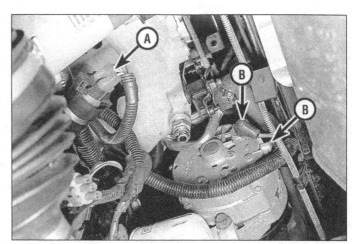

17.16 Disconnect the oil temperature sending unit connector (A) and the electrical leads to the alternator (B)

proper reinstallation. The important thing to remember is how the wire harness is routed because failure to reroute it properly will make it impossible to reattach all the connectors. If you are in doubt as to your ability to reroute the harness properly, a few sketches now will be helpful later at reinstallation time.
14 Unplug the engine wire harness connector at the right side of the engine compartment **(see illustrations)**, then disconnect the two fusible link connectors **(see illustration)**.
15 Disconnect the ground strap and the coolant temperature switch/
sender and connector on the left end of the cylinder head.

17.20 Remove the torque reaction strut rod through-bolts (arrows) and the strut rod

17.22 Disconnect the backup light switch connector (A) and the clutch slave cylinder bracket nuts (B) - wire the slave cylinder out of the way

17.31 Before disconnecting the struts, mark their alignment with the rear knuckles with paint to ensure proper realignment

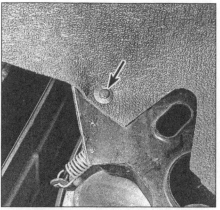

17.32 Though it isn't necessary to remove the inner fender splash shields during engine removal, be sure to disconnect the single fastener (arrow) that attaches the shield to the lower cradle or you will rip the shield when you raise the vehicle off the cradle

17.33 Remove the forward cradle through bolt nuts, but do not pull out the bolts yet

17.34 Put a sturdy support underneath the engine compartment to hold the engine/transaxle/cradle assembly off the floor

16   From underneath the vehicle, disconnect the oil temperature sending unit and the alternator leads **(see illustration)**.

17   If your vehicle is equipped with air conditioning, disconnect the electrical leads on the air conditioning compressor.

18   If your vehicle is equipped with an automatic transaxle, disconnect the cooler lines.

19   Disconnect the two leads on the starter solenoid.

20   Remove the strut rod **(see illustration)**. Work all three air conditioning wires out from between the air conditioning compressor and the compressor bracket by threading them around the front of the engine block and into the void behind the block.

21   Feed the main wire harness down between the intake manifold and the alternator bracket.

22   Disconnect the backup light switch connector and the clutch slave cylinder **(see illustration)**.

23   Disconnect the spark plug wire harness from the rocker arm cover, unplug the boots from the spark plugs and the distributor terminals and remove the spark plug wire harness. Disconnect the connector between the coil and the distributor and the ignition coil grounds.

24   Disconnect the radiator and heater hoses from the water pump.

25   Raise the vehicle and support it securely on jackstands.

26   Disconnect the air conditioning compressor line fitting bolt.

27   Disconnect the exhaust pipe bolts (Section 8).

28   Remove the rear wheels (Chapter 1).

29   Disconnect the parking brake cables.

30   Remove both rear brake calipers and hang them out of the way (Chapter 9). Do not disconnect the brake hoses. **Note:** *The caliper mounting bolts are Torx bolts, not Allen bolts.*

31   Mark the struts for realignment **(see illustration)** and remove the strut mounting bolts.

32   Remove the inner fender splash shields where they attach to the engine cradle **(see**

illustration).

33   Remove the nuts from the front cradle through-bolts **(see illustration)** but do not remove the through-bolts themselves at this time.

34   Slide a suitable support under the engine/transaxle/cradle assembly **(see illustration)**.

35   Lower the vehicle until the cradle is resting on the support. **Note:** *Unless your cradle support is higher than the axle line of the wheels, it will be necessary to remove the front wheels in order to lower the vehicle enough to lay the cradle on the support.*

36   Slide jackstands underneath the forward end of the cradle.

37   Remove the front cradle through-bolts.

38   Remove the rear cradle bolts **(see illustration)**. Disconnect the parking brake cable at the cradle.

39   Carefully raise the vehicle a few inches and check to make sure that everything is

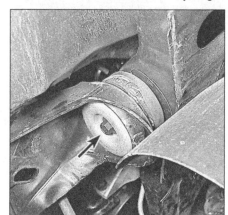

17.38 Remove the rear cradle retaining bolt (arrow) from each rear corner of the cradle

disconnected. If it is, raise the vehicle. The engine/transaxle/cradle assembly will remain resting on the support **(see illustration)**.

40   Remove the transaxle-to-block bolts and separate the engine from the transaxle by prying them apart.

17.39 Double check the engine compartment to make sure that everything is disconnected, then slowly raise the vehicle until it clears the engine/transaxle/cradle assembly, which will remain on its support

17.41 Carefully lift the engine out of the cradle after it is disengaged from the transaxle - make sure you have disconnected the exhaust manifold from the exhaust pipe

17.43 When installing the engine back into the cradle, use several guide bolts (arrows) to help align the block with the transaxle housing

41   Attach a hoist to the engine lifting brackets and raise the engine slightly **(see illustration)** until the front engine mount bolts are clear of the cradle.

42   Lift the engine out of the cradle.

Installation

43   Using the hoist to support the engine, carefully align the engine with the transaxle. **Caution:** *Do not try to force the engine and transaxle assemblies together or you might damage the splines on the transaxle input shaft or the diaphragm clutch fingers. Guide bolts will simplify this operation* **(see illustration)**.

44   Tighten the transaxle-to-block bolts to the specified torque.

45   Disconnect and remove the hoist.

46   Place the starter motor in position but do not install it at this time.

47   Install a new exhaust manifold gasket and fasten the exhaust manifold to the exhaust pipe.

48   Carefully lower the vehicle over the dolly. Make sure that the engine/transaxle/cradle assembly is aligned with the engine compartment well enough to allow the vehicle to be lowered almost all the way.

49   Position the hoist behind the vehicle in such a way that the boom arm is directly over the engine compartment. Reattach the hoist to the lifting points of the engine and raise the engine/transaxle/cradle assembly off its support until the front cradle mounting holes are aligned with the front cradle mounting brackets on the vehicle. Install the front cradle bolts finger tight. **Note:** *Have an assistant standing by to help align the struts with the knuckle mounting lugs.*

50   Install the rear cradle bolts and tighten them to the specified torque.

51   Tighten the front cradle bolts to the specified torque.

52   Raise the vehicle.

53   Align the paint marks on the struts with the marks on the knuckles. Install the strut mounting bolts and tighten them to the specified torque (see Chapter 10).

54   Install the calipers and the parking brake cable (Chapter 9). Adjust the parking brake cable as necessary.

55   Install the clutch cover dust shield and thread the cover bolts a few turns to hold the cover in place. Install the starter motor and tighten the starter mounting bolts. Tighten the dust cover mounting bolts securely.

56   Attach the lower radiator hose and the heater hose to the water pump and tighten the hose clamps securely.

57   If your vehicle is equipped with air conditioning, install the fitting for the air conditioning line and tighten the retaining bolt. Reconnect the air conditioning electrical connectors.

58   If your vehicle is equipped with a manual transaxle, install the shift cables and clutch slave cylinder. **Note:** *The linkage arms are slotted for readjustment, so leave the bolts on the ends of the cables loose at this time.*

59   If your vehicle is equipped with an automatic transaxle, install the cooler lines and shift linkage.

60   Connect the radiator hose to the thermostat housing.

61   Connect the heater hose to the intake manifold.

62   Connect the fuel lines at the throttle body and the fuel filter and tighten the fuel filter mounting bracket bolt.

63   Connect all vacuum lines to the throttle body assembly.

64   Reconnect the throttle cable at the throttle body bracket. Install the throttle cable bracket at the exhaust manifold and tighten the bolt securely.

65   Connect the ground straps.

66   Install the engine strut and tighten the mounting bolts to the specified torque.

67   Place the wire harness in position and reattach all electrical connectors.

68   Add oil (Chapter 1).

69   Add coolant (Chapter 1).

70   Install and adjust the drivebelts (Chapter 1).

71   If the engine has just been rebuilt, remove the spark plugs and crank the engine over until the oil pressure light goes out.

72   Reinstall the plugs and start the engine. Check for leaks.

73   Adjust the shift cables as necessary (Chapter 7).

74   Install the engine compartment lid.

---

18   Flywheel/driveplate and rear main bearing oil seal - removal and installation

---

Note: *The rear main bearing oil seal is a one-piece unit and can be replaced without removal of the oil pan or crankshaft.*

1   Remove the transaxle (Chapter 7). Follow all precautionary notes.

2   If your vehicle is equipped with a manual transaxle, remove the pressure plate and clutch disc (Chapter 8).

3   Remove the flywheel (manual transaxle) or driveplate (automatic transaxle) retaining bolts and separate it from the crankshaft.

4   Using a screwdriver or pry bar, carefully remove the rear main oil seal from the block.

5   Using solvent, thoroughly clean the crankshaft-to-seal and block mating surfaces.

6   Apply a light coat of engine oil to the outside surface of the new seal.

7   Using your fingertips, press the new seal evenly into position in the block.

8   Install the flywheel or driveplate and tighten the bolts to the specified torque.

9   If your vehicle is equipped with a manual transaxle, reinstall the clutch disc and pressure plate (Chapter 8).

10   Reinstall the transaxle (Chapter 7).

Chapter 2 Part B V6 engine

Contents

Specifications

Torque specifications

	Ft-lbs
Intermediate intake manifold bolts	15
Lower intake manifold nut	19
Upper intake manifold bolts	19
Camshaft sprocket bolts	15 to 20
Clutch or driveplate cover bolts	13 to 18
Cylinder head bolts (refer to illustration 9.10)	
1985	65 to 75
1986 and 1987	65 to 90
1988	
VIN W	
Step 1	33
Step 2	Turn additional 90 degrees
VIN S and 9	
Step 1	40
Step 2	Turn additional 90 degrees
Crankshaft pulley bolts	20 to 30
Crankshaft damper bolt	66 to 84
Distributor hold-down bolt	20 to 30
EGR valve mounting bolts	13 to 18
Engine mount-to-cradle nut	41
Engine mount-to-bracket nut	41
Engine strut-to-chassis nut and bolt	42
Engine strut-to-engine bracket bolt and nut	42
Support bracket nuts	35
Transaxle-to-engine bolts	48 to 63

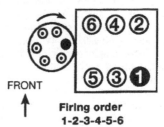

FRONT

**Firing order
1-2-3-4-5-6**

2.8L V6 ENGINE

**2.8L V6 ENGINE Cylinder
location and distributor rotation**

Torque specifications

	Ft-lbs (unless otherwise indicated)
Cradle bolts	
Rear	76
Front	66
Rear strut-to-knuckle bolts	140
Exhaust manifold	
Exhaust manifold bolt/stud lockwasher	18
Exhaust manifold bolt and lockwasher	18
Crossover pipe-to-exhaust manifold bolts	22
Muffler pipe-to-crossover pipe bolts	15
Timing cover mounting bolts	
Small	13 to 18
Large	20 to 30
Main bearing cap bolts	63 to 74
Oil pan mounting bolts	
Small	72 to 108 in-lbs
Large	14 to 22
Oil pump mounting bolt	26 to 35
Rocker arm cover bolts	72 to 108 in-lbs
Rocker arm studs	43 to 49
Starter motor mounting bolts	26 to 37
Water outlet housing bolts	20 to 30

1 General information

This part of Chapter 2 is devoted to in-vehicle repair procedures and removal and installation for the 2.8 liter V6 engine. All information concerning engine block and cylinder head servicing can be found in Part C of this Chapter.

The following repair procedures are based on the assumption that the engine is still installed in the vehicle. If you are referring to this part during a complete engine overhaul, that is with the engine already out of the vehicle and on a stand, many of the steps included here will be superfluous.

The specifications included in this Part of Chapter 2 apply to the 2.8L V6 engine. If you are looking for specifications regarding the 2.5L four cylinder engine, refer to Part A. For specifications related to cylinder head and block rebuilding procedures, refer to Part C of this Chapter.

2 Repair operations possible with the engine in the vehicle

Many major repair operations can be accomplished without removing the engine from the vehicle.

It is a good idea to clean the engine compartment and the exterior of the engine with some type of pressure washer before any work is begun. A clean engine will make the job easier and will prevent the possibility of getting dirt into internal areas of the engine.

Remove the engine compartment lid and cover the fenders to provide as much working room as possible and to prevent damage to the painted surfaces.

If oil or coolant leaks develop, indicating a need for gasket or seal replacement, repairs can generally be made with the engine in the vehicle. The oil pan gasket, the cylinder head gaskets, intake and exhaust manifold gaskets, timing cover gaskets and the front crankshaft oil seal are accessible with the engine in place.

Exterior engine components such as the water pump, starter motor, alternator, distributor, fuel pump and throttle body assembly, as well as the intake and exhaust manifolds, are easily removed for repair with the engine in place.

Since the cylinder heads can be removed without pulling the engine, valve component servicing can also be accomplished with the engine in the vehicle.

Replacement of, repairs to and inspection of the timing sprockets and chain and the oil pump are all possible with the engine in place.

In extreme cases caused by a lack of necessary equipment, repair or replacement of piston rings, pistons, connecting rods and rod bearings and reconditioning of the cylinder bores is possible with the engine in the vehicle. However, this practice is not recommended because of the cleaning and preparation work that must be done to the components involved.

Detailed removal, inspection and installation procedures for the above mentioned components can be found in the appropriate parts of Chapter 2 or in other Chapters in this manual.

3 Rocker arm covers - removal and installation

Front

1 Disconnect the cable from the negative battery terminal.

2 With the help of an assistant, remove the engine compartment lid. Then remove both side covers (Chapter 11).

3 Remove the vacuum boost line and tube.

4 Remove the throttle and downshift cables and bracket.

5 If your vehicle is equipped with cruise control, disconnect the cable and set it aside.

6 Disconnect the ground cable.

7 Remove the PCV valve from the rocker arm cover.

8 Remove the oil dipstick tube.

9 Remove the plug wires and bracket.

10 Remove the engine lift hook.

11 Remove the rocker arm cover bolts.

12 Remove the rocker arm cover. If the cover sticks to the cylinder head, shear it off by hitting the end of the rocker arm cover with the palm of your hand or by tapping it with a rubber mallet. If the cover still won't release, pry on it carefully until it comes loose. **Caution:** *Do not damage the sealing flange.*

13 Before reinstalling the cover, clean all dirt, oil and old gasket material from the sealing surfaces of the cover and cylinder head with a scraper and degreaser.

14 Apply a continuous 1/8-inch diameter bead of RTV-type sealant to the cover flange.

Be sure to apply the sealant inboard of the bolt holes.

15   Place the rocker arm cover gasket on the cylinder head, using care to line up the holes in the gasket with the bolt holes in the cylinder head.

16   Install the rocker arm cover on the cylinder head while the sealant is still wet and install the mounting bolts. Tighten the bolts a little at a time to the specified torque.

17   Connect the plug wires and bracket.

18   Attach the oil dipstick tube.

19   Install the PCV valve in the rocker arm cover.

20   Attach the ground cable.

21   Connect the cruise control cable (if applicable).

22   Connect the throttle and downshift cables and bracket.

23   Connect the vacuum boost line and tube.

24   Install the engine compartment lid and both side covers.

25   Attach the negative battery cable. **Note:** *If you are also removing the rear rocker arm cover, do not reattach the negative battery cable at this time.*

Rear

26   Disconnect the cable from the negative battery terminal.

27   Disconnect the torque reaction rod bolt at the cylinder head bracket.

28   Swing the torque reaction rod up and remove the bolt connecting the cylinder head bracket to the bracket at the front of the engine.

29   Loosen the lower bolt of the torque reaction rod bracket at the front of the engine.

30   Remove the upper two bolts of the torque reaction rod bracket at the front of the engine.

31   Remove the torque reaction rod bracket bolt at the cylinder head/exhaust manifold connection.

32   Remove the wiring harness (and covering sleeve) between the rocker arm cover and the lower plenum.

33   Remove the rocker arm cover bolts.

34   Remove the rocker arm cover. If it adheres to the cylinder head, shear it off by bumping the end of the cover with the palm of your hand or a rubber mallet. If the cover will not release, carefully pry on it until it comes loose. **Caution:** *Do not damage the rocker arm cover sealing flange.*

35   Clean the mating surfaces of the cylinder head and the rocker arm cover.

36   Apply a continuous 1/8-inch wide bead of RTV-type sealant to the cover flange. Be sure to apply the sealant inboard of the bolt holes.

37   Install the rocker arm cover gasket, using care to line up the holes in the gasket with the bolt holes in the cylinder head.

38   Install the rocker arm cover and bolts. Tighten the bolts to the specified torque.

39   Attach the wiring harness between the rocker arm cover and the lower plenum.

40   Install the torque reaction rod bracket bolt at the cylinder head/exhaust manifold connection and tighten it to the specified torque.

41   Install the upper two bolts of the torque reaction rod bracket at the front of the engine and tighten it to the specified torque.

42   Tighten the lower torque reaction rod bracket bolt at the front of the engine to the specified torque.

43   Install the bolt connecting the cylinder head bracket to the bracket at the front of the engine and tighten it to the specified torque.

44   Install the torque reaction rod bolt at the cylinder head bracket and tighten it to the specified torque.

45   Connect the negative battery cable.

4   Valve train components - replacement (cylinder head in place)

Refer to illustration 4.2

Note: *The following procedures apply to the rear cylinder head only - because of the front cylinder head's close proximity to the forward engine compartment bulkhead, removal of the valve train components from the front head is impossible unless the head is removed first.*

1   Remove the rocker arm cover as described in Section 3.

2   If only the pushrod is to be replaced, loosen the rocker arm nut to allow the rocker arm to be rotated away from the pushrod. Pull the pushrod out of the hole in the cylinder head **(see illustration)**.

3   If the rocker arm is to be removed, remove the rocker arm nut and pivot and lift off the rocker arm.

4   If the valve spring is to be removed, remove the spark plug from the affected cylinder.

5   Either of the two following procedures will enable you to hold the valve in place while the keepers are removed. If you have access to compressed air, install an air hose adapter in the spark plug hole. This adapter is also available at most auto parts stores. When air pressure is applied to the adapter, the valves will be held in place by the pressure.

6   If you do not have access to compressed air, bring the piston of the affected cylinder to approximately 45-degrees before top dead center (TDC) on the compression stroke. Feed a long piece of 1/4-inch cord through the spark plug hole until it fills the combustion chamber. Be sure to leave the end of the cord hanging out of the spark plug hole so it can be removed easily. Rotate the crankshaft with a wrench (in the normal direction of rotation) until slight resistance is felt.

7   Reinstall the rocker arm nut (without the rocker arm).

8   Insert the slotted end of a valve spring compression tool under the nut and com-

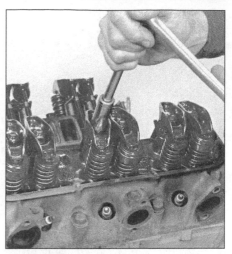

4.2 Loosen the rocker arm nuts

press the spring just enough to remove the spring keepers, then release the pressure on the tool.

9   Remove the retainer, cup shield, O-ring seal, spring, spring damper (if so equipped) and valve stem oil seal (if so equipped).

10   The rocker arm studs may be replaced by unscrewing the damaged one and replacing it with a new one. Be sure to reinstall the pushrod guide (if so equipped) under the stud nut and tighten the stud to the specified torque.

11   Inspection procedures for the hydraulic lifters are detailed in Section 6. Procedures regarding other valve train components are detailed in Chapter 2, Part C.

12   Installation of the valve train components is the reverse of the removal procedure. Always use new valve stem oil seals whenever the spring keepers have been disturbed. Before installing the rocker arms, coat the bearing surfaces of the arms and pivots with moly-base grease or engine assembly lube. Be sure to adjust the valve lash as detailed in Section 7.

5   Intake manifold - removal and installation

Refer to illustration 5.22

Removal

1   Relieve the fuel system pressure (see Chapter 4).

2   Disconnect the negative battery cable from the battery.

3   Remove both rocker arm covers (Section 3).

4   Drain the engine coolant (Chapter 1).

5   Disconnect and remove the intake hose between the throttle body and the elbow at the air cleaner canister.

6   Remove the distributor cap and the attached spark plug wires (Chapter 5).

7   Mark the position of the rotor and remove the distributor (Chapter 5).

5.22 Use a large screwdriver or prybar to break the intake manifold gasket seal

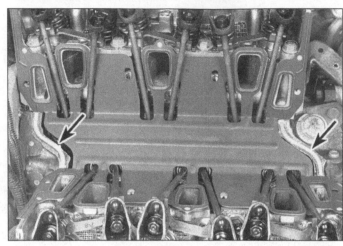

5.23 Apply a 3/16 inch bead of RTV sealant (arrows) to the front and rear ridges of the engine block (late models shown but earlier models require the RTV sealant in the same locations)

8   Disconnect the shift linkage, the throttle cable, the downshift cable and, if applicable, the cruise control cable.
9   Unbolt and remove the throttle body from the upper plenum.
10   Disconnect the radiator hose.
11   Disconnect the radiator fill inlet.
12   Disconnect the inlet and return heater hose and pipe from the throttle body.
13   Disconnect the wiring harness. Label all wires to simplify reinstallation.
14   Disconnect the heater hoses.
15   Disconnect all vacuum hoses. Label the hoses to insure correct installation.
16   Remove the brake booster pipe and bracket.
17   Remove the EGR pipe.
18   Remove the upper manifold plenum and the old gaskets.
19   Disconnect the fuel line inlet and return fittings at the fuel block.
20   Disconnect the wires from the fuel injectors.
21   Remove the intermediate intake manifold and old gasket. **Note:** *The intermediate intake manifold, fuel rail and injectors can be removed from the engine as an assembly.*
22   Remove the lower intake manifold and the old gaskets **(see illustration)**.

Installation

Refer to illustrations 5.23 and 5.24
23   Clean all gasket surfaces on the cylinder head and intake manifolds. Apply a smooth, continuous bead of sealant approximately 1/8-inch wide and 1/8-inch thick to the front and rear mating surfaces of the lower intake manifold and the top of the block **(see illustration)**. **Note:** *The bead configuration must completely seal in the oil. In order for it to be effective, the surface must be completely free of oil and dirt.*
24   Place the lower intake manifold and gasket in position. Tighten all bolts in the proper sequence to the specified torque **(see illustration)**.
25   Place the intermediate intake manifold

and gasket in position. Tighten all bolts in the proper sequence to the specified torque.
26   Connect the wires to the injectors and the fuel line fittings to the fuel metering block.
27   Place the upper manifold plenum and gaskets in position. Tighten all bolts in the proper sequence to the specified torque.
28   Bolt the throttle body to the upper plenum.
29   Install the EGR pipe.
30   Install the brake booster pipe and bracket.
31   Reattach all vacuum hoses.
32   Reattach the heater hoses.
33   Reconnect the wiring harness.
34   Install the radiator fill inlet.
35   Reattach the radiator hose.
36   Reconnect the inlet and return heater hose and pipe to the throttle body.
37   Hook up the shift linkage, the throttle cable, downshift cable and cruise control cable.
38   Install the distributor.
39   Reattach the intake hose between the throttle body and the elbow at the air cleaner. Tighten the hose clamps securely.
40   Fill the engine with coolant.
41   Install both rocker arm covers (Section 3).
42   Reattach the negative battery cable.
43   Check the ignition timing and coolant level (Chapter 1) and look for fluid leaks.

---

6   Hydraulic lifters - removal, inspection and installation

Refer to illustration 6.7
1   A noisy hydraulic lifter can be isolated while the engine is idling. Place a length of hose or tubing near the position of each valve while listening at the other end of the tube. Another method is to remove the rocker arm cover and, with the engine idling, place a finger on each of the valve spring retainers, one at a time. If a valve lifter is defective, it will be

evident from the shock felt at the retainer as the valve seats against the cylinder head.
2   Assuming that adjustment is correct, the most likely cause of a noisy valve lifter is a piece of dirt trapped between the plunger and the lifter body.
3   Remove the rocker arm covers (Section 3).
4   Remove the intake manifold (Section 5).
5   Loosen the rocker arm nut and rotate the rocker arm away from the pushrod (Section 4).
6   Remove the pushrod (Section 4).
7   To remove the lifter, a special hydraulic lifter removal tool should be used, or a scribe can be positioned at the inside top of the lifter and used to force the lifter up **(see illustration)**. Do not use pliers or other tools on the outside of the lifter body, as they will damage the finished surface and render the lifter useless.
8   The lifters should be kept in order for reinstallation in their original positions.
9   It is easier to simply replace a worn lifter with a new one than to repair a defective lifter. The internal components are not available separately - you must buy a lifter as a complete assembly. But sometimes disassembling and cleaning the internal components of a dirty lifter will restore normal operation. For complete inspection, disassembly and reassembly procedures, refer to Chapter 2C.
10   When installing the lifters, make sure they are replaced in their original bores. Coat them with moly-base grease or engine assembly lube.
11   The remaining installation steps are the reverse of removal.

---

7   Valve lash - adjustment

Refer to illustration 7.5
1   Disconnect the cable from the negative battery terminal.
2   Remove the rocker arm covers (Section 3).

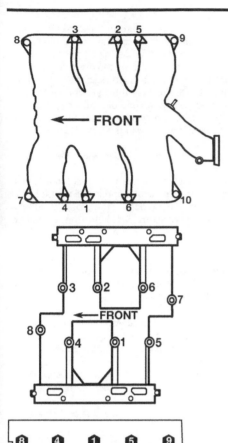

5.24 Each part of the intake manifold assembly has its own bolt tightening sequence

3   If the valve train components have just been serviced prior to this procedure, make sure that they are completely reassembled.

4   Rotate the crankshaft until the mark on the vibration damper lines up with the 0 mark on the timing tab, with the engine in the no. 1 firing position. To determine whether the no. 1 piston is at TDC, place your finger on the rocker arms for the no. 1 cylinder as the mark on the damper approaches the 0 mark. If the valves are not moving, the engine is in the no. 1 firing position. If they are moving, then the no. 4 piston is at TDC and you must rotate the crankshaft another 360 degrees.

5   Back out the rocker arm adjusting nut until play is felt at the pushrod, then turn it back in until all play is removed. This can be determined by rotating the pushrod while tightening the nut **(see illustration)**. All lash is removed the instant that you feel drag on the pushrod. Now turn the nut an additional 3/4-turn.

6   Adjust the number one, five and six cylinder intake valves and the number one, two and three cylinder exhaust valves with the crankshaft in this position, using the method just described.

7   Rotate the crankshaft one revolution (360 degrees) until the number four piston is at TDC. Adjust the number two, three and four cylinder intake valves and the number four, five and six cylinder exhaust valves.

8   Install the rocker arm covers (Section 3).

8   Exhaust manifolds - removal and installation

Front manifold

1   Disconnect the cable from the negative battery terminal.

2   Remove the rear compartment lid (Chapter 11).

3   Disconnect the brake vacuum hose.

4   Remove the bolts and detach the exhaust manifold heat shield.

5   Remove the crossover bolts.

6   Raise the vehicle and secure it on jack-stands.

7   Remove the front converter heat shield bolts and the heat shield.

8   Remove the lower manifold bolts.

9   Lower the vehicle.

10   Remove the upper manifold bolts and detach the exhaust manifold.

11   Clean the mating surfaces of the exhaust manifold and cylinder head thoroughly.

12   Position the exhaust manifold and gasket with the upper bolts.

13   Raise the vehicle.

14   Install the lower manifold bolts and tighten them to the specified torque.

15   Install the converter heat shield and tighten the bolts securely.

16   Lower the vehicle.

17   Tighten the upper manifold bolts to the specified torque.

18   Install the crossover bolts and tighten them to the specified torque.

19   Install the manifold heat shield and tighten the bolts securely.

20   Attach the brake vacuum hose.

21   Install the rear compartment lid.

22   Reattach the negative battery cable.

23   Inspect for exhaust and vacuum leaks.

Rear manifold

24   Remove the manifold-to-crossover bolts.

25   Remove the bolts and detach the exhaust manifold.

26   Clean the mating surfaces of the cylinder head and exhaust manifold thoroughly.

27   Position the exhaust manifold and gasket with the manifold bolts. Snug all the bolts before tightening them.

28   Tighten the exhaust manifold bolts to the specified torque.

29   Tighten the manifold-to-crossover bolts to the specified torque.

30   Check for exhaust and vacuum leaks.

Crossover pipe

31   Remove the rear compartment left side panel.

32   Remove the intake flex duct between

6.7 Use a scribe to remove the lifters

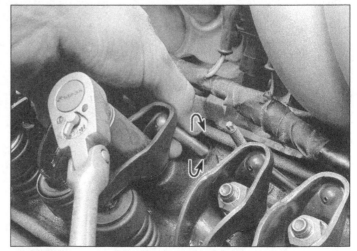

7.5 Twirling the pushrod with your thumb and index finger as you tighten the rocker arm adjuster nut will enable you to determine the point at which drag is imposed upon the pushrod

9.9a The rocker arm adjusting nuts must be loosened before you can swing the rocker arms out of the way to remove the pushrods

the throttle body and the elbow at the air cleaner housing.

33 Remove the EGR hose.

34 Remove the shift cables at the transaxle.

35 Remove the EGR tube between the exhaust crossover and the intake manifold.

36 Remove the front and rear crossover shields.

37 Detach the oxygen sensor connector.

38 Remove the bolts at the front and rear exhaust manifolds.

39 Raise the vehicle.

40 Remove the bolts at the catalytic converter.

41 Lower the vehicle.

42 Remove the crossover pipe.

43 Remove the oxygen sensor at the pipe.

44 Remove the EGR valve and adapter at the pipe.

45 Attach the oxygen sensor, EGR valve and adapter to the crossover pipe.

46 Install the crossover pipe and tighten the front and rear crossover-to-manifold bolts to the specified torque.

47 Raise the vehicle.

48 Install the catalytic converter bolts and tighten them securely.

49 Lower the vehicle.

50 Reconnect the oxygen sensor connec-

tor.

51 Install the front and rear crossover shields.

52 Install the EGR tube between the exhaust crossover and the intake manifold.

53 Reattach the shift cables at the transaxle.

54 Reconnect the EGR hose.

55 Install the intake flex hose between the throttle body and the elbow at the air cleaner canister.

56 Install the rear compartment left side panel.

9 Cylinder heads - removal and installation

Refer to illustrations 9.9a, 9.9b and 9.10

Front

1 Raise the vehicle and place it securely on jackstands.

2 Remove the engine block drain plug and drain the coolant (see Chapter 1).

3 Remove the jackstands and lower the vehicle.

4 Remove the intake manifold (Section 5).

5 Remove the exhaust crossover pipe (Section 8).

6 Remove the alternator and bracket (see Chapter 5).

7 Remove the oil level indicator tube.

8 Remove the rocker arm cover (Section 3).

9 Loosen the rocker arms and swing them out of the way **(see illustration)**. Remove the pushrods and store them separately to ensure reinstallation in their original positions **(see illustration)**.

10 Remove the cylinder head bolts by reversing the sequence that is used to tighten them **(see illustration)**.

11 Remove the cylinder head.

12 If a new cylinder head is being installed, transfer the various components, such as the exhaust manifold, brackets and coolant temperature sensor, from the old head.

13 Before installing the head, the gasket surfaces of both the head and the engine

block must be clean and free of nicks and scratches. Also, the threads in the block and on the head bolts must be completely clean, as any dirt or sealant in the threads will affect bolt torque. It's a good practice to recondition the bolts and the cylinder block bolt threads with a tap and die set.

14 Place the gasket in position over the locating dowels, with the note THIS SIDE UP visible.

15 Position the cylinder head over the gasket.

16 Coat the cylinder head bolt threads with an appropriate sealer and install the bolts.

17 Tighten the bolts in the proper sequence **(see illustration 9.10)** to the specified torque. Work up to the final torque in three steps.

18 Install the pushrods, making sure the lower ends are in the lifter seats.

19 Place the rocker arm ends over the pushrods and loosely install the rocker arm nuts.

20 Adjust the valve lash (Section 7).

21 Install the intake manifold and gaskets. Tighten all bolts to the specified torque (Section 5).

22 Attach the oil level indicator tube bracket to the head and tighten the bolt securely.

23 Install the heat stove pipe and air supply pipe.

24 Attach the alternator bracket and stud.

25 Install the exhaust pipe and the exhaust crossover pipe (Section 8).

26 Fill the cooling system with coolant. **Note:** *If you are also removing and installing the rear cylinder head, do not add coolant at this time.*

Rear

27 Raise the vehicle and place it securely on jackstands.

28 If you haven't already done so, remove the engine block drain plug and drain the coolant from the block (Chapter 1).

29 Disconnect the exhaust pipe from the exhaust manifold (Section 8).

30 Remove the jackstands and lower the vehicle.

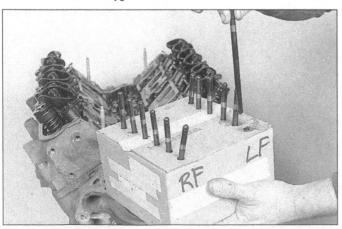

9.9b Always store the pushrods in a systematic way so they will be installed in their original positions

9.10 Tightening sequence for the cylinder head bolts

11.2 To detach the pump and strainer, remove the oil pump-to-rear main bearing cap bolt

12.3 Carefully pry the old seal out of the bore with a small screwdriver or equivalent – but don't scratch the seal bore or the crankshaft sealing surface

31   Remove the cruise control servo bracket, if equipped.

32   Remove the intake manifold (Section 5).

33   Disconnect the exhaust crossover pipe (Section 8).

34   Loosen the rocker arm nuts sufficiently to allow removal of the pushrods (Section 4).

35   Remove the pushrods (Section 4).

36   Loosen the head bolts by reversing the tightening sequence **(see illustration 9.10)**.

37   Remove the cylinder head.

38   If a new cylinder head is being installed, transfer all components from the old head.

39   Before reinstalling the cylinder head, clean all gasket surfaces on the head, cylinder block and intake manifold. Clean the cylinder block threaded holes and the cylinder head bolt threads with a tap and die set.

40   Place the new gasket in position over the dowel pins with the note THIS SIDE UP showing.

41   Place the cylinder head in position over the gasket.

42   Coat the cylinder head bolt threads with thread sealer or equivalent and install the bolts.

43   Tighten the cylinder head bolts in the proper sequence **(see illustration 9.10)** to the specified torque. Work up to the final torque in three steps.

44   Install the pushrods, making sure that the lower ends are in the lifter seats.

45   Place the rocker arm ends over the pushrods and loosely install the rocker arm nuts.

46   Adjust the valve lash (Section 7).

47   Install the intake manifold and gaskets (Section 5).

48   Install the exhaust crossover pipe (Section 8).

49   Install the cruise control servo bracket.

50   Raise the vehicle and support it securely on jackstands.

51   Connect the exhaust manifold to the exhaust pipe.

52   Lower the vehicle.

53   Adjust the drivebelts and fill the cooling system (Chapter 1).

10   Oil pan - removal and installation

Removal

1   Disconnect the cable from the negative battery terminal.

2   Raise the vehicle and support it on jackstands.

3   Drain the engine oil.

4   Unbolt the exhaust system from the manifolds, then remove the exhaust system.

5   If your vehicle is equipped with a manual transmission, remove the clutch cover (Chapter 7A).

6   If your vehicle is equipped with an automatic transmission, remove the flywheel shield (Chapter 7B).

7   Remove the starter (Chapter 5).

8   Loosen the alternator mounting bolts, remove the drivebelt, then pivot the alternator out of the way.

9   Support the engine with a hoist or an engine support fixture, then remove the right rear engine mount.

10   Remove the oil pan bolts. Note the different sizes used and their locations.

11   Remove the oil pan. Do not pry between the block and the pan as damage to the sealing surfaces may result.

Installation

12   Before installing the pan, make sure that the mating surfaces on the pan sealing flange, the block and the timing cover are clean and free of oil. Make sure that the rear main bearing cap and all threaded holes are clean and free of debris. If the original pan is being reinstalled, make sure that all sealant has been removed from the pan sealing flange.

13   With all the sealing surfaces clean, apply a 1/8-inch bead of RTV-type sealant to the oil pan sealing flange.

14   Position the pan and install the bolts finger tight.

15   Tighten all bolts to the specified torque. There is no specific order for torquing the bolts, but it is a good idea to tighten the end bolts first.

16   The remainder of installation is the reverse of removal.

17   Lower the vehicle.

18   Fill the engine with oil.

19   Attach the negative battery cable, then start and run the engine.

20   Inspect the oil pan for leaks.

11   Oil pump - removal and installation

Refer to illustration 11.2

1   Remove the oil pan (Section 10).

2   Remove the pump-to-rear main bearing cap bolt **(see illustration)** and separate the pump and extension shaft from the engine. **Note:** *Refer to Part C of this Chapter for the inspection and overhaul procedures for the oil pump.*

3   To install the pump, move it into position and align the top end of the hexagonal extension shaft with the hexagonal socket in the lower end of the distributor drive gear. The distributor drives the oil pump, so it is essential that this alignment is correct.

4   Install the oil pump-to-rear main bearing cap bolt and tighten it to the specified torque.

5   Reinstall the oil pan.

6   Fill the engine with oil.

7   Check the oil level and oil pressure and look for leaks.

12   Rear main bearing oil seal - replacement (engine in vehicle)

Refer to illustrations 12.3 and 12.12

Note: *A 360 degree (one piece) lip-type seal is utilized, which allows the oil pan to remain in place when performing this procedure.*

1   Remove the transaxle (see Chapter 7).

2   Remove the driveplate (if equipped) or clutch assembly and flywheel (see Chapter 8).

3   Using a screwdriver, pry out the old seal being very careful not to scratch the crankshaft sealing surface or the seal bore **(see illustration)**.

12.12 If the special tool is not available, tap around the seal with a blunt tool, slowly working it into position

13.5 Once the starter ring gear on the flywheel is immobilized with a screwdriver, you can break loose the bolts on the crankshaft pulley

4   Thoroughly clean the seal bore and the crankshaft sealing surface.

5   Inspect the crankshaft for scratches, burrs or nicks on the sealing surface. Even small imperfections on the crankshaft surface can damage the seal lip and cause oil leaks. If slight damage is evident on the crankshaft, clean the surface with crocus cloth. If the damage is too deep to clean up with crocus cloth, the crankshaft may have to be replaced.

6   A special seal installation tool is recommended to properly seat the seal in the bore without damaging the seat.

7   Lubricate the seal bore, seal lip and the sealing surface of the crankshaft with clean engine oil.

8   Position the seal over the mandril of the tool until the dust lip on the seal bottoms squarely against the collar of the tool.

9   Position the tool and the seal onto the crankshaft, mating the dowel pin on the tool with the alignment hole in the crankshaft.

10   Secure the tool to the crankshaft with the bolts provided with the tool.

11   Install the seal by rotating the T-handle on the tool until the collar bottoms against the block, seating the seal in the bore. After the seal is properly installed, remove the tool.

12   If the special tool is not available, carefully work the seal lip over the end of the crankshaft and tap the seal in with a hammer and a blunt drift until the seal is properly seated in the bore **(see illustration)**.

13   Reinstall the remaining components in the reverse order of removal.

14   Start the engine and check for oil leaks.

13   Vibration damper - removal and installation

Refer to illustrations 13.5 and 13.7
Caution: *The inertia weight section of the vibration damper is assembled to the hub with a rubber sleeve. The removal and installation procedures outlined in this section must be followed or the movement of the*

inertia weight section of the hub will destroy the tuning of the damper and the engine timing reference.

1   Disconnect the negative cable at the battery.

2   Loosen the accessory drivebelt adjusting bolts as necessary, then remove the drivebelts, tagging each one as it is removed to simplify reinstallation (Chapter 1).

3   Raise the vehicle and support it securely on jackstands.

4   Remove the right side inner fender splash shield for access.

5   Lock the starter ring gear with a large screwdriver and remove the bolts from the accessory drivebelt pulley **(see illustration)**, then remove the pulley.

6   With the ring gear still immobilized, remove the damper retaining bolt.

7   Attach a puller to the damper. Draw the damper off the crankshaft, being careful not to drop it as it breaks free **(see illustration)**. A clawtype gear puller should not be used to draw the damper off, as it may separate the outer portion of the damper from the hub. Use only a puller which bolts to the hub.
Note: *The vibration damper has three timing notches on the inertia ring. The number 1 cylinder timing reference mark will be identified by a dab of white paint in production. If a new damper assembly is installed, mark the new assembly in the same location for future reference. The number 1 cylinder reference is the first mark clockwise from the keyway when viewing the engine from the front.*

8   Before installing the damper, coat the front cover seal contact area of the damper with engine oil.

9   Apply sealant to the key and the keyway.

10   Place the damper in position over the key on the crankshaft. Make sure the damper keyway lines up with the key.

11   Using a damper installation tool, push the damper onto the crankshaft. The special tool distributes the pressure evenly around the hub. Use a large socket, washer and the damper retaining bolt if the tool is unavail-

able. Place the damper in position, insert the bolt through the socket and washer and thread the bolt into the end of the crankshaft. As the washer is tightened against the socket, the socket will press the damper onto the nose of the crankshaft.

12   Remove the installation tool, or the socket, washer and bolt, and install the damper retaining bolt. Tighten the bolt to the specified torque.

13   Install the accessory drive pulley and retaining bolts. Tighten them to the specified torque.

14   Install the right inner fender splash shield (Chapter 11).

15   Lower the vehicle.

16   Install the accessory drivebelts and adjust them (Chapter 1).

17   Attach the negative battery cable to the battery.

14   Crankcase front cover - removal and installation

Refer to illustrations 14.7a and 14.7b

1   If your vehicle is equipped with air conditioning, remove the compressor and mounting bracket. **Caution:** *Do not disconnect any of the air conditioning system hoses without having the system depressurized by a GM dealer or air conditioning technician.*

2   Remove the water pump as described in Chapter 3.

3   Remove the vibration damper (Section 13).

4   Raise the vehicle and support it securely on jackstands.

5   Remove the oil pan-to-front cover bolts (Section 10).

6   Lower the vehicle.

7   Remove the water pump and front cover mounting bolts and separate the cover from the engine. Note that some of the timing cover bolts pass through to the oil pan **(see illustrations)**.

8   Clean all oil, dirt and old gasket material from the sealing surfaces of the front cover

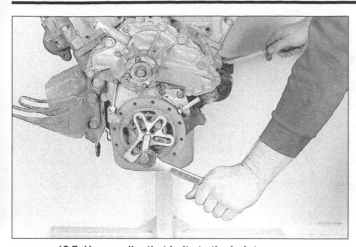

13.7 Use a puller that bolts to the hub to remove the vibration damper

14.7a Locations of the remaining front cover mounting bolts (the bottom two bolts and the top two are not visible)

and block. Replace the front cover oil seal as described in Section 15.

9   Install a new gasket, making sure that you don't damage any of the sealing surfaces.

10   Apply a 1/8-inch bead of RTV-type sealant to the oil pan contact surface of the cover.

11   Place the front cover in position on the engine block and install the mounting bolts. Tighten the bolts to the specified torque within five minutes of applying the RTV sealant. Note that the specified torque for the large bolts is not the same as for the smaller bolts.

12   Install the water pump (Chapter 3) and tighten the water pump bolts and nut to the specified torque. Note that there are three different size water pump bolts and each size bolt has a different torque specification.

13   Raise the vehicle.

14   Install the oil pan-to-cover bolts and tighten them to the specified torque.

15   Install the vibration damper (Section 13).

16   Lower the vehicle.

17   Install the air conditioning compressor

and bracket (Chapter 3).

18   Install and adjust the accessory drivebelts (Chapter 1).

19   Fill the cooling system with coolant (Chapter 1).

20   Attach the negative battery cable to the battery.

---

15   Front cover oil seal - replacement

Refer to illustrations 15.9 and 15.11

With front cover installed on engine

1   Remove the vibration damper (Section 13).

2   Pry out the old seal from the crankcase front cover with a large screwdriver. Be careful not to damage the surface of the crankshaft.

3   Lubricate the new seal with clean engine oil.

4   Place the seal in position with the lip

(open side) facing the engine.

5   Drive the seal into the cover until it is seated. A special tool is available for this purpose. These tools are designed to exert even pressure around the entire circumference of the seal as it is hammered into place. A section of large-diameter pipe or a large socket can also be used.

6   Install the vibration damper (Section 13).

7   Start the engine and check for leaks.

With front cover removed from engine

Note: *This method is preferable to the above method. Because the cover can be supported while the old seal is removed and the new one is installed, there is less possibility of damaging the cover and no chance of damaging the crankshaft.*

8   Remove the crankcase front cover (Section 14).

9   Using a large screwdriver, pry the old seal out of the front cover. Alternatively, support the cover and drive the seal out from the rear **(see illustration)**. Be careful not to damage the cover.

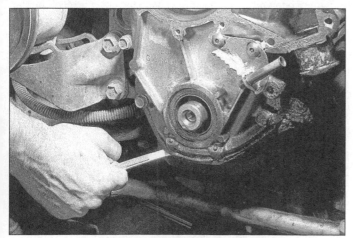

14.7b Some of the timing cover bolts pass through to the oil pan

15.9 The front cover oil seal can be pried out of the timing cover while the cover is still on the engine, but it's much easier to drive the old seal out with a hammer and punch while the cover is on a workbench

15.11 Drive the new oil seal into the front cover with a block of wood and a hammer

10   With the front of the cover facing up, place the new seal in position with the lip (open end of the seal) facing toward the inside of the cover.

11   Using a wooden block and hammer, drive the new seal into the cover until it is completely seated **(see illustration)**.

12   Install the cover (Section 14).

16   Timing chain and sprockets - inspection, removal and installation

Refer to illustrations 16.8, 16.9, 16.11 and 16.13

1   Disconnect the cable from the negative battery terminal.

2   Remove the vibration damper (Section 13).

3   Remove the crankcase front cover (Section 14).

Inspection

4   Before removing the chain and sprockets, visually inspect the teeth on the sprockets for signs of wear and check the chain for looseness.

5   If either or both sprockets show any

16.8 Turn the crankshaft until the marks (arrows) on the camshaft and crankshaft timing gears are in perfect alignment

signs of wear (edges on the teeth of the camshaft sprocket not square, bright or blue areas on the teeth of either sprocket, chipping, pitting, etc.), they should be replaced with new ones. Wear in these areas is very common.

6   If the timing chain has not been replaced recently, or if the engine has over 25,000 miles on it, the chain is almost certainly in need of replacement. Failure to replace a worn timing chain can result in erratic engine performance, loss of power and lowered gas mileage.

7   If either the camshaft sprocket, the crankshaft sprocket or the timing chain is worn, all three components must be replaced as a set. If you determine that the timing components require replacement, proceed as follows.

Removal

8   Turn the crankshaft until the marks on the camshaft and crankshaft are in exact alignment with the mark on the camshaft sprocket at 6 o'clock and the mark on the crankshaft sprocket at 12 o'clock **(see illustration)**. At this point, the number one and four pistons are at top dead center with the

number four piston in the firing position (you can verify this by checking the position of the rotor in the distributor). **Caution:** *Do not attempt to remove either sprocket or the timing chain until this step is completed.*

9   Remove the three camshaft sprocket retaining bolts by inserting a screwdriver into the sprocket holes and loosening the bolts with a breaker bar **(see illustration)**. If necessary, tap the sprocket with a softfaced hammer to dislodge it.

10   If you are pulling the camshaft rather than replacing the timing gear set, you may be able to remove the camshaft sprocket from the end of the camshaft without removing the entire gear set. Try to slide the sprocket off until it is free of the dowel pin that positions it on the end of the camshaft. If you are unable to do so, you will also have to remove the crankshaft sprocket.

11   Remove the crankshaft sprocket from the nose of the crankshaft with a puller **(see illustration)**. **Caution:** *Do not turn the crankshaft or camshaft after the sprockets and timing chain are removed.*

Installation

12   Push the crankshaft sprocket onto the nose of the crankshaft, aligning it with the key, until it seats against the shoulder.

13   Lubricate the thrust (rear) surface of the camshaft sprocket with moly-base grease or engine assembly lube **(see illustration)**.

14   Making sure that the timing marks are aligned as described in Step 8, slip the chain over the teeth of the camshaft and crankshaft sprockets. Then place the camshaft sprocket in place on the end of the camshaft and draw it into place by slowly tightening the three retaining bolts. **Caution:** *Do not hammer or attempt to drive the camshaft sprocket into place.*

15   With the chain and both sprockets in place, check again to ensure that the timing marks on the two sprockets are properly aligned. If they aren't, remove the cam sprocket and timing chain, turn the camshaft enough to change the chain position on the crankshaft sprocket one tooth, reinstall the

16.9 Insert a screwdriver into one of the camshaft sprocket holes to prevent it from turning, then remove each bolt

16.11 The crankshaft sprocket must be removed with a puller

16.13 Lubricate the thrust surface of the camshaft timing sprocket with moly base grease or engine assembly lube

chain and camshaft sprocket and recheck the alignment of the timing marks. Repeat this procedure as many times as necessary until the marks are in exact alignment.
16 Lubricate the chain with engine oil and install the remaining components in the reverse order of removal.

17 Engine mounts - replacement with engine in vehicle

Refer to illustrations 17.5 and 17.14
Note: *The engine and transaxle mounts are non-adjustable and seldom require service. However, mounts can eventually deteriorate. Broken or split mounts should be replaced immediately, because the additional strain placed on other mounts and driveline components will cause further damage. To determine the condition of the mounts on your vehicle, raise the engine with a floor jack and a block of wood under the pan. Raise it just enough to take the weight off the mounts and to place the mount rubber under slight ten-*

sion. If you detect a hard rubber surface covered with heat cracks, if the rubber has separated from its metal backing plate or if the rubber has split through the center, replace it.

Engine mount
1 Remove the engine compartment lid and side cover panels (Chapter 11). **Note:** *Do not remove the torsion rod retaining bolts.*
2 Remove the torque reaction strut bolt.
3 Raise the vehicle and support it securely on jackstands.
4 Using a hoist or a jack with a wood block under the oil pan, raise the engine slightly. **Note:** *Raise the engine just enough so that the old mount can be slipped out and the new mount can be slipped into place.*
5 Remove the engine mount-to-cradle nuts (**see illustration**).
6 Remove the upper mount-to-engine support bracket nuts.
7 Remove the old engine mount.
8 Install the new mount.
9 Tighten the engine mount-to-cradle nuts to the specified torque.
10 Tighten the engine mount-to-bracket nuts to the specified torque.
11 Lower the vehicle.
12 Install the torque reaction strut bolt and tighten it to the specified torque.
13 Remove the engine support.
14 Inspect the transaxle mounts for proper alignment (**see illustration**). If window A is not properly located, loosen the mount-to-cradle retaining nuts to permit the mount to reposition itself. **Caution:** *If the transaxle mounts are allowed to remain out of position, drive train component failure could occur.*
15 Install the engine compartment lid and side cover panels.

Torque reaction strut
16 Remove the strut-to-engine bracket bolt and nut.
17 Remove the strut-to-chassis bolt and nut.

18 Remove the strut.
19 Install the new strut.
20 Install the strut-to-chassis bolt and nut and tighten it to the specified torque.
21 Install the strut-to-engine bracket bolt and nut and tighten it to the specified torque.

Front transaxle mount
22 Remove the engine compartment lid and side cover panels (Chapter 11). **Note:** *Do not remove the torsion rod retaining bolts.*
23 Support the engine and transaxle.
24 Raise the vehicle and support it securely on jackstands.
25 Remove the mount-to-cradle nuts and support bracket.
26 Remove the forward transaxle mount and shield.
27 Position the shield and new mount.
28 Tighten the mount-to-cradle nuts to the specified torque.
29 Tighten the support bracket nuts to the specified torque.
30 Lower the vehicle.
31 Install the engine compartment lid and side cover panels (Chapter 11).

Rear transaxle mount
32 Remove the engine compartment lid and side cover panels (Chapter 11). **Note:** *Do not remove the torsion rod retaining bolts.*
33 Raise the vehicle and support it securely on jackstands.
34 Support the engine with a jack and block of wood.
35 Remove the mount-to-cradle nuts.
36 Remove the support bracket nuts.
37 Position the new rear transaxle mount.
38 Tighten the mount-to-cradle nuts to the specified torque.
39 Tighten the mount-to-bracket nuts to the specified torque.
40 Lower the vehicle.
41 Install the engine compartment lid and side cover panels.

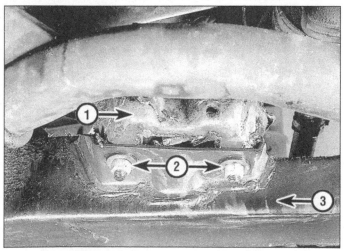

17.5 Typical engine mount-to-cradle mounting details

1 Engine mount
2 Mounting nuts
3 Cradle

17.14 When the forward mount is properly installed there should be equal space on each side of the lower mount bracket (A) in relationship to the upper bracket (B)

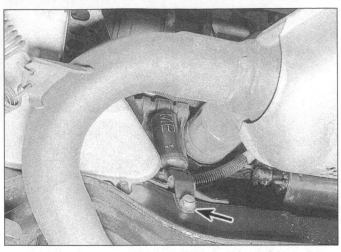

19.18 Remove the lower engine strut bolt (arrow) from the cradle crossmember

19.23 Remove the front engine mount nuts from the right cradle frame rail

18   Engine removal - methods and precautions

If you decide that the engine must be removed for overhaul or major repair work, the following preliminary steps will simplify the job.

Locate a suitable work area. Adequate work space for your vehicle, as well as a suitable storage space during the overhaul, is essential. A shop or garage, of course, is the ideal place to work. If one isn't available, at least try to secure the use of a flat, level, clean work surface made of concrete or asphalt.

Cleaning the engine compartment and engine prior to removal will help keep tools clean and organized.

Plan the entire operation ahead of time. Obtain all of the tools and equipment you will need prior to beginning the job. A vehicle hoist is a necessity.

Besides a hoist, you will a number of tools to perform a safe, relatively easy engine removal and installation job. A basic list would include complete sets of wrenches and sockets (described in the front of this manual), wooden blocks and plenty of rags and cleaning solvent for mopping up spills.

Plan for the vehicle to be out of use for a considerable amount of time. A machine shop will be required to perform some of the work which the do-it-yourselfer cannot accomplish due to a lack of special equipment. These shops often have a busy schedule, so it would be wise to consult them before removing the engine in order to accurately estimate the amount of time required to rebuild or repair components that may need work.

If you are a novice at engine removal, have a helper standing by when you remove the motor. Advice and assistance from someone with more experience is also helpful. One person cannot perform the multitude of tasks necessary to remove the engine from this vehicle.

Always use extreme caution when removing and installing the engine. Careless actions can cause serious injury. Take your time and plan ahead. Rehearse the next step before doing it. Although it's a major undertaking, a job of this nature can be successfully accomplished.

19   Engine - removal and installation

Refer to illustrations 19.18, 19.23, 19.28, 19.37a and 19.37b

Note: *If your vehicle is equipped with air conditioning, have it discharged by a dealer or automotive A/C specialist before you start removing the engine.*

Note: *Refer to the illustrations in Chapter 2A, Section 17 for additional information concerning the engine removal procedure.*

Removal

1   If the engine is not already warm, warm it up and drain the oil (Chapter 1).

2   Disconnect the negative battery cable and the ground strap from the right front corner of the cylinder head (near the dipstick), then disconnect the cable from the positive battery terminal. Disconnect the engine harness-to-junction block (both terminals are on the power distribution block). Disconnect the fusible links just below the junction block.

3   Allow the engine sufficient time to cool down, then drain the engine coolant. Disconnect the upper radiator hose from the thermostat housing.

4   Remove the louvered side covers from above the air cleaner and vapor canister on the left side and above the battery on the right side.

5   Disconnect the rear compartment lid ground strap. Mark the alignment of the lid mounting bolts and remove them. Remove the engine compartment lid (Chapter 11).

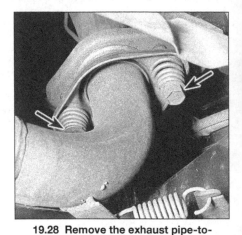

19.28 Remove the exhaust pipe-to-exhaust crossover pipe flange bolts (arrows) and springs

Caution: *Do not remove the torsion rod retaining bolts.*

6   Remove the torque reaction strut bolt that attaches the strut to the engine. Loosen the rear bolt and tilt the strut out of the way.

7   Remove the intake flex ducting between the throttle body and the elbow at the air cleaner housing.

8   Disconnect the throttle cable from the throttle body and set it aside. If your vehicle is equipped with the optional cruise control, disconnect the cruise control cable from the throttle linkage.

9   Remove all vacuum hoses connected to components mounted on the engine. Label all vacuum hoses to simplify reinstallation.

10   Disconnect the fuel line fittings between the filter and the fuel rail metering block and remove the fuel lines (Chapter 4).

11   Disconnect the main engine wire harness electrical connectors from the fuel pump relay, the coil, the distributor, the throttle position sensor, the idle air control valve, the oxygen sensor, the coolant temperature sensor, the ground between the engine and the cradle on the front left corner of the forward cylinder head, the oil pressure sending

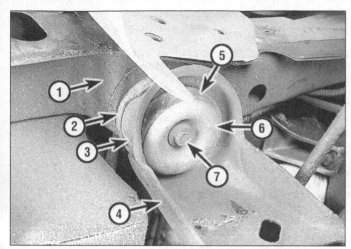

19.37a Typical rear cradle-to-body mounting details (right rear shown)

1	Right rear body rail	5	Lower cushion
2	Spacer	6	Retainer
3	Upper cushion	7	Right rear cradle
4	Frame assembly (cradle)		mounting bolt

19.37b Typical right forward cradle-to-body mounting details (right rear shown)

1	Right forward cradle	4	Forward frame mounting
	assembly		bolt
2	Bushing	5	Nut
3	Right side forward body rail		

unit, the alternator, the starter solenoid, the A/C compressor (if equipped) and all other electrical connections. Look carefully for additional grounds and disconnect them if necessary. **Note:** *Although most of the main engine wire harness connectors are distinctive enough in shape to make it unnecessary to label every wire, it's a good idea to sketch a diagram of the proper routing for the harness before removing the wire harness to facilitate proper reinstallation.*

12   If your vehicle is equipped with an automatic transmission, disconnect the shift cable, Neutral safety/backup lamp switch, speedometer sensor connector, TV cable, TCC connector and the transaxle cooler lines (Chapter 7B).

13   If your vehicle is equipped with a manual transaxle, remove the clutch slave cylinder and shield, the shifter cables, the backup light switch connector and the speedometer sensor connector (Chapter 7A).

14   Disconnect the lower radiator hose and the heater hose from the water pump.

15   Remove the rear heat shield between the muffler and the front trunk well bulkhead.

16   Disconnect, plug and set aside the air conditioning lines at the A/C compressor. If you have not already done so, disconnect the air conditioning wiring harness.

17   Remove the rear console and pull the ECM harness forward, through the grommet in the bulkhead panel, until it is out of the way.

18   Remove the lower engine strut front bolt **(see illustration).**

19   Raise the vehicle on a hydraulic hoist.

20   Remove the rear wheels. Then remove the inner fender splash shields where they attach to the engine cradle.

21   Disconnect the parking brake cables. Remove the brake calipers and support the calipers out of the way (Chapter 9). **Note:** *Do*

not disconnect the brake hoses.

22   Mark the suspension struts for realignment and remove the strut to-knuckle through bolts (see Chapter 10).

23   Remove the nuts from the front engine mount-to-cradle through studs **(see illustration).** These nuts must be removed before the cradle is lowered onto a dolly or pallet.

24   Loosen but do not remove the cradle bolts. **Warning:** *If the rear bolts are removed at this time, the unsupported cradle will fall, resulting in serious injury to anyone underneath.*

25   Slide a dolly or a pallet underneath the engine/transaxle/cradle assembly, then lower the vehicle until the engine/transaxle/cradle assembly is resting on the dolly or pallet. **Note:** *Unless the cradle support (pallet or dolly) is higher than the axle line of the wheels, it will be necessary to remove the front wheels in order to lower the vehicle enough to lay the cradle on the support.*

26   Slide jackstands underneath the forward edge of the cradle and remove the front and rear cradle bolts.

27   Make sure that the engine/transaxle/cradle assembly is resting firmly on the dolly or pallet. Raise the vehicle slightly and check to make sure that everything is disconnected. Then raise the vehicle, leaving the engine/transaxle/cradle assembly on the dolly or pallet.

28   Disconnect the exhaust pipe-to-exhaust crossover bolts and springs **(see illustration).**

29   Remove the starter motor bolts, then remove the starter motor and set it aside. Remove the transaxle-to-block bolts and separate the engine from the transaxle by prying them apart.

30   Attach a hoist to the engine lifting brackets and raise the engine slowly until the front engine mount studs are clear of the cradle. Then lift the engine out of the cradle.

Installation

31   Carefully lower the engine assembly into the cradle. Using the hoist to support the engine while you guide it into place, align the engine with the transaxle. **Caution:** *Do not attempt to force the engine and transaxle assemblies together or you might damage the splines on the transaxle input shaft or the diaphragm clutch fingers. Guide bolts inserted through several bolt holes between the engine and transaxle can simplify this operation.*

32   Install the transaxle-to-engine bolts and tighten them to the specified torque.

33   Make sure that the front engine mount studs fit through their holes in the right cradle rail, then disconnect and remove the hoist.

34   Place the starter motor in position but do not install it at this time (the starter motor is easier to position if you do it now instead of after the cradle is bolted to the frame).

35   Install a new exhaust pipe-to-exhaust crossover pipe gasket and tighten the exhaust pipe-to-exhaust crossover bolts securely. Don't forget to install the springs.

36   Carefully lower the vehicle over the engine/transaxle/cradle assembly. If the dolly or pallet has been moved since the engine was removed, make sure that it is realigned with the engine compartment so that the vehicle can be lowered almost all the way.

37   Position the engine hoist behind the vehicle in such a way that the boom is directly over the engine compartment. Attach the hoist to the lifting brackets on the engine and carefully raise the engine/transaxle/cradle assembly the final few inches into place. Align the front cradle mounting holes with the holes in the brackets protruding from the body and install the front cradle bolts finger tight. Align the rear cradle bolt holes with the holes in the body and tighten the rear bolts to the specified torque. Then tighten the front bolts to the specified torque **(see illustrations).**

38   Raise the vehicle. Install the front engine mount-to-cradle through stud nuts and tighten them to the specified torque. Also install the engine strut front bolt and tighten it securely.

39   Install the front wheels and snug the wheel lug nuts finger tight.

40   Align the paint marks on the struts with the marks on the knuckles. Install the strut mounting bolts and tighten them to the specified torque.

41   Install the calipers. Tighten the caliper bolts to the specified torque. Install the parking brake cable and adjust as necessary. Install the rear wheels and snug the wheel lug nuts finger tight.

42   If your vehicle is equipped with a manual transaxle, place the clutch dust shield in position and thread in the cover bolts a few turns to hold the cover in place. Push the starter motor into position until the starter pinion engages with the flywheel ring gear and tighten the starter motor bolts. Tighten the dust cover mounting bolts securely.

43   Attach the lower radiator hose and the heater hose to the water pump and tighten the hose clamps securely.

44   If your vehicle is equipped with air conditioning, install the line fitting bolt to the back of the compressor and tighten it securely. Also attach the electrical connectors to the compressor.

45   Feed the ECM harness back through the bulkhead panel and plug it into the main engine harness, then install the rear console pad (Chapter 11).

46   If your vehicle is equipped with an automatic transaxle, attach the cooler fittings and the TV cable (Chapter 7B).

47   Install the rear heat shield between the muffler and the trunk well forward bulkhead.

48   Connect the fuel line to the fuel filter and the return line to the small rubber hose at the rear of the fuel tank.

49   Lower the vehicle and tighten the lug nuts on all four wheels to the specified torque.

50   Connect the radiator hose to the thermostat housing.

51   Reattach the electrical connectors of the main engine harness to their respective components, including the fuel pump relay, the coil, the distributor, the throttle position sensor, the idle air control valve, the oxygen sensor, the coolant temperature sensor, the ground between the engine and the cradle on the front left corner of the forward cylinder head, the oil pressure sending unit, the alternator, the starter solenoid, the A/C compressor (if equipped) and all other electrical devices. Make sure that all ground connectors are attached properly.

52   Plug the engine harness connector into the junction block and connect the fusible links below the junction box.

53   Attach the throttle cable to the throttle body linkage and, if equipped, the cruise control cable to the throttle body linkage.

54   If your vehicle is equipped with an automatic transaxle, attach the speedometer sensor connector, the TCC switch connector, the Neutral safety/backup lamp switch and the shift cable and bracket (Chapter 7B).

55   If your vehicle is equipped with a manual transaxle, install the speedometer sensor connector, the backup light switch connector, the shifter cables and the clutch slave cylinder and shield (Chapter 7A).

56   Connect the fuel feed and return lines to the fuel metering block at the fuel rail.

57   Connect all vacuum hoses.

58   Install the throttle body-to-elbow air intake hose.

59   Add fresh engine oil and coolant.

60   Connect the battery cables and the plastic battery protector.

61   Install the rear compartment lid and side panels.

62   Install and adjust the drivebelts (Chapter 1).

63   If the engine has just been rebuilt, remove the spark plugs and crank the engine over until the oil pressure light goes out.

64   Reinstall the spark plugs and start the engine. Check for leaks.

65   Adjust the shift cables as necessary (Chapter 7).

66   Install the engine compartment lid.

Chapter 2 Part C
General engine overhaul procedures

Contents

Specifications

2.5 liter four-cylinder engine

General data

Displacement	151 cu in
Bore and stroke	4.00 x 3.00 in
Compression ratio	
1984 through 1986	9.0 to 1
1987 and 1988	8.3 to 1
Cylinder numbers	1-2-3-4 (right-to-left - transaxle is on left)
Firing order	1-3-4-2
Distributor rotation	Clockwise
Oil pressure	36 to 41 psi at 2,000 rpm

Valve and related components

Valve face angle	45 degrees
Valve seat angle	46 degrees
Valve margin	0.025 in minimum
Valve stem diameter	
1984 through 1986	0.342 to 0.343 in
1987 and 1988	
Intake	0.313 to 0.314 in
Exhaust	0.312 to 0.313 in
Stem-to-guide clearance	0.001 to 0.003 in
Valve seat width	
Intake	0.035 to 0.075 in
Exhaust	
1984 through 1986	0.058 to 0.097 in
1987 and 1988	0.058 to 0.105 in
Valve spring free length	
1984 through 1986	2.08 in
1987 and 1988	1.78 in

2.5 liter four-cylinder engine (continued)

Valve and related components (continued)

Valve spring installed height
    1984 through 1986 .... 1.69 in
    1987 and 1988 .... 1.44 in
Valve spring pressure and length
    Valve closed
        1984 through 1986 .... 78 to 86 lbs at 1.66 in
        1987 and 1988 .... 71 to 78 lbs at 1.26 in
    Valve opened
        1984 through 1986 .... 170 to 180 lbs at 1.26 in
        1987 and 1988 .... 158 to 170 lbs at 1.04 in
Valve lifter leak down rate .... 12 to 90 sec with 50 lb load
Valve lifter body diameter .... 0.8420 to 0.8427 in
Valve lifter bore diameter .... 0.8435 to 0.8445 in
Clearance in valve lifter bore .... 0.0025 in
Valve lifter plunger travel .... 0.125 in
Pushrod length
    1984 .... 9.754 in
    1985 and 1986 .... 8.3996 in
    1987 and 1988 .... 8.299 in
Cylinder head warpage limit .... 0.006 in

Camshaft

Lobe lift (intake and exhaust) .... 0.398 in
Bearing journal diameter .... 1.869 in
Bearing oil clearance .... 0.0007 to 0.0027 in
Camshaft end play .... 0.0015 to 0.0050 in

Crankshaft and connecting rods

Crankshaft end play .... 0.0035 to 0.0085 in
Connecting rod side center .... 0.006 to 0.022 in
Main journal diameter .... 2.300 in
Main journal oil clearance .... 0.0005 to 0.0022 in
Main journal taper limit .... 0.0005 in
Main journal out-of-round limit .... 0.0005 in
Crankpin diameter .... 2.000 in
Crankpin oil clearance .... 0.0005 to 0.0026 in
Crankpin taper limit .... 0.0005 in
Crankpin out-of-round limit .... 0.0005 in

Cylinder bore and piston

Cylinder bore diameter .... 4.000 in
Out-of-round limit .... 0.001 in
Taper limit .... 0.005 in
Piston-to-bore clearance .... 0.0014 to 0.0022 in (measured 1.800 in down from piston top)
Piston ring gap
    Top compression .... 0.010 to 0.020 in
    Second compression .... 0.010 to 0.020 in
    Oil control .... 0.020 to 0.060 in
Piston ring side clearance
    Top compression .... 0.002 to 0.003 in
    Second compression .... 0.001 to 0.003 in
    Oil control .... 0.015 to 0.055 in

Oil pump

Gear lash .... 0.009 to 0.015 in
Gear pocket
    1984 through 1987
        Depth .... 0.995 to 0.998 in
        Diameter .... 1.503 to 1.506 in
    1988
        Depth .... 0.514 to 0.516 in
        Diameter .... NA
Gear
    Diameter
        Drive (inner) gear .... 1.496 to 1.500 in
        Idler (outer) gear .... NA
    Side clearance .... 0.004 in maximum
End clearance .... 0.002 to 0.005 in

Torque specifications

Ft-lbs (Unless otherwise indicated)

Rocker arm nuts/bolts
    1984 and 1985 20
    1986 through 1988 24
Main bearing cap bolts 70
Connecting rod cap nuts 32
Cylinder head bolts **(refer to illustration 9.19 in Chapter 2A)**
    1984 and 1985 92
    1986 and 1987
        Step 1 18
        Step 2
            All but bolt 9 22
            Bolt 9 29
        Step 3
            All but bolt 9 .... Turn an additional 120 degrees
            Bolt 9 .... Turn an additional 90 degrees
    1988
        Step 1 18
        Step 2
            All but bolt 9 26
            Bolt 9 18
        Step 3 (all bolts) .... Turn an additional 90 degrees
Oil pump driveshaft retainer plate bolts .... 120 in-lbs
Oil pan-to-block retaining bolts
    1984 .... 75 in-lbs
    1985 .... 54 in-lbs
    1986 through 1988 .... 90 in-lbs
Oil pan drain plug 25
Oil screen support nut 37
Oil pump-to-block bolt 22
Oil pump cover bolt .... 120 in-lbs
Pushrod cover-to-block nut .... 90 in-lbs
Roller lifter guide retainer-to-block stud .... 90 in-lbs
Harmonic balancer bolt
    1984 and 1985 200
    1986 through 1988 162
Flywheel/driveplate-to-crankshaft bolts
    1984 through 1986 44
    1987 and 1988
        Automatic 55
        Manual 69
Clutch pressure plate bolts 15
Intake manifold-to-cylinder head bolt **(refer to illustration 5.20 in Chapter 2A)**
    1984 and 1985 29
    1986
        Bolt 1 25
        Bolt 2 37
        Bolt 3 28
    1987 and 1988 25
Exhaust manifold-to-cylinder head bolts **(refer to illustration 8.13 in Chapter 2A)**
    1984 through 1986 (all bolts)... 44
    1987 and 1988
        Bolts 1, 2, 6 and 7 31
        Bolts 3, 4 and 5 37
Force balancer assembly-to-block bolts
    Short bolts
        Step 1 .... 108 in-lbs
        Step 2 .... Turn an additional 75 degrees
    Long bolts
        Step1 .... 108 in-lbs
        Step 2 .... Turn an additional 90 degrees
Distributor hold-down clamp bolt
    1984 and 1985 22
    1986 through 1988 15
EGR valve-to-manifold bolt
    1984 and 1985 .... 120 in-lbs
    1986 through 1988 16
Water outlet housing bolt 20
Thermostat housing bolt... 20

Torque specifications (continued)

Ft-lbs (Unless otherwise indicated)

Water pump-to-block bolt	25
Timing gear cover-to-block bolt	90 in-lbs
Rocker arm bolt	
1984 and 1985	20
1986 through 1988	24
Rocker arm cover bolt	
1984 and 1985	72 in-lbs
1986 through 1988	45 in-lbs
Camshaft thrust plate-to-block bolt	
1984 and 1985	84 in-lbs
1986 through 1988	90 in-lbs

2.8 liter V6 engine

General data

Displacement	170.8 cu in
Bore and stroke	3.51 x 2.99 in
Compression ratio	8.5:1
Cylinder numbers (right-to-left - transaxle is on left)	
Rear bank (trunk)	1-3-5
Front bank (bulkhead)	2-4-6
Firing order	1-2-3-4-5-6
Distributor rotation	Clockwise

Valves and related components

Valve face angle	45 degrees
Valve seat angle	46 degrees
Valve seat runout limit	0.002 in
Stem-to-guide clearance	0.001 to 0.003 in
Valve seat width	
1984 through 1987	
Intake	0.049 to 0.059 in
Exhaust	0.063 to 0.075 in
1988	
Intake	0.061 to 0.073 in
Exhaust	0.067 to 0.079 in
Valve margin	0.025 in minimum
Valve spring installed height	1.58 in
Valve spring free length	1.91 in
Valve spring pressure/length	
1984 through 1987	
Valve closed	87.9 lbs at 1.58 in
Valve open	195.1 lbs at 1.18 in
1988	
Valve closed	90 lbs at 1.70 in
Valve open	215 lbs at 1.30 in
Valve lash	3/4-turn from zero lash
Damper free length	1.86 in
Cylinder head warpage limit	0.006 in

Camshaft

Lobe lift	
Intake	0.231 to 0.263 in
Exhaust	0.263 to 0.273 in
Journal diameter	1.869 to 1.871 in
Journal oil clearance	0.001 to 0.004 in
Crankshaft and connecting rods	
Crankshaft end play	0.002 to 0.008 in
Main journal diameter	2.649 to 2.650 in
Main journal taper limit	0.0002 in
Main journal out-of-round limit	0.0002 in
Main bearing oil clearance	0.0016 to 0.0032 in
Main thrust bearing clearance	0.0021 to 0.0033 in
Crank pin diameter	2.000 to 2.001 in
Crank pin taper limit	0.0002 in
Crank pin out-of-round limit	0.0002 in
Connecting rod bearing oil clearance	0.0014 to 0.0037 in
Connecting rod side clearance	0.006 to 0.017 in

Cylinder bore and piston

Cylinder bore diameter	3.506 to 3.470 in
Out-of-round limit	0.0008 in
Taper limit	0.0008 in
Compression rings	
Groove clearance	
Top ring	0.001 to 0.003 in
Second ring	0.002 to 0.004 in
Gap (both compression rings)	0.010 to 0.020 in
Oil ring	
Groove clearance	0.001 in max
Gap	0.020 to 0.055 in

Oil pump

Gear lash	0.009 to 0.015 in
Gear pocket	
Depth	1.195 to 1.198 in
Diameter	NA
Gear diameter	1.498 to 1.500 in
Side clearance	0.003 to 0.004 in
End clearance	0.002 to 0.005 in

Torque specifications

	Ft-lbs (unless otherwise indicated)
Camshaft sprocket	15 to 20
Camshaft cover (rear)	72 to 108 in-lbs
Cylinder head bolts (refer to illustration 9.10 in Chapter 2B)	
1985	65 to 75
1986 and 1987	65 to 90
1988	
VIN W	
Step 1	33
Step 2	Turn additional 90 degrees
VIN S and 9	
Step 1	40
Step 2	Turn additional 90 degrees
Connecting rod cap bolt	34 to 40
Crankshaft pulley bolts	20 to 30
Crankshaft damper bolt	66 to 84
Distributor hold-down bolt	20 to 30
Driveplate-to-torque converter bolts	25 to 35
EGR valve	13 to 18
Exhaust manifold	
Exhaust manifold bolt/stud	18
Exhaust manifold bolt	18
Crossover pipe-to-exhaust manifold bolts	22
Muffler pipe-to-crossover pipe bolts	15
Flywheel bolts	45 to 55
Intermediate intake manifold bolts	15
Lower intake manifold nut	19
Upper intake manifold bolts	19
Main bearing cap bolts	63 to 74
Oil pan mounting bolts	
Small	72 to 108 in-lbs
Large	14 to 22
Oil pump mounting bolt	26 to 35
Oil pump cover	72 to 108 in-lbs
Oil pressure switch	48 to 60 in-lbs
Oil drain plug	15 to 20
Rocker arm cover bolts	72 to 108 in-lbs
Rocker arm studs	43 to 49
Spark plugs	84 to 180 in-lbs
Starter motor bolts	26 to 37
Transmission-to-engine block	48 to 63
Water outlet bolts	20 to 30
Water pump pulley bolts	13 to 18

1   General information

Included in this portion of Chapter 2 are the general overhaul procedures for the cylinder head and internal engine components. The information ranges from advice concerning preparation for an overhaul and the purchase of replacement parts to detailed, step-by-step procedures covering removal and installation of internal engine components and the inspection of parts.

The following Sections have been written based on the assumption that the engine has been removed from the vehicle. For information concerning in-vehicle engine repair, as well as removal and installation of the external components necessary for the overhaul, see Part A or B of Chapter 2 (depending on engine type) and Section 8 of this Part.

The specifications included here are only those necessary for the inspection and overhaul procedures which follow. Refer to Part A or B for additional specifications related to the various engines covered in this manual.

2   Compression check

Refer to illustration 2.4

1   A compression check will tell you what mechanical condition the engine is in. Specifically, it can tell you if the compression is down due to leakage caused by worn piston rings, defective valves and seats or a blown head gasket. **Note:** *The engine must be at normal operating temperature and the battery must be fully charged for this check.*

2   Begin by cleaning the area around the spark plugs before you remove them. This will keep dirt from falling into the cylinders while you are performing the compression test.

3   Disconnect the coil wire from the distributor (see Chapter 5). Remove the FUEL PUMP fuse from the fuse block (see Chapters 4 and 12).

4   With the compression gauge in the number one spark plug hole, crank the engine over at least four compression strokes and observe the gauge **(see illustration)**. The compression should build up quickly in a healthy engine. Low compression on the first stroke, which does not build up during successive strokes, indicates leaking valves, a blown head gasket or a cracked head. Record the highest gauge reading obtained.

5   Repeat the procedure for the remaining cylinders. The lowest compression reading should not be less than 70% of the highest reading. No reading should be less than 100 psi.

6   If the readings are below normal, pour a couple of teaspoons of engine oil (a squirt can works great for this) into each cylinder, through the spark plug hole, and repeat the test.

7   If the compression increases after the oil

is added, the piston rings are worn. If the compression does not increase significantly, the leakage is occurring at the valves or head gasket. Leakage past the valves may be caused by burned valve seats or faces or warped, cracked or bent valves.

8   If two adjacent cylinders have equally low compression, there is a strong possibility that the head gasket between them is blown. The appearance of coolant in the combustion chambers or the crankcase would verify this condition.

9   If the compression is higher than normal, the combustion chambers are probably coated with carbon deposits. If that it the case, the cylinder head(s) should be removed and decarbonized.

10   If compression is down or varies greatly between cylinders, it would be a good idea to have a leak-down test performed by an automotive repair shop. This test will pinpoint exactly where the leakage is occurring and how severe it is.

3   Top Dead Center (TDC) for number one piston - locating

Note: *The following procedure is based on the assumption that the distributor is correctly installed. If you are trying to locate TDC to install the distributor correctly, piston position must be determined by feeling for compression at the number one spark plug hole, then aligning the ignition timing marks as described in Step 6.*

1   Top Dead Center (TDC) is the highest point in the cylinder that each piston reaches as it travels up-and-down when the crankshaft turns. Each piston reaches TDC on the compression stroke and again on the exhaust stroke, but TDC generally refers to piston position on the compression stroke.

2   Positioning the piston(s) at TDC is an essential part of many procedures such as valve train component removal and distributor removal.

3   Before beginning this procedure, be sure to place the transaxle in Neutral and apply the parking brake or block the rear wheels. Also, disable the ignition system by detaching the coil wire from the center terminal of the distributor cap and grounding it on the block with a jumper wire (models with a distributor) or disconnecting the small-wire electrical connector from the coil pack (models with direct ignition). Remove the spark plugs (see Chapter 1).

4   When looking at the drivebelt end of the engine, normal crankshaft rotation is clockwise. In order to bring any piston to TDC, the crankshaft must be turned with a socket and ratchet attached to the bolt threaded into the center of the lower drivebelt pulley (vibration damper) on the crankshaft.

5   Have an assistant turn the crankshaft with a socket and ratchet as described above while you hold a finger over the number one spark plug hole. **Note:** *See the Specifications*

2.4 Using a compression gauge to check cylinder compression

for the engine you are working on for the number one cylinder location.

6   When the piston approaches TDC, air pressure will be felt at the spark plug hole. Have your assistant stop turning the crankshaft when the timing marks at the crankshaft pulley are aligned. **Note:** *On models with no distributor (direct ignition systems), there may be no timing marks visible. After you feel pressure, stop turning the crankshaft, then insert a plastic pen into the spark plug hole. As the piston rises, the pen will be pushed out. Note where the pen stops moving out – this is TDC.*

7   If the timing marks are bypassed, turn the crankshaft two complete revolutions clockwise until the timing marks are properly aligned.

8   After the number one piston has been positioned at TDC on the compression stroke, TDC for any of the remaining pistons can be located by turning the crankshaft one-half turn (180-degrees) on four-cylinder engines or one-third turn (120-degrees) on V6 engines to get to TDC for the next cylinder in the firing order.

4   Engine overhaul - general information

Refer to illustration 4.4

It is not always easy to determine when, or if, an engine should be completely overhauled, as a number of factors must be considered.

High mileage is not necessarily an indication that an overhaul is needed, while low mileage does not preclude the need for an overhaul. Frequency of servicing is probably the most important consideration. An engine that has had regular and frequent oil and filter changes, as well as other required maintenance, will most likely give many thousands of miles of reliable service. Conversely, a neglected engine may require an overhaul very early in its life.

Excessive oil consumption is an indication that piston rings and/or valve guides are

4.4 On later 2.8L V6 engines, the oil pressure sending unit is located just behind the oil filter adapter (arrow)

in need of attention. Make sure, however, that oil leaks are not responsible before deciding that the rings and guides are bad. Have a compression or leak-down test performed by an experienced tune-up mechanic to determine the extent of the work required.

If the engine is making obvious knocking or rumbling noises, the connecting rod and/or main bearings are probably at fault. Check the oil pressure with a gauge, installed in place of the oil pressure sending unit, and compare it to the Specifications. **Note:** *The oil pressure sending unit on most models is near the oil filter* **(see illustration).** If it is extremely low, the bearings and/or oil pump are probably worn out.

Loss of power, rough running, excessive valve train noise and high fuel consumption rates may also point to the need for an overhaul, especially if they are all present at the same time. If a complete tune-up does not remedy the situation, major mechanical work is the only solution.

An engine overhaul involves restoring the internal parts to the specifications of a new engine. During an overhaul, the piston rings are replaced and the cylinder walls are reconditioned (rebored or honed). If a rebore is done, new pistons are also required. The main and connecting rod bearings are replaced with new ones and, if necessary, the crankshaft may be reground to restore the journals. Generally, the valves are serviced as well, since they are usually in less-than-perfect condition at this point. While the engine is being overhauled, other components, such as the carburetor, distributor, starter and alternator can be rebuilt as well. The end result should be a like-new engine that will give many thousands of trouble-free miles.

Before beginning the engine overhaul, read through the entire procedure to familiarize yourself with the scope and requirements of the job. Overhauling an engine is not difficult, but it is time consuming. Plan on the vehicle being tied up for a minimum of two weeks, especially if parts must be taken to an automotive machine shop for repair or reconditioning. Check on availability of parts and make sure that any necessary special tools and equipment are obtained in advance. Most work can be done with typical hand tools, although a number of precision

measuring tools are required for inspecting parts to determine if they must be replaced. Often an automotive machine shop will handle the inspection of parts and offer advice concerning reconditioning and replacement. **Note:** *Always wait until the engine has been completely disassembled and all components, especially the engine block, have been inspected before deciding what service and repair operations must be performed by an automotive machine shop.* Since the block's condition will be the major factor to consider when determining whether to overhaul the original engine or buy a rebuilt one, never purchase parts or have machine work done on other components until the block has been thoroughly inspected. As a general rule, time is the primary cost of an overhaul, so it does not pay to install worn or sub-standard parts.

As a final note, to ensure maximum life and minimum trouble from a rebuilt engine, everything must be assembled with care in a spotlessly clean environment.

5   Engine rebuilding alternatives

The do-it-yourselfer is faced with a number of options when performing an engine overhaul. The decision to replace the engine block, piston/connecting rod assemblies and crankshaft depends on a number of factors, with the primary consideration being the condition of the block. Other considerations are cost, access to machine shop facilities, parts availability, time required to complete the project and experience.

Some of the rebuilding alternatives include:

Individual parts - If the inspection procedures reveal that the engine block and most engine components are in reusable condition, purchasing individual parts may be the most economical alternative. The block, crankshaft and piston/connecting rod assemblies should all be inspected carefully. Even if the block shows little wear, the cylinder bores should receive a finish hone; a job for an automotive machine shop.

Master kit (crankshaft kit) - This rebuild package usually consists of a reground crankshaft and a matched set of pistons, connecting rods and bearings. The

pistons will already be installed on the connecting rods. These kits are commonly available for standard cylinder bores, as well as for engine blocks which have been bored to a regular oversize.

Short block - A short block consists of an engine block with a crankshaft, camshaft and piston/connecting rod assemblies already installed. All new bearings are incorporated and all clearances will be correct. Depending on where the short block is purchased, a guarantee may be included. The existing valve train components, cylinder head and external parts can be bolted to the short block with little or no machine shop work necessary.

Long block - A long block consists of a short block plus an oil pump, oil pan, cylinder head and valve train components, timing sprockets and chain and timing chain cover. All components are installed with new bearings, seals and gaskets incorporated throughout. The installation of manifolds and external parts is all that is necessary. Some form of guarantee is usually included with the purchase.

Give careful thought to which alternative is best for you and discuss the situation with local automotive machine shops, auto parts dealers or dealership parts personnel before ordering or purchasing replacement parts.

6   Engine removal - methods and precautions

If it has been decided that an engine must be removed for overhaul or major repair work, certain preliminary steps should be taken.

Locating a suitable work area is extremely important. A shop is, of course, the most desirable place to work. Adequate work space, along with storage space for the vehicle, is very important. If a shop or garage is not available, at the very least a flat, level, clean work surface made of concrete or asphalt is required.

Cleaning the engine compartment and engine prior to removal will help keep tools clean and organized.

A vehicle hoist will also be necessary.

If the engine is being removed by a novice, a helper should be available. Advice and aid from someone more experienced would also be helpful. There are many instances when one person cannot simultaneously perform all of the operations required when lifting the engine out of the vehicle.

Plan the operation ahead of time. Arrange for or obtain all the tools and equipment you will need prior to beginning the job. Some of the equipment necessary to perform engine removal and installation safely and with relative ease are, in addition to a vehicle hoist, complete sets of wrenches and sockets as described in the front of this manual, wooden blocks and plenty of rags and cleaning solvent for mopping up the inevitable spills.

Plan for the vehicle to be out of use for a considerable amount of time. A machine shop will be required to perform some of the work which the do-it-yourselfer cannot accomplish due to a lack of special equipment. These shops often have a busy schedule, so it would be wise to consult them before removing the engine in order to accurately estimate the amount of time required to rebuild or repair components that may need work.

Always use extreme caution when removing and installing the engine. Serious injury can result from careless actions. Plan ahead. Take your time and a job of this nature, although major, can be accomplished successfully.

7　Engine overhaul disassembly sequence

1　It is much easier to disassemble and work on the engine if it is mounted on a portable engine stand. These stands can often be rented for a reasonable fee from an equipment rental yard. Before the engine is mounted on a stand, the flywheel/driveplate should be removed from the engine (refer to Chapter 8).

2　If a stand is not available, it is possible to disassemble the engine with it blocked up on a sturdy workbench or on the floor. Be careful not to tip or drop the engine when working without a stand.

3　If you are going to obtain a rebuilt engine, all external components must come off your old engine to be transferred to the replacement engine, just as they will if you are doing a complete engine overhaul yourself. These include:

　　Alternator and brackets
　　Emissions control components
　　Distributor or coil pack, spark plug wires
　　　and spark plugs
　　Thermostat and housing cover
　　Water pump
　　Fuel injection components
　　Intake and exhaust manifolds
　　Oil filter
　　Fuel pump
　　Engine mounts
　　Flywheel or driveplate

Note: *When removing the external components from the engine, pay close attention to details that may be helpful or important during installation. Note the installed position of gaskets, seals, spacers, pins, washers, bolts and other small items.*

4　If you are obtaining a short block, which consists of the engine block, crankshaft, pistons and connecting rods all assembled, then the cylinder head, oil pan and oil pump will have to be removed as well. See *Engine rebuilding alternatives* for additional information regarding the different possibilities to be considered.

5　If you are planning a complete overhaul, the engine must be disassembled and the

8.3　The dial indicator plunger must be positioned directly above and in-line with the pushrod (use a short length of vacuum hose to hold the plunger over the pushrod end, if you encounter difficulty keeping the plunger on the pushrod)

components removed in the following order:

　　Valve cover
　　Intake and Exhaust manifolds
　　Rocker arms and pushrods
　　Valve lifters
　　Cylinder head
　　Timing chain cover
　　Timing chain/sprockets/gears
　　Camshaft
　　Oil pan
　　Oil pump
　　Piston/connecting rod assemblies
　　Crankshaft

6　Before beginning the disassembly and overhaul procedures, make sure the following items are available:

　　Common hand tools
　　Small cardboard boxes or plastic bags
　　　for storing parts
　　Gasket scraper
　　Ridge reamer
　　Vibration damper puller
　　Micrometers
　　Small hole gauges
　　Telescoping gauges
　　Dial indicator set
　　Valve spring compressor
　　Cylinder surfacing hone
　　Piston ring groove cleaning tool
　　Electric drill motor
　　Tap and die set
　　Wire brushes
　　Cleaning solvent

8　Camshaft - removal, inspection and installation

Camshaft lobe lift check

With cylinder head(s) installed
Refer to illustration 8.3

1　In order to determine the extent of cam lobe wear, the lobe lift should be checked

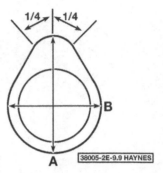

8.9　To verify camshaft lobe lift, measure the major (A) and minor (B) diameters of each lobe with a micrometer or vernier caliper - subtract each minor diameter from the major diameter to arrive at the lobe lift

prior to camshaft removal. Refer to Part A or B and remove the valve cover(s).

2　Position the number one piston at TDC on the compression stroke (see Section 3).

3　Beginning with the number one cylinder valves, loosen the rocker arm nuts and pivot the rocker arms sideways. Mount a dial indicator on the engine and position the plunger against the top of the first pushrod. The plunger must be directly in line with the pushrod **(see illustration)**.

4　Zero the dial indicator, then very slowly turn the crankshaft in the normal direction of rotation (clockwise) until the indicator needle stops and begins to move in the opposite direction. The point at which it stops indicates maximum cam lobe lift.

5　Record this figure for future reference, then reposition the piston at TDC on the compression stroke.

6　Move the dial indicator to the remaining number one cylinder rocker arm and repeat the check. Be sure to record the results for each valve.

7　Repeat the check for the remaining valves. Since each piston must be at TDC on the compression stroke for this procedure, work from cylinder-to-cylinder, following the firing order sequence.

8　After the check is complete, compare the results to this Chapter's Specifications. If camshaft lobe lift is less than specified, cam lobe wear has occurred and a new camshaft should be installed.

With cylinder head(s) removed
Refer to illustration 8.9

9　If the cylinder heads have already been removed, an alternate method of lobe measurement can be used. Remove the camshaft, as described below. Using a micrometer, measure the lobe at its highest point. Then measure the base circle perpendicular (90-degrees) to the lobe **(see illustration)**. Do this for each lobe and record the results.

10　Subtract the base circle measurement from the lobe height. The difference is the lobe lift. See Step 8 above.

8.13 On the 2.5L four-cylinder engine, turn the crankshaft until the holes in the gear are aligned with the thrust plate bolts, then remove them with a ratchet and socket

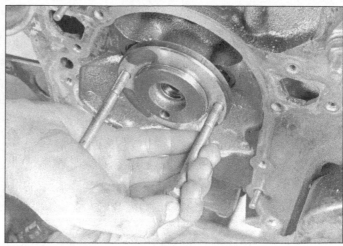

8.15 On all except 2.5L four-cylinder engines, thread long bolts into the sprocket bolt holes to use as a handle when removing and installing the camshaft

Removal

Refer to illustrations 8.13, 8.15 and 8.17

11   Refer to the appropriate Sections in Part A or B and remove the timing chain/gear cover, chain and sprockets (if equipped), hydraulic lifters and pushrods.

12   Remove the distributor (see Chapter 5).

13   Remove the camshaft thrust plate-to-block bolts. On 2.5L models, you can access these bolts through the holes in the camshaft gear **(see illustration)**

14   On 2.5L models, slide out the camshaft and gear together as an assembly.

15   On 2.8L models, thread long bolts into the camshaft sprocket bolt holes to use as a handle when removing the camshaft from the block **(see illustration)**.

16   Carefully pull the camshaft out. Support the cam near the block so the lobes don't nick or gouge the bearings as it's withdrawn.

17   When reinstalling the camshaft, be sure to coat the bearing journals and lobes, as well as the wear surfaces of all other valve train components, with moly-base grease or engine assembly lube. Again, install the camshaft very carefully to avoid damaging the camshaft bearings. The remainder of installation is the reverse of removal. On 2.5L four-cylinder engines, be sure to align the timing gear marks **(see illustration)**.

Inspection

Camshaft

Refer to illustration 8.19

18   After the camshaft has been removed from the engine, cleaned with solvent and dried, inspect the bearing journals for uneven wear, pitting and evidence of seizure. If the journals are damaged, the bearing inserts in the block are probably damaged as well. Both the camshaft and bearings will have to be replaced.

19   Measure the bearing journals with a micrometer **(see illustration)** to determine if they're excessively worn or out-of-round.

20   Check the camshaft lobes for heat discoloration, score marks, chipped areas, pitting and uneven wear. If the lobes are in good condition, and if the lobe lift measurements are as specified, the camshaft can be reused.

Camshaft bearings

21   Check the bearings in the block for wear and damage. Look for galling, pitting and discolored areas.

22   The inside diameter of each bearing can be determined with a telescoping gauge and outside micrometer or an inside micrometer. Subtract the camshaft bearing journal diameters from the corresponding bearing inside diameters to obtain the bearing oil clearance. If it's excessive, new bearings will be required regardless of the condition of the originals.

23   Camshaft bearing replacement requires special tools and expertise that place it outside the scope of the home mechanic. Take the block to an automotive machine shop to ensure the job is done correctly.

8.17 Align the timing marks as shown here when installing the camshaft on the 2.5L four-cylinder engine

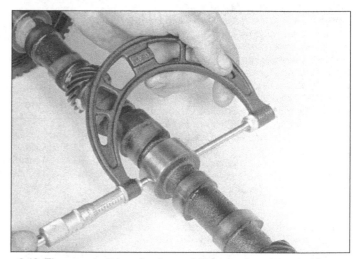

8.19 The camshaft bearing journal diameter is subtracted from the bearing inside diameter to obtain the oil clearance, which must be as listed in this Chapter's Specifications

9.3a Use a valve spring compressor to compress the springs, then remove the keepers from the valve stem with a magnet or small needle-nose pliers

9.3b If you can't pull the valve through the guide, deburr the edge of the stem end and the area around the top of the keeper groove with a file or whetstone

10.12 Check the cylinder head gasket surface for warpage by trying to slip a feeler gauge under the straightedge (see this Chapter's Specifications for the maximum warpage allowed and use a feeler gauge of that thickness)

Gears (2.5L engine only)

24   Check the camshaft drive and driven gears for cracks, missing teeth and excessive wear. If the teeth are highly polished, pitted and galled, or if the outer hardened surface of the teeth is flaking off, new parts will be required. If one gear is worn or damaged, replace both gears as a set. Never install one new and one used gear. **Note:** *We recommend replacing the gears as a set whenever they are removed. The camshaft gear must be pressed off the camshaft by an automotive machine shop.*

25   Check the end clearance with a feeler gauge and compare it to the Specifications. If it's less than the minimum specified, the spacer ring should be replaced. If it's excessive, the thrust plate must be replaced. In either case, the gear will have to be pressed off the camshaft, so take the parts to an automotive machine shop.

9   Cylinder head - disassembly

Refer to illustrations 9.3a and 9.3b

Note: *New and rebuilt cylinder heads are commonly available for most engines at dealerships and auto parts stores. Due to the fact that some specialized tools are necessary for the disassembly and inspection procedures, and replacement parts may not be readily available, it may be more practical and economical for the home mechanic to purchase a replacement head rather than taking the time to disassemble, inspect and recondition the original head.*

1   Cylinder head disassembly involves removal of the intake and exhaust valves and their related components. If they are still in place, remove the nuts or bolts and pivot balls, then separate the rocker arms and/or shafts from the cylinder head. Label the parts or store them separately so they can be reinstalled in their original locations.

2   Before the valves are removed, arrange to label and store them, along with their related components, so they can be kept

separate and reinstalled in the same valve guides. Measure the valve spring installed height (see Section 13) for each valve and compare it to the Specifications. If it is greater than specified, the valves will require servicing. Tell the automotive machine shop who does this work about this out-of-spec condition.

3   Compress the valve spring with a spring compressor and remove the keepers **(see illustration)**. Carefully release the valve spring compressor and remove the retainer, the shield (if so equipped), the springs, the valve guide seal and/or O-ring seal, any spring seat shims and the valve from the head. If the valve binds in the guide (won't pull through), push it back into the head and deburr the area around the keeper groove with a fine file or whetstone **(see illustration)**.

4   Repeat the procedure for the remaining valves. Remember to keep together all the parts for each valve so they can be reinstalled in the same locations.

5   Once the valves have been removed and safely stored, the head should be thoroughly cleaned and inspected. If a complete engine overhaul is being done, finish the engine disassembly procedures before beginning the cylinder head cleaning and inspection process.

10   Cylinder head - cleaning and inspection

1   Thorough cleaning of the cylinder head and related valve train components, followed by a detailed inspection, will enable you to decide how much valve service work must be done during the engine overhaul.

Cleaning

2   Scrape away all traces of old gasket material and sealing compound from the head gasket, intake manifold and exhaust manifold sealing surfaces.

3   Remove any built-up scale around the coolant passages.

4   Run a stiff wire brush through the oil holes to remove any deposits that may have formed in them.

5   It is a good idea to run a tap into each of the threaded holes to remove any corrosion and thread sealant that may be present. If compressed air is available, use it to clear the holes of the debris produced by this operation.

6   Clean the exhaust and intake manifold stud threads with a die. Clean the rocker arm pivot bolt or stud threads with a wire brush.

7   Clean the cylinder head with solvent and dry it thoroughly. Compressed air will speed the drying process and ensure that all holes and recessed areas are clean. **Note:** *Decarbonizing chemicals are available and may prove very useful when cleaning cylinder heads and valve train components. They are very caustic and should be used with caution. Be sure to follow the instructions on the container.*

8   Clean the rocker arms, pivot balls and pushrods with solvent and dry them thoroughly. Compressed air will speed the drying process and can be used to clean out the oil passages.

9   Clean all the valve springs, keepers, retainers, shields and spring seat shims with solvent and dry them thoroughly. Do the components from one valve at a time to avoid mixing up the parts.

10   Scrape off any heavy deposits that may have formed on the valves, then use a motorized wire brush to remove deposits from the valve heads and stems. Again, make sure the valves do not get mixed up.

Inspection

Refer to illustrations 10.12, 10.14a, 10.14b, 10.19, 10.20, 10.21a and 10.21b

Cylinder head

11   Inspect the head very carefully for cracks, evidence of coolant leakage or other damage. If cracks are found, a new cylinder head should be obtained.

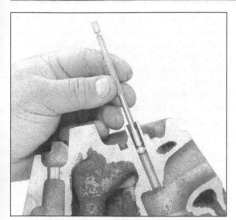

10.14a Use a small-hole gauge to determine the inside diameter of the valve guides (the gauge is then measured with a micrometer)

10.14b A dial indicator can also be used to determine the valve stem-to-guide clearance (the reading must be divided by two to obtain the actual clearance

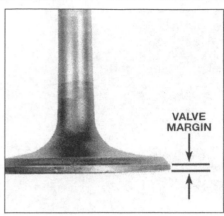

10.19 The margin width on each valve must be as specified (if no margin exists, the valve must be replaced)

12   Using a straightedge and feeler gauge, check the head gasket mating surface for warpage **(see illustration)**. If the warpage exceeds 0.006-inch over the length of the head, it can be resurfaced at an automotive machine shop.

13   Examine the valve seats in each of the combustion chambers. If they are pitted, cracked or burned, the head will require valve service that is beyond the scope of the home mechanic.

14   Measure the inside diameter of the valve guides (at both ends and the center of each guide) with a small hole gauge and a 0-to-1-inch micrometer **(see illustration)**. Record the measurements for future reference. These measurements, along with the valve stem diameter measurements, will enable you to compute the valve stem-to-guide clearances. These clearances, when compared to the Specifications, will be one factor that will determine the extent of valve service work required. The guides are measured at the ends and at the center to determine if they are worn in a bellmouth pattern (more wear at the ends). If they are, guide reconditioning or replacement is necessary. As an alternative to using a small-hole gauge and micrometer, use a dial indicator to measure

the lateral movement of each valve stem with the valve in the guide and approximately 1/16-inch off the seat **(see illustration)**.

Rocker arm components

15   Check the rocker arm faces, where they contact the pushrod ends and valve stems, for pits, wear and rough spots. Check the pivot contact areas as well.

16   Inspect the pushrod ends for scuffing and excessive wear. Roll the pushrod on a flat surface, such as a piece of glass, to determine if it is bent.

17   Any damaged or excessively worn parts must be replaced with new ones.

Valves

18   Carefully inspect each valve face for cracks, pits and burned spots. Check the valve stem and neck for cracks. Rotate the valve and check for any obvious indication that it is bent. Check the end of the stem for pits and excessive wear. The presence of any of these conditions indicates the need for valve service by a properly equipped shop.

19   Measure the width of the valve margin on each valve and compare it to Specifications. Any valve with a margin narrower than specified will have to be replaced with a new

one **(see illustration)**.

20   Measure the valve stem diameter **(see illustration)**. **Note:** *The exhaust valves used in the 2.5L four-cylinder engine have tapered stems and are approximately 0.001-inch larger at the tip end than at the head end.* By subtracting the stem diameter from the corresponding valve guide diameter, the valve stem-to-guide clearance is obtained. Compare the results to the Specifications. If the stem-to-guide clearance is greater than specified, the guides will have to be reconditioned and new valves may have to be installed, depending on the condition of the old valves.

Valve components

21   Check each valve spring for wear and pits on the ends. Measure the free length and compare it to the Specifications **(see illustration)**. Any springs that are shorter than specified have sagged and should not be reused. Stand the spring on a flat surface and check it for squareness **(see illustration)**.

22   Check the spring retainers and keepers for obvious wear and cracks. Any questionable parts should be replaced with new ones, as extensive damage will occur in the event of failure during engine operation.

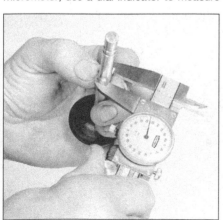

10.20 Measure the valve stem diameter at three points

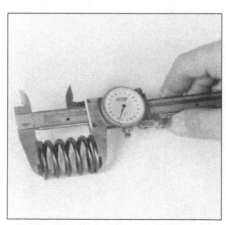

10.21a Measure the free length of each valve spring with a dial or vernier caliper

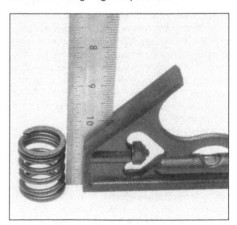

10.21b Check each valve spring for squareness

**12.6 Measuring valve spring
installed height**

**12.7 Checking the valve stem seals for
leakage (models with O-ring type seals)**

23   If the inspection process indicates that the valve components are in generally poor condition and worn beyond the limits specified, which is usually the case in an engine that is being overhauled, reassemble the valves in the cylinder head and refer to Section 12 for valve servicing recommendations.

24   If the inspection turns up no excessively worn parts, and if the valve faces and seats are in good condition, the valve train components can be reinstalled in the cylinder head without major servicing. Refer to the appropriate Section for cylinder head reassembly procedures.

11   Valves - servicing

1   Because of the complex nature of the job and the special tools and equipment needed, servicing of the valves, the valve seats and the valve guides (commonly known as a "valve job") is best left to a professional.

2   The home mechanic can remove and disassemble the head, do the initial cleaning and inspection, then reassemble and deliver the head to a dealer service department or an automotive machine shop for the actual valve servicing.

3   The dealer service department, or automotive machine shop, will remove the valves and springs, recondition or replace the valves and valve seats, recondition the valve guides, check and replace the valve springs, spring retainers and keepers (as necessary), replace the valve seals with new ones, reassemble the valve components and make sure the installed spring height is correct. The cylinder head gasket surface will also be resurfaced if it is warped.

4   After the valve job has been performed by a professional, the head will be in like-new condition. When the head is returned, be sure to clean it again to remove any metal particles and abrasive grit that may still be present from the valve service or head resurfacing operations. Use compressed air, if available, to blow out all the oil holes and passages.

12   Cylinder head - reassembly

Refer to illustrations 12.6 and 12.7

1   Regardless of whether or not the head was sent to an automotive repair shop for valve servicing, make sure it is clean before beginning reassembly.

2   If the head was sent out for valve servicing, the valves and related components will already be in place. Begin the reassembly procedure with Step 6.

3   Install new seals on each of the valve guides. Using a hammer and a deep socket, gently tap each seal into place until it is properly seated on the guide. Do not twist or cock the seals during installation or they will not seal properly on the valve stems.

4   Install the valves, taking care not to damage the new valve stem oil seals, drop the valve spring shim(s) around the valve guide boss and set the valve spring, cap and retainer in place.

5   Compress the spring with a valve compressor tool and install the valve locks. Release the compressor tool, making sure the locks are seated properly in the valve stem upper groove. If necessary, grease can be used to hold the locks in place while the compressor tool is released.

6   Double-check the installed valve spring

**13.2 A special tool is required to remove
the ridge from the top of each cylinder
(do it before removing the piston)**

height. If it was correct before reassembly it should still be within the specified limits. If it is not, install an additional valve spring seat shims (available from your dealer) to bring the height to within the specified limit **(see illustration)**.

7   On models equipped with O-ring-type seals, check the seals with a vacuum pump and adapter **(see illustration)**. A properly installed oil seal should not leak vacuum.

8   Install the rocker arms and tighten the nuts to the specified torque. Be sure to lubricate the ball pivots with moly-base grease or engine assembly lube.

13   Piston/connecting rod assembly - removal

Refer to illustrations 13.2, 13.6 and 13.8

1   Prior to removal of the piston/connecting rod assemblies, the engine should be positioned upright.

2   Using a ridge reamer, completely remove the ridge at the top of each cylinder. Follow the manufacturer's instructions provided with the ridge reaming tool **(see illustration)**. Failure to remove the ridge before attempting to remove the piston/connecting rod assemblies will result in piston breakage.

3   After the cylinder wear ridges have been removed, turn the engine upside-down.

4   Before the connecting rods are removed, check the endplay. Mount a dial indicator with its stem in line with the crankshaft and touching the side of the number one connecting rod cap.

5   Push the connecting rod backward, as far as possible, and zero the dial indicator. Next, push the connecting rod all the way to the front and check the reading on the dial indicator. The distance that it moves is the endplay. If the endplay exceeds the service limit, a new connecting rod will be required. Repeat the procedure for the remaining connecting rods.

6   An alternative method is to slip feeler gauges between the connecting rod and the crankshaft throw until the play is removed **(see illustration)**. The endplay is then equal

**13.6 Checking connecting rod endplay
with a feeler gauge**

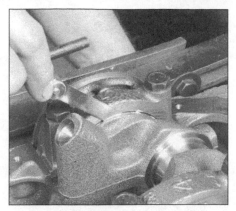

13.8 To prevent damage to the crankshaft journals and cylinder walls, slip sections of hose over the rod bolts before removing the pistons

14.1 Checking crankshaft endplay with a dial indicator

14.3 Checking crankshaft endplay with a feeler gauge

to the total thickness of the feeler gauges.

7   Check the connecting rods and connecting rod caps for identification marks. If they are not plainly marked, identify each rod and cap, using a small punch to make the appropriate number of indentations to indicate the cylinders they are associated with.

8   Loosen each of the connecting rod cap nuts approximately 1/2-turn. Remove the number one connecting rod cap and bearing insert. Do not drop the bearing insert out of the cap. Slip a short length of plastic or rubber hose over each connecting rod cap bolt to protect the crankshaft journal and cylinder wall when the piston is removed **(see illustration)** and push the connecting rod/piston assembly out through the top of the engine. Use a wooden tool to push on the upper bearing insert in the connecting rod. If resistance is felt, double-check to make sure that all of the ridge was removed from the cylinder.

9   Repeat the procedure for the remaining cylinders. After removal, reassemble the connecting rod caps and bearing inserts in their respective connecting rods and install the cap nuts finger-tight. Leaving the old bearing inserts in place until reassembly will help prevent the connecting rod bearing surfaces from being accidentally nicked or gouged.

14   Crankshaft - removal

Refer to illustrations 14.1, 14.3 and 14.4

1   Before the crankshaft is removed check the endplay. Mount a dial indicator with the stem in line with the crankshaft and just touching one of the crank throws **(see illustration)**.

2   Push the crankshaft all the way to the rear and zero the dial indicator. Next, pry the crankshaft to the front as far as possible and check the reading on the dial indicator. The distance that it moves is the endplay. If it is greater than specified, check the crankshaft thrust surfaces for wear. If no wear is apparent, new main bearings should correct the endplay.

3   If a dial indicator is not available, feeler

gauges can be used. Gently pry or push the crankshaft all the way to the front of the engine. Slip feeler gauges between the crankshaft and the front face of the thrust main bearing **(see illustration)** to determine the clearance, which is equivalent to crankshaft endplay.

4   Loosen each of the main bearing cap bolts 1/4-turn at a time, until they can be removed by hand. Check the main bearing caps to see if they are marked as to their locations. They are usually numbered consecutively from the front of the engine to the rear. If they are not, mark them with number stamping dies or a center-punch **(see illustration)**. Most main bearing caps have a cast-in arrow, which points to the front of the engine.

5   Gently tap the caps with a soft-face hammer, then separate them from the engine block. If necessary, use the main bearing cap bolts as levers to remove the caps. Try not to drop the bearing insert if it comes out with the cap.

6   Carefully lift the crankshaft out of the engine. It is a good idea to have an assistant available, since the crankshaft is quite heavy. With the bearing inserts in place in the engine block and in the main bearing caps, return the caps to their respective locations on the engine block and tighten the bolts finger-tight.

15.1a A hammer and large punch can be used to drive the soft plugs into the block

14.4 Use a centerpunch or number stamping dies to mark the main bearing caps to ensure installation in their original locations on the block - make the punch marks near one of the bolt heads

15   Engine block - cleaning

Refer to illustrations 15.1a, 15.1b and 15.10

1   Remove the soft plugs from the engine block. To do this, knock the plugs into the block, using a hammer and punch, then grasp them with large pliers and pull them back through the holes **(see illustrations)**.

15.1b Using pliers to remove a soft plug from the block

15.10 A large socket on an extension can be used to force the new soft plugs into their bores

2 Using a gasket scraper, remove all traces of gasket material from the engine block. Be very careful not to nick or gouge the gasket sealing surfaces.

3 Remove the main bearing caps and separate the bearing inserts from the caps and the engine block. Tag the bearings according to which cylinder they are removed from and whether they were in the cap or the block, then set them aside.

4 Remove the threaded oil gallery plugs from the front and back of the block.

5 If the engine is extremely dirty, it should be taken to an automotive machine shop to be steam cleaned or hot tanked. Any bearings left in the block, such as the camshaft bearings, will be damaged by the cleaning process, so plan on having new ones installed while the block is at the machine shop.

6 After the block is returned, clean all oil holes and oil galleries one more time. Brushes for cleaning oil holes and galleries are available at most auto parts stores. Flush the passages with warm water until the water runs clear, dry the block thoroughly and wipe all machined surfaces with a light, rust preventative oil. If you have access to compressed air, use it to speed the drying process and to

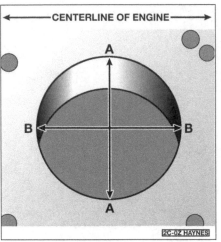

16.4a Measure the diameter of each cylinder at a right angle to engine centerline (A), and parallel to the engine centerline (B) - out-of-round is the difference between A and B; taper is the difference between A and B at the top of the cylinder and A and B at the bottom of the cylinder

blow out all the oil holes and galleries.

7 If the block is not extremely dirty or sludged up, you can do an adequate cleaning job with warm soapy water and a stiff brush. Take plenty of time and do a thorough job. Regardless of the cleaning method used, be very sure to thoroughly clean all oil holes and galleries, dry the block completely and coat all machined surfaces with light oil.

8 The threaded holes in the block must be clean to ensure accurate torque readings during reassembly. Run the proper-size tap into each of the holes to remove any rust, corrosion, thread sealant or sludge and to restore any damaged threads. If possible, use compressed air to clear the holes of debris produced by this operation. Now is a good time to thoroughly clean the threads on the head bolts and the main bearing cap bolts as well.

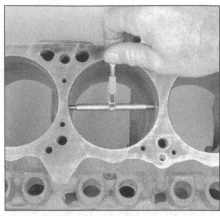

16.4b The ability to "feel" when the telescoping gauge is at the correct point will be developed over time, so work slowly and repeat the check until you're satisfied the bore measurement is accurate

9 Reinstall the main bearing caps and tighten the bolts finger-tight.

10 After coating the sealing surfaces of the new soft plugs with a non-hardening gasket sealant, such as Permatex no. 2, install them in the engine block **(see illustration)**. Make sure they are driven in straight and seated properly or leakage could result. Special tools are available for this purpose, but equally good results can be obtained using a socket with an outside diameter that will just slip into the soft plug and a hammer.

11 If the engine is not going to be reassembled right away, cover it with a large plastic trash bag to keep it clean.

16 Engine block - inspection

Refer to illustrations 16.4a, 16.4b, 16.4c, 16.7a and 16.7b

1 Thoroughly clean the engine block as described in Section 16 and double-check to

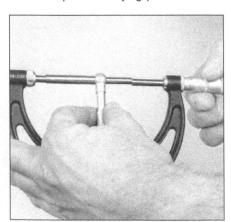

16.4c The gauge is then measured with a micrometer to determine the bore size

16.7a If this is the first time you've ever honed cylinders, you'll get better results with a "bottle brush" hone than you will with a traditional spring-loaded hone

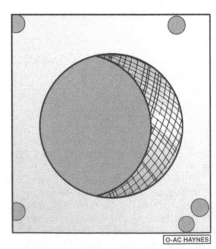

16.7b The cylinder hone should leave a smooth, crosshatch pattern with the lines intersecting at approximately a 60-degree angle

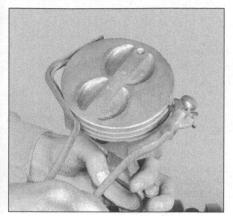

17.4 The piston ring grooves can be cleaned with a special tool, as shown here, or a piece of a broken piston ring

17.10 Check the piston ring side clearance with a feeler gauge at several points around the groove

17.11 Measure the piston diameter at a 90-degree angle to the piston pin at the specified point on the skirt (see the Specifications)

make sure that the ridge at the top of each cylinder has been completely removed.

2   Visually check the block for cracks, rust and corrosion. Look for stripped threads in the threaded holes. It is also a good idea to have the block checked for hidden cracks by an automotive machine shop that has the special equipment to do this type of work. If defects are found, have the block repaired, if possible, or replaced.

3   Check the cylinder bores for scuffing and scoring.

4   Using an expansion gauge and micrometer, measure each cylinder's diameter at the top (just under the ridge), center and bottom of the cylinder bore, parallel to the crankshaft axis **(see illustrations)**. Next, measure each cylinder's diameter at the same three locations across the crankshaft axis. Compare the results to the Specifications. If the cylinder walls are badly scuffed or scored, or if they are out-of-round or tapered beyond the limits given in the Specifications, have the engine block rebored and honed at an automotive machine shop. If a rebore is done, oversize pistons and rings will be required.

5   If the cylinders are in reasonably good condition and not worn to the outside of the limits, and if the piston-to-cylinder clearances can be maintained properly, then they do not have to be rebored. Honing is all that is necessary.

6   Before honing the cylinders, install the main bearing caps, without the bearings, and tighten the bolts to the specified torque.

7   To perform the honing operation you will need the proper-size hone (with fine stones), plenty of light oil or honing oil, some rags and an electric drill motor. Mount the hone in the drill motor, compress the stones and slip the hone into the first cylinder **(see illustration)**. Lubricate the cylinder thoroughly, turn on the drill and move the hone up and down in the cylinder at a pace which will produce a fine crosshatch pattern on the cylinder walls, with the crosshatch lines intersecting at approximately a 60-degree angle **(see illustration)**. Be sure to use plenty of lubricant and do not take off any more material than is absolutely

necessary to produce the desired finish. Do not withdraw the hone from the cylinder while it is running. Instead, shut off the drill and continue moving the hone up and down in the cylinder until it comes to a complete stop, then compress the stones and withdraw the hone. Wipe the oil out of the cylinder and repeat the procedure on the remaining cylinders. If you do not have the tools or do not desire to perform the honing operation, most automotive machine shops will do it for a reasonable fee.

8   After the honing job is complete, chamfer the top edges of the cylinder bores with a small file so the rings will not catch when the pistons are installed.

9   The entire engine block must be thoroughly washed again with warm, soapy water to remove all traces of the abrasive grit produced during the honing operation. Be sure to run a brush through all oil holes and galleries and flush them with running water. After rinsing, dry the block and apply a coat of light rust preventative oil to all machined surfaces. Wrap the block in a plastic trash bag to keep it clean and set it aside until reassembly.

17 Piston/connecting rod assembly - inspection

Refer to illustrations 17.4, 17.10 and 17.11

1   Before the inspection process can be carried out, the piston/connecting rod assemblies must be cleaned and the original piston rings removed from the pistons. **Note:** *Always use new piston rings when the engine is reassembled.*

2   Using a piston ring installation tool, carefully remove the rings from the pistons. Do not nick or gouge the pistons in the process.

3   Scrape all traces of carbon from the top (or crown) of the piston. A hand-held wire brush or a piece of fine emery cloth can be used once the majority of the deposits have been scraped away. Do not, under any circumstances, use a wire brush mounted in a

drill motor to remove deposits from the pistons. The piston material is soft and will be eroded away by the wire brush.

4   Use a piston ring groove cleaning tool to remove any carbon deposits from the ring grooves. If a tool is not available, a piece broken off an old ring will do the job. Be very careful to remove only the carbon deposits. Do not remove any metal and do not nick or scratch the sides of the ring grooves **(see illustration)**.

5   Once the deposits have been removed, clean the piston/rod assemblies with solvent and dry them thoroughly. Make sure that the oil return holes in the back sides of the ring grooves are clear.

6   If the pistons are not damaged or worn excessively, and if the engine block is not rebored, new pistons will not be necessary. Normal piston wear appears as even, vertical wear on the piston thrust surfaces and slight looseness of the top ring in its groove. New piston rings, on the other hand, should always be used when an engine is rebuilt.

7   Carefully inspect each piston for cracks around the skirt, at the pin bosses and at the ring lands.

8   Look for scoring and scuffing on the thrust faces of the skirt, holes in the piston crown and burned areas at the edge of the crown. If the skirt is scored or scuffed, the engine may have been suffering from overheating or abnormal combustion, which caused excessively high operating temperatures. The cooling and lubrication systems should be checked thoroughly. A hole in the piston crown is an indication that abnormal combustion (preignition) was occurring. Burned areas at the edge of the piston crown are usually evidence of spark knock (detonation). If any of the above problems exist, the causes must be corrected or the damage will occur again.

9   Corrosion of the piston (evidenced by pitting) indicates that coolant is leaking into the combustion chamber or the crankcase. Again, the cause must be corrected or the problem may persist in the rebuilt engine.

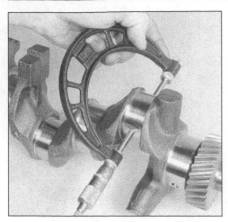

18.2 Measure the diameter of each crankshaft journal at several points to detect taper and out-of-round conditions

10   Measure the piston ring side clearance by laying a new piston ring in each ring groove and slipping a feeler gauge between the ring and the edge of the ring groove **(see illustration)**. Check the clearance at four locations around each groove. Be sure to use the correct ring for each groove; they are different. If the side clearance is greater than specified, new pistons will have to be used.
11   Check the piston-to-bore clearance by measuring the bore (see Section 16) and the piston diameter. Make sure that the pistons

and bores are correctly matched. Measure the piston across the skirt **(see illustration)**. Subtract the piston diameter from the bore diameter to obtain the clearance. If it is greater than specified, the block will have to be rebored and new pistons and rings installed. Check the piston-to-rod clearance by twisting the piston and rod in opposite directions. Any noticeable play indicates that there is excessive wear, which must be corrected. The piston/connecting rod assemblies should be taken to an automotive machine shop to have new piston pins installed and the pistons and connecting rods rebored.
12   If the pistons must be removed from the connecting rods, such as when new pistons must be installed, or if the piston pins have too much play in them, they should be taken to an automotive machine shop. While they are there, have the connecting rods checked for bend and twist, as automotive machine shops have special equipment for this purpose. Unless new pistons or connecting rods must be installed, do not disassemble the pistons from the connecting rods.
13   Check the connecting rods for cracks and other damage. Temporarily remove the rod caps, lift out the old bearing inserts, wipe the rod and cap bearing surfaces clean and inspect them for nicks, gouges or scratches. After checking the rods, replace the old bearings, slip the caps into place and tighten the nuts finger tight.

18   Crankshaft - inspection

Refer to illustration 18.2
1   Clean the crankshaft with solvent and dry it thoroughly. Be sure to clean the oil holes with a stiff brush and flush them with solvent. Check the main and connecting rod bearing journals for uneven wear, scoring, pitting or cracks. Check the remainder of the crankshaft for cracks and damage. Automotive machine shops are equipped with Magnaflux machines to check the crankshaft for cracks that may not be visible to the eye.
2   Using a micrometer, measure the diameter of the main and connecting rod journals **(see illustration)** and compare the results to the Specifications. By measuring the diameter at a number of points around the journal's circumference you will be able to determine whether or not the journal is out of round. Take the measurement at each end of the journal, near the crank counterweights, to determine whether the journal is tapered.
3   If the crankshaft journals are damaged, tapered, out-of-round or worn beyond the limits given in the Specifications, have the crankshaft reground by a reputable automotive machine shop. Be sure to use the correct undersize bearing inserts if the crankshaft is reconditioned.

19   Main and connecting rod bearings - inspection

Refer to illustration 19.3
1   Even though the main and connecting rod bearings should be replaced with new ones during the engine overhaul, the old bearings should be retained for close examination, as they may reveal valuable information about the condition of the engine.
2   Bearing failure occurs primarily because of lack of lubrication, the presence of dirt or other foreign particles, overloading the engine and corrosion. Regardless of the cause of bearing failure, it must be corrected before the engine is reassembled to prevent it from happening again.
3   When examining the bearings, remove them from the engine block, the main bearing caps, the connecting rods and the rod caps and lay them out on a clean surface in the same general position as their location in the engine. This will enable you to match any bearing problems with the corresponding crankshaft journal **(see illustration)**.
4   Dirt and other foreign particles get into the engine in a variety of ways. It may be left in the engine during assembly, or it may pass through filters or breathers. It may get into the oil, and from there into the bearings. Metal chips from machining operations and normal engine wear are often present. Abrasives are sometimes left in engine components after reconditioning, especially when parts are not thoroughly cleaned using the proper cleaning methods. Whatever the

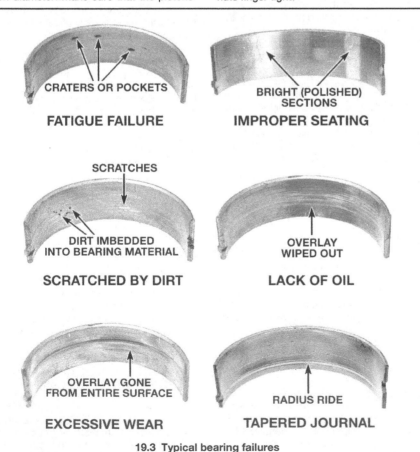

CRATERS OR POCKETS

FATIGUE FAILURE

BRIGHT (POLISHED) SECTIONS

IMPROPER SEATING

SCRATCHES

DIRT IMBEDDED INTO BEARING MATERIAL

SCRATCHED BY DIRT

OVERLAY WIPED OUT

LACK OF OIL

OVERLAY GONE FROM ENTIRE SURFACE

EXCESSIVE WEAR

RADIUS RIDE

TAPERED JOURNAL

19.3 Typical bearing failures

20.3a Use the piston to square up the ring in the cylinder prior to checking the ring end gap

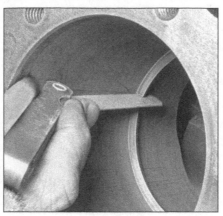

20.3b Measure the ring end gap with a feeler gauge

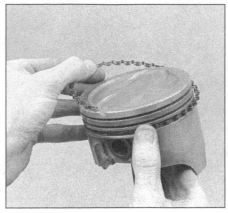

20.9a Installing the spacer/expander in the oil control ring groove

source, these foreign objects often end up embedded in the soft bearing material and are easily recognized. Large particles will not embed in the bearing and will score or gouge the bearing and shaft. The best prevention for this cause of bearing failure is to clean all parts thoroughly and keep everything spotlessly clean during engine assembly. Frequent and regular engine oil and filter changes are also recommended.

5   Lack of lubrication (or lubrication breakdown) has a number of interrelated causes. Excessive heat (which thins the oil), overloading (which squeezes the oil from the bearing face) and oil leakage or throwoff (from excessive bearing clearances, worn oil pump or high engine speeds) all contribute to lubrication breakdown. Blocked oil passages, which usually are the result of misaligned oil holes in a bearing shell, will also oil-starve a bearing and destroy it. When lack of lubrication is the cause of bearing failure, the bearing material is wiped or extruded from the steel backing of the bearing. Temperatures may increase to the point where the steel backing turns blue from overheating.

6   Driving habits can have a definite effect on bearing life. Full-throttle, low-speed operation (or *lugging* the engine) puts very high loads on bearings, which tends to squeeze out the oil film. These loads cause the bearings to flex, which produces fine cracks in the bearing face (fatigue failure). Eventually the bearing material will loosen in pieces and tear away from the steel backing. Short-trip driving leads to corrosion of bearings because insufficient engine heat is produced to drive off the condensed water and corrosive gases. These products collect in the engine oil, forming acid and sludge. As the oil is carried to the engine bearings, the acid attacks and corrodes the bearing material.

7   Incorrect bearing installation during engine assembly will lead to bearing failure as well. Tight-fitting bearings leave insufficient bearing oil clearance and will result in oil starvation. Dirt or foreign particles trapped behind a bearing insert result in high spots on the bearing which can lead to failure.

20 Piston rings - installation

Refer to illustrations 20.3a, 20.3b, 20.9a, 20.9b and 20.12

1   Before installing the new piston rings, the ring end gaps must be checked. It is assumed that the piston ring side clearance has been checked and verified correct (Section 18).

2   Lay out the piston/connecting rod assemblies and the new ring sets so the ring sets will be matched with the same piston and cylinder during the end gap measurement and engine assembly.

3   Insert the top ring into the cylinder and square it up with the cylinder walls by pushing it in with the top of the piston, until it is near the bottom of ring travel in the cylinder. To measure the end gap, slip a feeler gauge between the ends of the ring **(see illustrations)**. Compare the measurement to the Specifications.

4   If the gap is larger or smaller than specified, double-check to make sure that you have the correct rings before proceeding.

5   If the gap is too small, it must be enlarged or the ring ends may come in contact with each other during engine operation, which can cause serious damage to the engine. The end gap can be increased by filing the ring ends very carefully with a fine file. Mount the file in a vise equipped with soft jaws, slip the ring over the file with the ends contacting the file face and slowly move the ring to remove material from the ends. When performing this operation, file only from the outside in.

6   Excess end gap is not critical unless it is greater than 0.040-inch. Again, double-check to make sure you have the correct rings for your engine.

7   Repeat the procedure for the rest of the rings. Remember to keep rings, pistons and cylinders matched up.

8   Once the ring end gaps have been checked, the rings can be installed on the pistons.

9   The oil control ring (lowest one on the piston) is installed first. It is composed of

three separate components. Slip the spacer/expander into the groove **(see illustration)**, then install the lower side rail **(see illustration)**. Do not use a piston ring installation tool on the oil ring side rails, as they may be damaged. Instead, place one end of the side rail into the groove between the spacer/expander and the ring land, hold it firmly in place and slide a finger around the piston while pushing the rail into the groove. Next, install the upper side rail in the same manner.

10   After the three oil ring components have been installed, check to make sure that both the upper and lower side rails can be turned smoothly in the ring groove.

11   The number two (middle) ring is installed next. It should be stamped with a mark so it can be readily distinguished from the top ring. **Note:** *Always follow the instructions printed on the ring package or box - different manufacturers may require different approaches.* Do not mix up the top and middle rings, as they have different cross sections.

12   Use a piston ring installation tool and make sure that the identification mark is facing the top of the piston, then slip the ring into the middle groove on the piston **(see**

20.9b Do not use a piston ring tool when installing the oil ring side rails

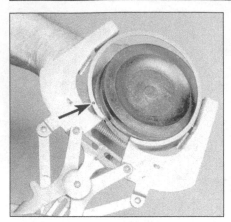

20.12 Installing the compression rings with a ring expander - the mark (arrow) must face up

21.10 Lay the Plastigage strips (arrow) on the main bearing journals, parallel to the crankshaft centerline

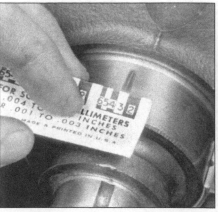

21.14 Compare the width of the crushed Plastigage to the scale on the envelope to determine the main bearing oil clearance (always take the measurement at the widest point of the Plastigage); be sure to use the correct scale - standard and metric ones are included

illustration). Do not expand the ring any more than is necessary to slide it over the piston.

13   Install the number one (top) ring in the same manner. Make sure the identifying mark is facing up.

14   Repeat the procedure for the remaining pistons and rings. Be careful not to confuse the number one and number two rings.

21   Crankshaft - installation and main bearing oil clearance check

Refer to illustrations 21.10 and 21.14

1   Crankshaft installation is generally one of the first steps in engine reassembly. It is assumed at this point that the engine block and crankshaft have been cleaned, inspected and repaired or reconditioned.

2   Position the engine with the bottom facing up.

3   Remove the main bearing cap bolts and lift out the caps. Lay them out in the proper order to help ensure that they are installed correctly.

4   If they are still in place, remove the old bearing inserts from the block and the main bearing caps. Wipe the main bearing surfaces of the block and caps with a clean, lint-free cloth. They must be kept spotlessly clean.

5   Clean the back sides of the new main bearing inserts and lay one bearing half in each main bearing saddle in the block. Lay the other bearing half from each bearing set in the corresponding main bearing cap. Make sure the tab on the bearing insert fits into the recess in the block or cap. Do not hammer the bearing into place and do not nick or gouge the bearing faces. No lubrication should be used at this time.

6   The flanged thrust bearing must be installed in the number three cap and saddle (2.8L V6 engines) or the number five cap and saddle (2.5L four-cylinder engines).

7   Clean the faces of the bearings in the

block and the crankshaft main bearing journals with a clean, lint-free cloth. Check or clean the oil holes in the crankshaft, as any dirt here can go only one way - straight into the new bearings.

8   Once you are certain that the crankshaft is clean, carefully lay it in position (an assistant would be very helpful here) in the main bearings.

9   Before the crankshaft can be permanently installed, the main bearing oil clearance must be checked.

10   Trim several pieces of the appropriate size of Plastigage slightly shorter than the width of the main bearings and place one piece on each crankshaft main bearing journal, parallel with the journal axis (see illustration).

11   Clean the faces of the bearings in the caps and install the caps in their respective positions (do not mix them up) with the arrows pointing toward the front of the engine. Do not disturb the Plastigage.

12   Starting with the center main and working out toward the ends, tighten the main bearing cap bolts, in three steps, to the torque listed in this Chapter's Specifications. Do not rotate the crankshaft at any time during this operation.

13   Remove the bolts and carefully lift off the main bearing caps. Keep them in order. Do not disturb the Plastigage or rotate the crankshaft. If any of the main bearing caps are difficult to remove, tap them gently from side-to-side with a soft-face hammer to loosen them.

14   Compare the width of the crushed Plastigage on each journal to the scale printed on the Plastigage container to obtain the main bearing oil clearance (see illustration). Check the Specifications to make sure it is correct.

15   If the clearance is not correct, double-check to make sure you have the right size bearing inserts. Also, make sure that no dirt or oil is between the bearing inserts and the main bearing caps or the block when the clearance was measured.

16   Carefully scrape all traces of the Plastigage material off the main bearing journals and the bearing faces. Do not nick or scratch the bearing faces.

17   Carefully lift the crankshaft out of the engine. Clean the bearing faces in the block, then apply a thin, uniform layer of moly-base grease or engine assembly lube to each of the bearing surfaces. Be sure to coat the thrust flange faces as well as the journal face of the thrust bearing.

18   Lubricate the rear main bearing oil seal where it contacts the crankshaft with moly-base grease or engine assembly lube. Note that on four-cylinder engines and V6 engines a 360 degree lip-type seal is utilized and is installed after the crankshaft is in place (refer to Parts A or B of this Chapter).

19   Rotate the crankshaft a number of times by hand and check for any obvious binding.

20   Check the crankshaft endplay with a feeler gauge or a dial indicator as described in Section 15.

22   Piston/connecting rod assembly - installation and bearing oil clearance check

Refer to illustrations 22.5, 22.8, 22.10, 22.12 and 22.14

1   Before installing the piston/connecting rod assemblies the cylinder walls must be perfectly clean, the top edge of each cylinder must be chamfered, and the crankshaft must be in place.

2   Remove the connecting rod cap from the end of the number one connecting rod. Remove the old bearing inserts and wipe the bearing surfaces of the connecting rod and cap with a clean, lint-free cloth (they must be kept spotlessly clean). Slip pieces of rubber hose over the connecting rod bolts to prevent crankshaft damage.

3   Clean the back side of the new upper

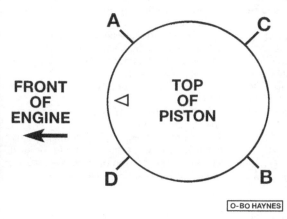

22.5 Piston end gap positions

A   Oil ring side rail gap - lower
B   Oil ring side rail gap - upper
A   Top compression ring gap
C   Second compression ring
    gap and oil ring spacer gap

FRONT
OF
ENGINE

TOP
OF
PISTON

A   C

D   B

O-BO HAYNES

22.8 The notch or arrow on each piston must face the front (drivebelt end) of the engine

bearing half, then lay it in place in the connecting rod. Make sure that the tab on the bearing fits into the recess in the rod. Do not hammer the bearing insert into place and be very careful not to nick or gouge the bearing face. Do not lubricate the bearing at this time.

4   Clean the back side of the other bearing insert and install it in the rod cap. Again, make sure the tab on the bearing fits into the recess in the cap, and do not apply any lubricant. It is critically important that the mating surfaces of the bearing and connecting rod are perfectly clean and oil-free when they are assembled.

5   Position the piston ring gaps as shown, (see illustration), then slip a section of plastic or rubber hose over the connecting rod cap bolts.

6   Lubricate the piston and rings with clean engine oil and attach a piston ring compressor to the piston. Leave the skirt protruding about 1/4-inch to guide the piston into the cylinder. The rings must be compressed as far as possible.

7   Rotate the crankshaft until the number one connecting rod journal is as far from the number one cylinder as possible (bottom dead center), and apply a coat of engine oil to the cylinder walls.

8   With the notch on top of the piston facing to the front of the engine (see illustration), slip the piston/connecting rod assembly into the number one cylinder bore and rest the bottom edge of the ring compressor on the engine block. Tap the top edge of the ring compressor to make sure it is contacting the block around its entire circumference.

9   Clean the number one connecting rod journal on the crankshaft and the bearing faces in the rod.

10   Carefully tap on the top of the piston with the end of a wooden hammer handle (see illustration) while guiding the end of the connecting rod into place on the crankshaft journal. The piston rings may try to pop out of the ring compressor just before entering the cylinder bore, so keep some pressure on the ring compressor. Work slowly, and if any resistance is felt as the piston enters the cylinder, stop immediately. Find out what is hanging up and fix it before proceeding. Do not, for any reason, force the piston into the cylinder, as you will break a ring and/or the piston.

11   Once the piston/connecting rod assembly is installed, the connecting rod bearing oil clearance must be checked before the rod cap is permanently bolted in place.

12   Cut a piece of the appropriate size Plastigage slightly shorter than the width of the connecting rod bearing and lay it in place on the number one connecting rod journal, parallel with the journal axis. It must not cross the oil hole in the journal (see illustration).

13   Clean the connecting rod cap bearing face, remove the protective hoses from the connecting rod bolts and install the rod cap in place. Make sure the mating mark on the cap is on the same side as the mark on the connecting rod. Install the nuts and tighten them to the torque listed in this Chapter's Specifications, working up to it in three steps. Do not rotate the crankshaft at any time during this operation.

14   Remove the rod cap, being careful not to disturb the Plastigage. Compare the width of the crushed Plastigage to the scale printed on the Plastigage container to obtain the oil clearance (see illustration). Compare it to the Specifications to make sure the clearance is correct. If the clearance is not correct, double-check to make sure that you have the correct-size bearing inserts. Also, recheck the crankshaft connecting rod journal diameter

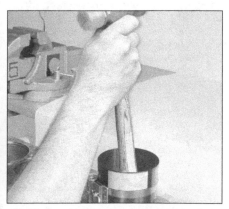

22.10 If resistance is encountered when tapping the piston/connecting rod assembly into the block, stop immediately and make sure the rings are fully compressed

22.12 Position the Plastigage strip on the bearing journal, parallel to the journal axis

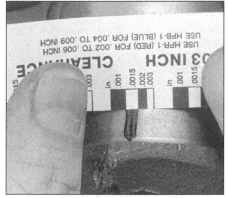

22.14 The crushed Plastigage is compared to the scale printed on the container to obtain the bearing oil clearance

and make sure that no dirt or oil was between the bearing inserts and the connecting rod or cap when the clearance was measured.

15   Carefully scrape all traces of the Plastigage material off the rod journal and bearing face. Be very careful not to scratch the bearing - use your fingernail or a piece of hardwood. Make sure the bearing faces are perfectly clean, then apply a uniform layer of moly-base grease or engine assembly lube to both of them. You will have to push the piston into the cylinder to expose the face of the bearing insert in the connecting rod. Be sure to slip the protective hoses over the rod bolts first.

16   Slide the connecting rod back into place on the journal, remove the protective hoses from the rod cap bolts, install the rod cap and tighten the nuts to the torque listed in this Chapter's Specifications. Again, work up to the torque in three steps.

17   Repeat the entire procedure for the remaining piston/connecting rod assemblies. Keep the back sides of the bearing inserts and the inside of the connecting rod and cap perfectly clean when assembling them. Make sure you have the correct piston for the cylinder and that the notch on the piston faces to the front of the engine when the piston is installed. Remember, use plenty of oil to lubricate the piston before installing the ring compressor. Also, when installing the rod caps for the final time, be sure to lubricate the bearing faces adequately.

18   After all the piston/connecting rod assemblies have been properly installed, rotate the crankshaft a number of times by hand and check for any obvious binding.

19   As a final step, the connecting rod endplay must be checked. Refer to Section 14 for this procedure. Compare the measured endplay to the Specifications to make sure it is correct.

---

23 Force balancer/oil pump assembly (1988 2.5L engines) - removal and installation

Removal

Note: *To remove the oil pump only, it is not necessary to remove the entire force balancer/oil pump assembly. To remove only the oil pump, see Chapter 2A.*

1   1988 model 2.5L engines are equipped with a force balancer. The assembly consists of two eccentrically weighted shafts and gears which are counter-rotated by a concentric gear on the crankshaft at twice crankshaft speed, dampening engine vibration. The oil filter, a pick-up screen and gerotor type oil pump are also integral parts of the assembly, so the oil pump removal procedure in Section 11 does not apply. The oil pump is driven from the back side of one of the balancers.

2   The balancer-equipped engine can be distinguished by the element-type oil filter located in the oil pan.

3   The force balancer/oil pump assembly must be removed to disassemble the engine for overhaul.

4   Remove the oil pan (see Chapter 2A).

5   Position the number 1 piston at TDC on the compression stroke (see Section 3).

6   Unbolt the balancer assembly. and remove it from the engine. **Warning:** *The assembly is heavy, so support it carefully before removing the bolts and be careful not to drop it!*

Installation

7   Position the crankshaft by measuring from the engine block to the first cut of the double notch on the reluctor ring. The distance should be 1-11/16 inches. If it isn't, turn the crankshaft until it is. Position the measuring ruler on the backside of the reluctor ring and measure from the first notch to the engine block.

8   Mount the balancer with the counterweights parallel and pointing AWAY from the crankshaft. Tighten the bolts to the torque listed in this Chapter's Specifications.

9   Rotate the crankshaft four times and check for clearance between the fourth counterweight and the balancer weights.

10   Install the oil pan.

11   Install a new oil filter and add oil (see Chapter 1). Run the engine and check for leaks.

---

24 Oil pump/pressure regulator valve (1988 2.5L engines) - removal, inspection and installation

Removal

Note: *It isn't necessary to remove the force balancer assembly to service the oil pump or pressure regulator valve.*

1   Remove the oil pan (see Section 10).

2   Remove the restrictor (if equipped).

3   Remove the oil pump cover assembly and oil pump gears or, on later models, the gerotor assembly.

4   **Warning:** *The pressure regulator valve is under pressure. Exercise caution when unscrewing the plug or removing the pin, as bodily injury may result.* Remove the pressure regulator valve plug (or pin) and spring, then remove the valve itself. If the valve is stuck, clean the valve and pump housing with carburetor cleaner or solvent.

5   Remove any sludge, oil or varnish from the parts with carburetor cleaner or solvent. If the varnish on any of the parts is difficult to remove, allow them to soak for awhile.

Inspection

6   Inspect all parts for the presence of foreign material. If you find evidence of contamination, determine its source.

7   Inspect the oil pump pocket and oil pump cover assembly for cracks, scoring, and casting imperfections.

8   Inspect the pressure regulator valve for scoring and sticking. Remove burrs with a fine oil stone.

9   Inspect the pressure regulator valve spring for distortion and loss of tension. If you have any doubt regarding the condition of the spring, replace it.

10   Clean the screen assembly and inspect it for damage.

11   Inspect the pump gears for chipping, galling, and wear.

12   We recommend replacing the oil pump gears (or gerotor assembly on later models) whenever they are removed, since they are so critical to proper engine lubrication.

Installation

13   Lubricate all internal parts with engine oil.

14   To assure priming and avoid engine damage, pack all pump cavities with petroleum jelly.

15   Install the oil pump gears.

16   Install the oil pump cover assembly.

17   Install the pressure regulator valve and spring.

18   Install the pressure regulator plug or pin. Make sure it's properly secured.

19   Install the oil pump cover assembly and tighten the bolts securely.

20   Install the restrictor (if equipped) and a new filter.

21   Install the oil pan.

22   Fill the crankcase to the correct level with clean engine oil.

23   Remove the oil pressure sending unit and install an oil pressure gauge in its place.

24   Start the engine and note the oil pressure. If it doesn't build up quickly, remove the oil pan and examine the pump. If necessary, disassemble the pump and repack all cavities with petroleum jelly. Running the engine without oil pressure will cause extensive damage.

---

25 Engine overhaul - reassembly sequence

Before beginning engine reassembly, make sure you have all the necessary new parts, gaskets and seals as well as the following items on hand:

Common hand tools
A 1/2-inch drive torque wrench
Piston ring installation tool
Piston ring compressor
Short lengths of rubber hose to fit over rod bolts
Plastigage
Feeler gauges
A fine-tooth file
New engine oil
Engine assembly lube or moly-base grease
RTV-type gasket sealant
Anaerobic-type gasket sealant
Thread locking compound

2   In order to save time and avoid problems, engine reassembly must be done in the

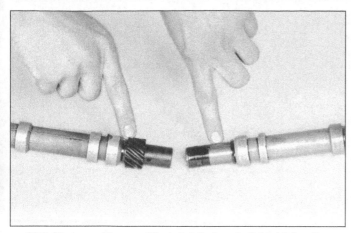

26.3 The pre-oiling modified distributor shaft (right) has the gear and advance weights ground off

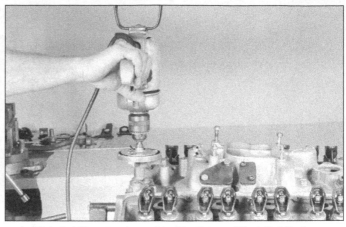

26.4 A drill motor connected to the modified distributor drives the oil pump

following order.

> Rear main oil seal (two-piece seal only)
> Crankshaft and main bearings
> Piston rings
> Piston/connecting rod assemblies
> Oil pump
> Oil pan
> Camshaft
> Timing chain/sprockets or gears
> Timing chain/gear cover
> Valve lifters
> Cylinder heads and pushrods
> Intake and exhaust manifolds
> Oil filter
> Pre-oil the engine (2.8 liter V6 engines
>   only - Section 26)
> Valve covers
> Fuel pump
> Water pump
> Rear main oil seal (one-piece seals)
> Flywheel/driveplate
> Fuel injection components
> Thermostat and housing cover
> Distributor (if equipped), spark plug
>   wires and spark plugs
> Emissions control components
> Alternator

26 Pre-oiling the engine after overhaul (2.8L V6 engines only)

Refer to illustrations 26.3, 26.4 and 26.5

1   After an overhaul it is a good idea to pre-oil the engine before it is installed and initially started. This will reveal any problems with the lubrication system at a time when corrections can be made easily and without major engine damage. Pre-oiling the engine will also allow the parts to be lubricated thoroughly in a normal fashion, but without the heavy loads associated with the combustion process placed upon them.

2   The engine should be assembled completely with the exception of the distributor and the valve covers.

3   A modified distributor will be needed for this procedure. This pre-oil tool is a distributor body with the bottom gear ground off and

the advance weight assembly removed from the top of the shaft (**see illustration**).

4   Place the pre-oiler into the distributor shaft access hole at the rear of the intake manifold and make sure the bottom of the shaft mates with the oil pump. Clamp the modified distributor into place just as you would an ordinary distributor. Now attach an electric drill motor to the top of the shaft (**see illustration**).

5   With the oil filter installed, all oil ways plugged (oil-pressure sending unit at rear of block) and the crankcase full of oil as shown on the dipstick, rotate the pre-oiler with the drill. Make sure the rotation is in a clockwise direction. Soon, oil should start to flow from the rocker arms, signifying that the oil pump and lubrication system are functioning (**see illustration**). It may take two or three minutes for oil to flow to all of the rocker arms. Allow the oil to circulate through the engine for a few minutes, then shut off the drill motor.

6   Check for oil leaks at the filter and all gasket and seal locations.

7   Remove the pre-oil tool, then install the distributor and rocker arm covers.

27 Initial start-up and break-in after overhaul

1   Once the engine has been properly installed in the vehicle, double check the engine oil and coolant levels.

2   With the spark plugs out of the engine and the coil high-tension lead grounded to the engine block, crank the engine over until oil pressure registers on the gauge (if so equipped) or until the oil light goes off.

3   Install the spark plugs, hook up the plug wires and the coil high tension lead.

4   Make sure the carburetor choke plate is closed, then start the engine. It may take a few moments for the gasoline to reach the carburetor, but the engine should start without a great deal of effort.

5   As soon as the engine starts it should be set at a fast idle to ensure proper oil circulation and allowed to warm up to normal oper-

ating temperature. While the engine is warming up, make a thorough check for oil and coolant leaks.

6   Shut the engine off and recheck the engine oil and coolant levels. Restart the engine and check the ignition timing and the engine idle speed (refer to Chapter 1). Make any necessary adjustments.

7   Drive the vehicle to an area with minimum traffic, accelerate at full throttle from 30 to 50 mph, then allow the vehicle to slow to 30 mph with the throttle closed. Repeat the procedure 10 or 12 times. This will load the piston rings and cause them to seat properly against the cylinder walls. Check again for oil and coolant leaks.

8   Drive the vehicle gently for the first 500 miles (no sustained high speeds) and keep a constant check on the oil level. It is not unusual for an engine to use oil during the break-in period.

9   At approximately 500 to 600 miles, change the oil and filter, retorque the cylinder head bolts and recheck the valve clearances (if applicable).

10   For the next few hundred miles, drive the vehicle normally. Do not either pamper it or abuse it.

11   After 2000 miles, change the oil and filter again and consider the engine fully broken in.

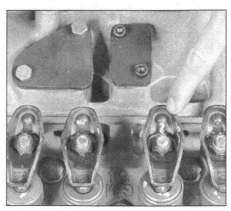

26.5 Oil will flow out of the holes in the rocker arms if the lubrication system is functioning properly

Notes

Chapter 3
Cooling, heating and air conditioning systems

Contents

Specifications

Radiator cap pressure rating	15 psi
Thermostat rating	195 degrees F

Torque specifications | **Ft-lbs** (unless otherwise indicated)

Fan frame-to-radiator support bolts	84 in-lbs
Transaxle oil cooler fittings	20
Radiator-to-radiator support bolts	84 in-lbs
Coolant recovery reservoir bolts	27 in-lbs
Coolant hose clamps at radiator	18 in-lbs
Water pump retaining bolts (4-cylinder engine)	22
Water pump-to-timing chain cover bolts (2.8 liter V6 engine)	
8 mm bolt	13 to 22
10 mm bolt	20 to 35
Thermostat housing bolts	20

Torque specifications (continued)
Ft-lbs (unless otherwise indicated)

Air conditioning system fittings

Coupled hose assembly at compressor	36 in-lbs
Coupled hose assembly at outlet tube	17
Coupled hose assembly at inlet tube...	30
Inlet tube at compressor hose assembly	30
Inlet tube at accumulator	30
Rear compressor outlet tube-to-front compressor outlet tube..	17
Compressor outlet tube at compressor hose assembly	17
Evaporator-to-condenser tube...	17
Evaporator core tube-to-accumulator	30
Inlet tube assembly at condenser	13
Outlet tube assembly at condenser	17
Compressor hose assembly at compressor	36 in-lbs
Compressor retaining bolts..	37

1   General information

The cooling system consists of a radiator, coolant recovery reservoir, thermostatically controlled cooling fan, thermostat and housing, engine driven water pump and a drivebelt.

The radiator is a cross-flow design. Radiators on automatic transaxle equipped vehicles are fitted with additional inlet and outlet fittings which circulate transaxle fluid through an oil cooler inside the left side tank. Vehicles equipped with air conditioning have a condenser mounted directly in front of the radiator.

A pressure cap on the cross-flow radiator allows a buildup of 15 psi in the cooling system. This pressure raises the boiling point of the coolant to approximately 262°F at sea level.

A plastic recovery bottle is located in the front compartment immediately behind the radiator and is connected to the radiator filler neck by a rubber hose. As the vehicle is driven, the coolant expands and flows from the radiator into the recovery bottle. When the engine is stopped, the coolant is drawn back into the radiator, maintaining the proper level of coolant in the radiator at all times.

The fan is driven by an electric motor, activated by a coolant temperature fan switch.

The water pump is mounted on the front of the engine block and is driven by a belt from the crankshaft pulley. The pump draws coolant from the radiator and circulates it through water jackets in the engine block, intake manifold and cylinder head(s). Coolant travels between the radiator and the engine through stainless steel pipes running underneath the vehicle. The right side pipe carries coolant from the radiator to the water pump; the left side pipe transports hot coolant from the thermostat housing back to the radiator.

Some of the hot coolant is diverted through the heater core located under the dashboard on the right side of the passenger compartment. A fan blows the heated air through several heating ducts located in the dash.

Air conditioning is available as an option. The compressor is mounted on the front of the engine. All other components - the condenser, the evaporator, the accumulator, etc. - are located in the front compartment.

2   Antifreeze - general information

Warning 1: *Do not allow antifreeze to come in contact with your skin or painted surfaces of the vehicle. Flush contaminated areas immediately with plenty of water. Do not store new coolant or leave old coolant lying around where it is easily accessible to children and pets, because they are attracted by its sweet taste. Ingestion of even a small amount can be fatal. Wipe up the garage floor and drip pan coolant spills immediately. Keep antifreeze containers covered and repair leaks in your cooling system immediately.*
Warning 2: *DO NOT remove the radiator cap or the coolant reservoir cap while the cooling system is hot, as escaping steam could cause injury.*

The cooling system should be filled with a water/ethylene glycol based antifreeze solution which will prevent freezing down to at least -20°F. It also provides protection against corrosion and increases the coolant boiling point.

The cooling system should be drained, flushed and refilled at least every other year (see Chapter 1). The use of antifreeze solutions for periods of longer than two years is likely to cause damage and encourage the formation of rust and scale in the system.

Before adding antifreeze to the system, check all hose connections and retorque the cylinder head bolts, because antifreeze tends to leak through very minute openings.

The exact mixture of antifreeze to water which you should use depends on the relative weather conditions. The mixture should contain at least 50 percent antifreeze, but should never contain more than 70 percent antifreeze.

3   Thermostat and housing - removal and installation

Refer to illustration 3.2
Warning: *The engine must be completely cool before beginning this procedure.*
1   The thermostat housing is located on the left upper end of the cylinder head on the 2.5L four-cylinder engine and on the upper right end, between the cylinder heads, of the 2.8L V6 engine. The removal procedures are the same for both. Remove the thermostat housing cap by pushing down firmly and twisting it clockwise.
2   Grasp the handle of the thermostat and gently pull it out of the housing **(see illustration)**.
3   Wipe out the thermostat housing with a clean rag.
4   Wipe off the thermostat O-ring and lubricate it with white grease.
5   Check the thermostat (Section 4) to determine whether it must be replaced.
6   Install the thermostat by pushing it down into the housing until it is properly seated.
7   Install the thermostat housing cap.
8   If you are replacing a damaged thermostat housing or a leaking thermostat housing gasket, repeats Steps 1 and 2, then disconnect the radiator hose from the housing.
Note: This is the highest point in the cooling system, but some coolant will be spilled

3.2 Pull the thermostat from the thermostat housing by lifting straight up on the wire handle

when removing the hose. It's a good idea to have some rags ready to mop up the spilled coolant.

9   Disconnect the coolant temperature sensor wire from the sensor.

10 If you are replacing the gasket, leave the coolant temperature sensor attached to the thermostat housing. If you are replacing the thermostat housing, remove the coolant temperature sensor from the housing. **Caution:** *Care must be taken when handling the coolant sensor. Damage to this sensor will affect the operation of the fuel injection system.*

11 Unbolt and remove the thermostat housing.

12 Scrape all old gasket material from the mating surface of the cylinder head. Inspect the housing for damage on the outside and corrosion on the inside.

13 Apply a thin coat of RTV sealant to both sides of the new gasket and position it on the thermostat housing flange.

14 Install the thermostat housing and gasket and tighten the bolts securely.

15 Wrap the threads of the coolant temperature sensor with Teflon tape, install it in the thermostat housing and tighten it securely. Reconnect the coolant temperature sensor wire.

16 Top up the cooling system with fresh coolant (Chapter 1).

17 Install the thermostat and the thermostat housing pressure cap.

18 Start the engine and check the new thermostat housing gasket for leaks.

4   Thermostat - check

1   If you suspect a malfunctioning thermostat, replace it. It's an inexpensive item and the replacement procedure only takes a few minutes. If you are unsure of the thermostat's condition, it's easy to check.

2   Remove the thermostat (Section 3).

3   Inspect the thermostat for excessive corrosion and damage. Replace it with a new one if either of these conditions is noted.

4   Place the thermostat in a container filled with water. You will need a cooking thermometer to measure water temperature.

5   Heat the water to the temperature stamped on the bottom of the thermostat (probably 195°F). If the thermostat in your vehicle is rated at some temperature other than 195°F, heat the water to that temperature. When submerged in hot water heated to its rated operating temperature, the valve should open all the way.

6   Allow the water to cool until it's about 10 degrees below the temperature indicated on the thermostat (about 185 degrees). At this temperature, the thermostat valve should close completely.

7   Reinstall the thermostat if it operates properly. If it does not, purchase a new thermostat of the same temperature rating.

5   Recovery bottle - removal and installation

1   Disconnect the overflow hose at the filler neck of the radiator by prying the pinch clamp loose with a small screwdriver. **Caution:** *If there is any coolant in the bottle, or if the radiator is filled to the rim of the filler neck, coolant may spill onto the vehicle when the overflow hose is disconnected. To minimize spillage, place several rags under the filler neck outlet for the overflow hose before disconnecting it.*

2   Remove the recovery bottle mounting bolts and detach the bottle.

3   Drain the coolant from the bottle and rinse it out.

4   Inspect the bottle for cracks and swelling. Check the cap to make sure it is sealing properly. If the bottle is damaged, replace it.

5   Installation is the reverse of removal.

6   Electric cooling fan motor circuit - testing

Circuit operation

1   The cooling fan motor is operated by the fan relay. When the coolant temperature reaches 235°F, the coolant temperature switch closes, turning the fan on.

2   Vehicles equipped with air conditioning have an additional relay which causes the fan motor to operate whenever the air conditioning is in use.

System test

3   Run the engine at a fast idle until the fan comes on. The fan should begin to operate before the coolant warning light in the instrument panel comes on.

4   If your vehicle is equipped with a heavy duty cooling fan, the fan will run at a higher speed if the engine is allowed to reach a higher temperature.

5   If the fan does not come on when it

should, check the fan fuse and the fusible link. Refer to the wiring diagrams for the fuse and fusible link locations, as well as the wiring associated with the various cooling fan systems. Different engine and accessory combinations utilize different fan operating systems, and although all are essentially similar, slight differences in wiring and switching are used.

6   If the fan runs with the ignition switch in the Off position, check the fan relay for stuck contacts.

7   If the fan does not run, disconnect the coolant temperature switch wire and check it for power. If there is no power to the switch, the problem is an open circuit or blown fuse or fusible link.

8   If there is power to the switch, use a jumper cable to ground the wire. If the fan motor runs, replace the coolant temperature switch. **Note:** *Be sure to apply Teflon tape to the threads of the new switch.*

9   If the fan motor does not run, check the fan relay with a test lamp. If there is power into the relay, but not out, replace the relay.

10 If the relay is passing current, check the fan motor connector. If there is power to the fan motor, but it does not run, replace the fan motor.

11 On models with air conditioning, if the fan runs when the engine is hot, but does not run when the air conditioning is turned on, turn the air conditioning to Norm and the ignition switch to Run. With the engine cool, connect a jumper between terminals H and K of the air conditioning switch connector. If the coolant fan runs, replace the air conditioning control head.

7   Electric cooling fan - removal and installation

Refer to illustrations 7.2, 7.5a and 7.5b
Warning: *If the ignition switch is turned to the On position, the fan can start even when the engine is not operating. The negative battery cable should always be disconnected when you are working in the vicinity of the fan.*

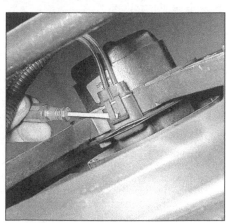

7.2 The fan motor electrical connector has a locking tab, which you can't see from above (it must be released to detach the connector)

7.5a Remove the retaining nut on the fan motor output shaft (arrow) to remove the fan

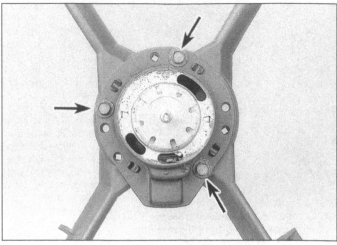

7.5b Remove the three mounting bolts (arrows) to remove the fan motor from the fan frame assembly

8.2a The radiator drain plug (arrow) is located at the lower right rear corner

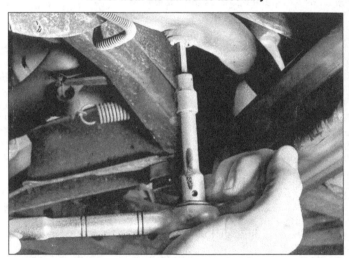

8.2b Both coolant pipes have a small Allen bolt drain plug

Removal

1   Disconnect the cable from the negative terminal of the battery.
2   Disconnect the wire harness at the fan motor **(see illustration)**.
3   Remove the bolts attaching the fan frame assembly to the radiator.
4   Lift the fan frame assembly from the engine compartment.
5   Remove the fan from the motor **(see illustration)**, then unbolt the motor from the fan frame **(see illustration)**. Remove the motor.

Installation

6   Install the new fan motor in the fan frame and tighten the fan motor bolts securely. Install the fan and tighten the nut.
7   Place the fan motor and frame assembly in position and tighten the four retaining bolts securely.
8   Plug in the electrical connector.
9   Connect the negative battery cable.
10   Check for proper operation as the engine warms to operating temperature.

8   Radiator - removal, servicing and installation

Refer to illustrations 8.2a, 8.2b, 8.3, 8.9 and 8.10
Warning: *The engine must be completely cool before beginning this procedure. Also, when working in the vicinity of the electric fan, disconnect the negative battery cable from the battery to prevent the fan from coming on accidentally.*

Removal

1   Disconnect the negative cable from the battery.
2   Drain the cooling system by opening the petcock at the bottom right corner of the radiator **(see illustration)** and the drain bolts in the coolant pipes **(see illustration)**.
3   Scribe alignment marks on the front compartment lid along the edges of the hinge brackets **(see illustration)**, remove the hinge bracket bolts and remove the lid.
4   Disconnect the overflow hose at the filler neck, remove the two coolant recovery tank

bolts and remove the coolant recovery tank.
5   Remove the fan frame assembly (Section 7).
6   Loosen the radiator inlet and outlet hose clamps and disconnect the hoses.
7   On automatic transaxle equipped

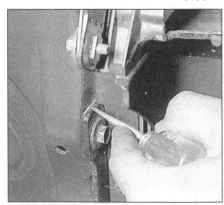

8.3 Be sure to scribe alignment marks along the edges of the front compartment lid mounting brackets with an awl or small screwdriver before removing the lid

8.9 Disconnect the air shroud from the radiator support bracket by cutting the plastic fasteners on each side near the top of the bracket, then remove the support bracket

8.10 Grasp the radiator by the side tanks and carefully remove it from the vehicle

models, disconnect and plug the transmission cooler lines.

8   Remove the bolts attaching the radiator to the radiator support brackets.

9   With a pair of diagonal cutters, snip off the heads of the plastic pop fasteners that anchor the air shroud to the top of the radiator **(see illustration)**. These fasteners are not reusable. You will need a couple of new ones for reinstallation.

10   Grasp the radiator by the side tanks and carefully lift it from the engine compartment **(see illustration)**.

Inspection and servicing

11   Carefully examine the radiator for evidence of leaks and damage. Any necessary repairs should be performed by a radiator repair shop.

12   With the radiator removed, brush accumulations of insects and leaves from the cooling fins.

13   Inspect the radiator hoses. If they have become cracked, swollen or otherwise deteriorated, replace them.

14   Flush the radiator as described in

Chapter 1.

15   Replace the radiator cap with a new one of the same rating. If the cap is relatively new, have it pressure tested.

Installation

16   If you are installing a new radiator, transfer the fittings from the old unit to the new one.

17   Lower the radiator into position. Make sure that the bottom of the radiator is properly located on the lower mounting pads.

18   Install the bolts attaching the radiator to the radiator support brackets and tighten them to the specified torque.

19   Install the transaxle oil cooler lines, if equipped.

20   Reattach the coolant hoses to the radiator fittings and tighten the hose clamps securely.

21   Place the fan frame assembly in position, install the fan assembly mounting bolts and tighten them securely.

22   Reconnect the fan motor electrical connector.

23   Place the coolant recovery tank in position and tighten the two mounting bolts securely.

24   Reconnect the overflow hose to the filler neck. Use a new clamp if the old one does

not securely pinch down the end of the hose.

25   Fill the cooling system with coolant (see Chapter 1).

9   Coolant and heater pipes - removal and installation

Refer to illustrations 9.6, 9.12, 9.28 and 9.29

Coolant pipes

Removal

1   Disconnect the cable from the negative terminal of the battery.

2   Drain the cooling system (Chapter 1).

3   Remove the spare tire, jack and jack storage bracket.

4   Remove the four mounting screws along the front edge and the plastic fasteners along the sides and back edge of the spare tire storage panel and remove it.

5   If you are replacing the left (radiator inlet) coolant pipe, loosen the left front wheel nuts. If you are replacing the right (radiator outlet) coolant pipe, loosen the right front wheel nuts.

6   Loosen the hose clamp attaching the rubber radiator hose to the stainless steel coolant pipe **(see illustration)**.

7   Remove the coolant pipe bracket screw. **Note:** *This is a self-tapping screw, so it may not back out easily. The best way to remove it is to push on it with a pry bar or screwdriver while backing it out. Usually you will need a new screw for reinstallation.*

8   Jack up the front of the vehicle and support it securely on jackstands.

9   Loosen the hose clamp attaching the coolant hose to the water pump or thermostat housing in the engine compartment.

10   Remove the bracket bolts and disconnect the front and rear brackets from the coolant pipe.

11   Remove the left or right front wheel.

12   Swing the coolant pipe out from the rear and then work it out from between the front shock and lower A-arm **(see illustration)**. Note how the pipe is positioned before removing it.

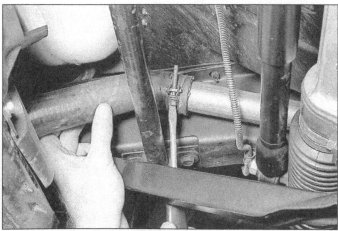

9.6 Remove the hose clamp attaching the coolant pipe to the radiator hose

9.12 Swing the coolant pipe out from the rear of the wheel housing and work it out from between the front shock absorber and the lower A-arm

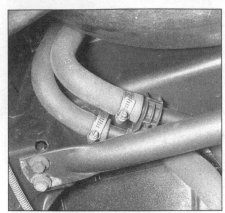

9.28 The heater hoses travel straight down from the front compartment to the underside of and just behind the spare tire recess - to remove either pipe loosen the hose clamp on the hose attached to the pipe which you intend to replace, pull the hose and pipe apart and remove the underbody bracket bolt and bracket

Installation

13   Work the new pipe into position at the front.
14   Swing it into position at the engine compartment and attach the water pump or thermostat housing hose. Tighten the hose clamp securely.
15   Attach the radiator inlet or outlet hose at the front compartment. Tighten the hose clamp securely.
16   Install the mounting brackets underneath the vehicle.
17   Install the front wheel and tighten the lug nuts to the specified torque.
18   Lower the vehicle.
19   Install the self-tapping screw in the slotted support bracket in the front compartment.
20   Install the spare tire panel.
21   Install the jack storage bracket, jack and spare tire.
22   Fill the cooling system with a fresh mixture of coolant and water.
23   Start the engine and check for leaks.

Heater pipes

Removal

24   Disconnect the cable from the negative terminal of the battery.
25   Drain the coolant (Chapter 1).
26   Raise the vehicle and secure it on jackstands.
27   Depending upon which pipe is being replaced, disconnect the heater hose at the water pump or the heater pipe on the underside of the intake manifold.
28   Remove the appropriate hose clamp for the pipe being replaced just in front of the underbody bracket underneath and to the right rear of the spare tire storage panel **(see illustration)**. Remove the retaining bolt for the bracket and remove the bracket.
29   Remove the main fuel tank support bracket and the two smaller heater pipe

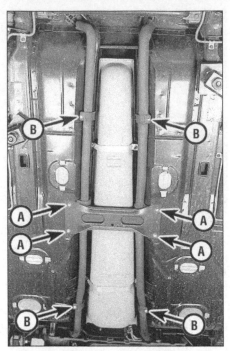

9.29 The main fuel tank support bracket bolts (A) and the heater pipe bracket bolts (B) must come off before either pipe can be removed

brackets from the underside of the vehicle **(see illustration)**.

Installation

30   Place the new heater pipe in position and install the two smaller heater pipe brackets. Tighten them securely.
31   Install the fuel tank main support bracket.
Note: *Be sure to coat the shoulders (between the threads and the flanges) of the four support bracket mounting bolts with sealant (see Fuel tank removal and installation in Chapter 4).*
32   Attach the pipe to the rubber hoses at both ends and tighten the hose clamps securely.
33   Install the bracket underneath the spare tire storage panel and tighten the bracket bolt securely.
34   Lower the vehicle.
35   Add coolant (Chapter 1).
36   Start the engine and allow it to warm up. Check for leaks.

---

10   Water pump - checking

Refer to illustration 10.4

1   A failure in the water pump can cause overheating and serious engine damage because a defective pump will not circulate coolant through the engine.
2   There are three ways to check the operation of the water pump while it is still installed on the engine. If any of the following checks indicate that the pump is defective, replaced it with either a new or rebuilt unit.
3   Squeeze the upper radiator hose while the engine is running at normal operating

10.4 The water pump weep hole (arrow) will drip coolant when the seal for the pump shaft fails

temperature. If the water pump is working properly, a pressure surge will be felt as the hose is released.
4   The water pump is equipped with a "weep" (vent) hole **(see illustration)**. If the pump seal fails, small amounts of coolant will leak through the weep hole. You will need to get underneath the water pump to see the weep hole, so raise the vehicle and place it on jackstands. Use a flashlight or a drop light to help you determine whether there is evidence of leakage from the weep hole.
5   If the water pump shaft bearing fails, it will usually emit a squealing sound while it is running. Do not confuse drivebelt slippage, which also makes a squealing sound, with water pump bearing failure. Even before the bearing actually fails, shaft wear can be detected by grasping the pulley firmly and moving it up and down. If excessive side play is noted, the shaft and/or the bearing are worn and the pump should be replaced.

---

11   Water pump - removal and installation

Warning: *The engine must be completely cool before beginning this procedure.*

Four-cylinder-engine

Removal

1   Disconnect the cable from the negative battery terminal.
2   Drain the cooling system.
3   Remove the water pump drivebelt.
4   Remove the lower radiator hose at the pump support housing.
5   Remove the water pump attaching bolts. Work the water pump forward out of the housing and lift it from the engine compartment.

Installation

6   Clean the water pump gasket surface on the pump support housing.
7   Run a 1/8-inch bead of RTV sealant around the mounting flange of the new water pump.

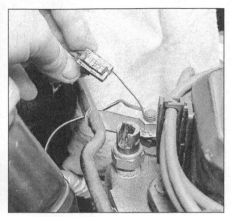

12.4 Ground the dark green wire of the coolant temperature sending unit connector with a short piece of wire

8   While the sealer is still wet, install the water pump on the housing and tighten the water pump bolts to the specified torque.

9   Install the lower radiator hose on the housing spigot.

10   Install the water pump drivebelt and adjust it to the proper tension.

11   Fill the cooling system with fresh coolant.

12   Attach the negative battery cable, start the engine and check for coolant leaks.

V6 engine

Removal

13   Remove the right side louvered cover over the engine compartment.

14   Disconnect the negative cable from the battery and remove the plastic battery shield.

15   Drain the cooling system.

16   Remove the nut securing the wiring harness to the junction block behind the battery and position the wires out of the way.

17   Remove the EGR vacuum solenoid and bracket, which is located just below the thermostat housing cap.

18   Loosen the right rear wheel lug nuts.

19   Raise the rear of the vehicle and support it securely on jackstands.

20   Remove the right rear wheel.

21   Loosen the alternator adjusting bolts and remove the drivebelt.

22   Remove the water pump bolts and nut.

23   Lower the vehicle and lift out the water pump from the top.

Installation

24   Clean the water pump mating surface on the block, making sure all the old sealant is removed.

25   Apply a 1/8-inch bead of RTV sealant to the new water pump mating surface.

26   Put the water pump in position, allowing it to hang on the upper mounting stud.

27   Raise the vehicle and support it securely on jackstands.

28   Install the water pump bolts and nut, tightening them to the specified torque. Note that the two bottom bolts have a different torque requirement than the other bolts and the nut.

29   The remainder of the installation is the reverse of the removal procedure.

30   Fill the cooling system with fresh antifreeze mixture.

31   Start the engine, bring it to normal operating temperature and check for coolant leaks.

12   Coolant temperature indicator circuit - check and switch replacement

Refer to illustration 12.4

1   The coolant temperature indicator system consists of a warning lamp mounted in the instrument panel and a coolant temperature switch located in the top left front corner of the cylinder head.

2   If the coolant temperature indicator comes on with the engine running, but not overheated, remove the connector from the coolant temperature switch. If the indicator goes out, install a new switch.

3   To install a new switch, remove the old unit and seal the threads of the new one with Teflon tape. Install the new switch and tighten it securely.

4   If the coolant temperature indicator does not come on before starting the engine, and the bulb is good, ground the dark green wire at the coolant temperature switch **(see illustration)**. If the indicator comes on, the circuit is good. Replace the switch.

13   Coolant temperature gauge - check and replacement

1   An optional coolant temperature gauge indicates coolant temperature between 100 and 260ºF. Vehicles equipped with the gauge have a transducer instead of a coolant temperature switch. The transducer is located in the same location as the coolant temperature switch.

2   If the coolant temperature gauge shows hot when the engine is cold, the sender circuit may be shorted. Remove the connector from the temperature sender switch.

3   If the gauge now shows cold, the wiring is good. Install a new coolant temperature switch. Be sure to seal the threads of the new unit with Teflon tape.

4   If the coolant temperature gauge remains hot, check for a short to ground in the dark green wire between the sender and the instrument panel.

5   If the coolant temperature gauge shows cold when the engine is hot or warm, the sender circuit may be open. Remove the connector from the temperature sender switch and connect terminal 8 (tan or dark green/white wire) of the connector to ground.

6   If the display now reads hot, the wiring is good. Install a new coolant temperature switch. Be sure to seal the threads of the new unit with Teflon tape.

7   If the display remains at cold, check for an open in the wire back to the instrument panel.

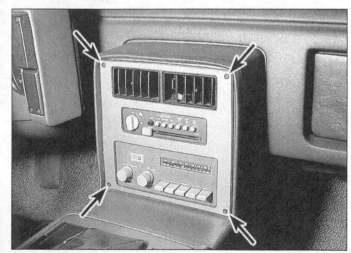

14.1 There are four screws (arrows) retaining the console fascia

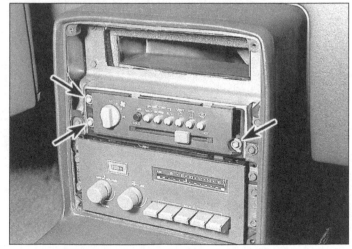

14.2 Remove these three screws (arrows) to pull out the heater control assembly

14.3a Unplug the electrical connector on the left . . .

14.3b . . . then pull the heater control assembly out a little further and unplug the big electrical connector

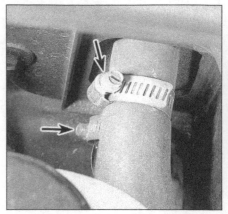

15.3 The heater hoses are located at the upper right side of the firewall under the hood (arrows)

14 Heater control cable - replacement

Refer to illustrations 14.1, 14.2, 14.3a and 14.3b

1   Remove the four Torx retaining bolts at the corners of the console fascia **(see illustration)** and remove the fascia.

2   Remove the three mounting screws from the heater/air conditioner control assembly **(see illustration)** and pull the assembly out of the console.

3   Unplug both electrical connectors from the backside of the heater/air conditioner control assembly **(see illustrations)**. **Note:** *The larger plug has a retaining clip on the bottom that must be pried away from the plug before it can be disconnected.*

4   Pop the clip off the end of the heater control cable pivot and disconnect the cable. **Note:** *To remove the clip, grasp it firmly at its circumference with needle nose pliers and pinch it slightly. The tangs gripping the pivot pin will release and the clip can be slid off.*

5   Remove the wire harness retaining

bracket attaching screws from the heater core access cover and the four mounting screws from the heater core cover.

6   Disconnect the cable brackets affixed to the heater module, disconnect the heater control cable at the other end and pull out the old cable. Note carefully the proper routing for the cable.

7   Thread in a new cable, connect it to the temperature valve shaft lever arm that actuates the temperature valve (door) between the heater core and the air conditioner evaporator.

8   Reconnect the cable brackets affixed to the heater module.

9   Install the heater core access cover and tighten the screws securely. **Note:** *There is a spring retaining clip affixed to the inside of the access cover that helps to support the heater core. Make sure that it is seated against the core properly before securing the cover.*

10   Reconnect the heater control cable to the heater control pivot and install the retaining clip.

11   Plug in the two electrical connectors to

the rear of the heater/air conditioner control assembly.

12   Slide the heater/air conditioner control assembly back into position in the console and tighten the three mounting screws securely.

13   Place the console fascia in position and tighten the four Torx screws securely.

14   Check the operation of the new cable by starting the engine, turning on the heater and air conditioner and making sure that the temperature valve (door) is routing hot or cold air properly.

15 Heater core - removal and installation

Refer to illustrations 15.3, 15.4 and 15.5

1   Drain the cooling system (Chapter 1).

2   Disconnect the electrical connector and the fluid tube from the windshield washer reservoir, remove the reservoir mounting bolts and remove the reservoir to gain access to the lower heater hose clamp.

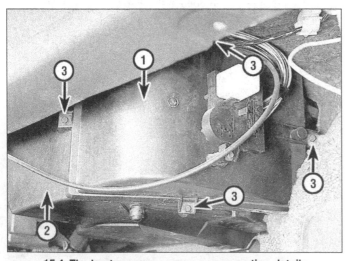

15.4 The heater core access cover mounting details

1	Heater core access cover	3	Mounting screws
2	Heater core case		

15.5 Heater core mounting details

1	Heater core	3	Retainer clip
2	Heater core case	4	Retainer clip screw

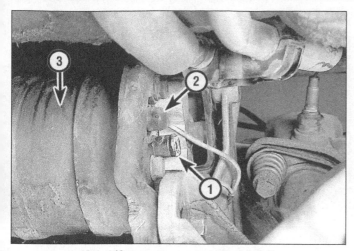

18.4 A/C compressor rear head details

1   High pressure relief valve   3   A/C compressor
2   High pressure cut-off switch

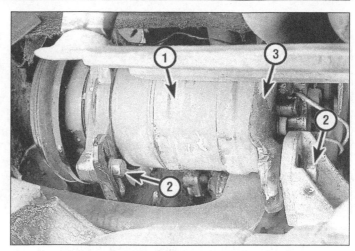

18.6 Lower A/C compressor mounting details

1   A/C compressor   3   Bracket
2   Pivot bolts

3   Loosen the inlet and outlet heater hose clamps and pull both hoses off the inlet and outlet pipes protruding forward through the right side of the heater/air conditioner module cover **(see illustration)**.
4   From inside the vehicle, remove the heater core access cover **(see illustration)**.
5   Remove the heater core retaining bracket bolts and the bracket and pull the heater core out of the heater module **(see illustration)**.
6   Install the new core.
7   Install the retaining bracket and tighten the bolts securely.
8   Install the heater core access cover.
Note: *There is a spring retaining clip affixed to the inside of the access cover that helps to support the heater core. Make sure that it is seated against the core properly before securing the cover.*
9   Slip the heater hoses onto the pipes of the new heater core and tighten the hose clamps securely.
10   Install the windshield washer reservoir, tighten the mounting bolts securely, plug in the electrical connector and connect the washer fluid tube to the reservoir.
11   Start the engine, run the heater and check the inlet and outlet pipe hose connections for leaks.

16   Heater blower motor - removal and installation

1   Disconnect the cable from the negative terminal of the battery.
2   Remove the blower motor cooling tube.
3   Disconnect the electrical connectors at the air conditioning accumulator and the power and ground wires to the blower motor.
4   Remove the five blower motor mounting bolts and remove the old blower.
5   Installation of a new blower motor is the reverse of removal.

17   Air conditioning system - servicing

Warning: *The air conditioning system is under high pressure. Do not remove any component of the system before taking the vehicle to a dealer or automotive air conditioning repair facility and having the air conditioner discharged.*
1   Regularly inspect the condenser fins, located ahead of the radiator, and brush away leaves and bugs.
2   Check the condition of the system hoses periodically. If there is any sign of deterioration or hardening, replace them or have them replaced by a dealer or repair shop.
3   At the recommended intervals, check and adjust the compressor drivebelt as described in Chapter 1.
4   Because of the special tools, equipment and skills required to diagnose and service air conditioning systems and because of the differences between the various systems that may be installed on various models, major air

18.7 The A/C compressor can be removed from above by lifting it out from the space between the engine and the firewall

conditioning service procedures cannot be covered in this manual.
5   However, once the system is depressurized by a professional shop, removal and installation of the major components of the air conditioning system can be performed by the home mechanic with regular hand tools. Therefore, performing the work at home before and after major component servicing will save you the labor costs for removing and installing damaged or worn out components.

18   Air conditioning compressor - removal and installation

Refer to illustrations 18.4, 18.6 and 18.7
Warning: *The air conditioning system must be discharged before performing this procedure.*
1   Disconnect the negative cable at the battery.
2   Remove the air conditioning drivebelt (Chapter 1).
3   Disconnect the inlet and outlet hose fittings from the rear of the compressor.
4   If your vehicle is equipped with a DA-6 compressor, unplug the clutch coil terminals on the top and the high pressure cutoff switch on the rear of the compressor **(see illustration)**.
5   If your vehicle is equipped with a V5 compressor, unplug the clutch coil terminals on the top and the high side low pressure and high side cut-off switches from the rear of the compressor.
6   Remove the compressor mounting bolts from the compressor bracket **(see illustration)**. **Note:** *It's easier to remove the pivot and tension adjusting bolts from underneath the vehicle. The other bolts can be reached from above.*
7   Remove the compressor by lifting out the rear end first **(see illustration)**.
8   Installation of a new or rebuilt compressor is the reverse of the removal procedure.

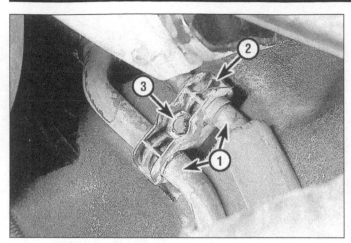

19.2 The A/C refrigerant lines mount to the body

| 1 | A/C lines | 2 | Clamp | 3 | Bolt |

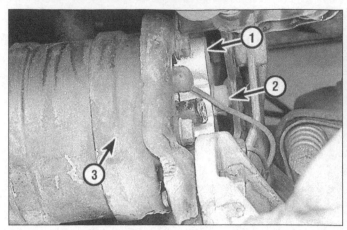

19.3 Location of the coupled hose fittings on the A/C compressor

1   Coupled hose assembly
2   Coupled hose assembly retaining bolt
3   A/C compressor

19 Air conditioning coupled hose assembly - replacement

Refer to illustrations 19.2 and 19.3
Warning: *The air conditioning system must be discharged before performing this procedure.*

Removal

1   Remove the cable from the negative terminal of the battery.
2   Remove the bolt from the hose assembly clamp **(see illustration)**.
3   Disconnect the coupled hose assembly at the compressor **(see illustration)**.
4   Disconnect the coupled hose assembly at the inlet tube.
5   Disconnect the coupled hose assembly at the outlet tube.
6   Remove the coupled hose assembly.

Installation

7   Install new O-rings lubricated with 525

viscosity refrigerant oil at all connector fittings. When replacing O-rings, the connection design should be carefully identified to assure installation of the correct O-ring. Some connections use a captive O-ring design connector. Assembly and tightening procedure is the same as for the standard O-ring design. However, the captive O-ring design uses different O-rings. When replacing the O-rings, it is important that the proper O-ring is used.
8   Install the coupled hose fitting bolt at the compressor. Tighten it to the specified torque.
9   Install and tighten securely the bolt at the coupled hose assembly clamp.
10   Install the coupled hose fitting at the inlet tube assembly and tighten it to the specified torque.
11   Install the coupled hose fitting at the outlet tube assembly and tighten it to the specified torque.
12   Take the vehicle to a dealer or automotive air conditioning shop and have the system recharged.

20 Air conditioning inlet and outlet tubes - removal and installation

Refer to illustrations 20.6 and 20.10
Warning: *The air conditioning system must be discharged before performing this procedure.*

Removal

1   Disconnect the cable from the negative terminal of the battery.
2   Remove the jack and spare tire.
3   Remove the jack storage bracket.
4   Remove the four bolts across the front of the spare tire storage panel and the plastic pop fasteners along the sides and rear lip of the panel. **Note:** *To extract each plastic pop fastener, grasp the head with a pair of pliers or pry it up with a screwdriver tip, then lift up about 3/8-inch and pull the fastener out.*
5   Remove the plastic spare tire storage panel.
6   If you are replacing the outlet tube and hose (low pressure) assembly, disconnect it from the threaded fitting on the top left side

20.6 If you are replacing the outlet (low pressure) pipe, disconnect the threaded fitting (arrow) at the top of the accumulator

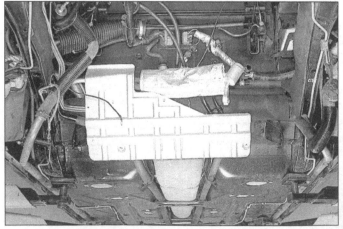

20.10 Remove the catalytic converter heat shield

21.5 Both of these threaded fittings (arrows) must be disconnected before the condenser can be removed

21.7 Removing the condenser from underneath the vehicle

of the accumulator **(see illustration)**. If you are replacing the inlet (high pressure) tube, disconnect it at the threaded fitting of the condenser inlet pipe, just below and in front of the brake booster.

7   If you are replacing the outlet tube and hose, remove the bolt and support bracket from the lower rear bulkhead of the front compartment, just below the blower motor.

8   Unbolt the compressor hose fitting in the engine compartment and disconnect the inlet and outlet hoses from the fitting.

9   Raise the vehicle and support it securely on jackstands.

10   Remove the catalytic converter heat shield **(see illustration)**.

11   Remove the three support brackets underneath the vehicle.

12   Unbolt and remove the fuel tank reinforcement brace.

13   Remove the inlet or outlet tube.

Installation

14   Place the new inlet or outlet tube in position.

15   Install the fuel tank reinforcement brace and tighten the bolts securely.

16   Install the three underbody support brackets and tighten the bolts securely.

17   Install the catalytic converter heat shield and tighten the bolts securely.

18   Lower the vehicle.

19   If you are replacing the inlet tube and hose assembly, install the inlet hose support bracket on the lower rear wall of the front compartment and tighten securely.

20   Connect the inlet and outlet lines to the compressor hose fitting in the engine compartment and tighten the bolt to the specified torque.

21   Connect the inlet tube to the accumulator and tighten it to the specified torque.

22   Set the plastic spare tire panel in place. Install the plastic pop fasteners and the four small bolts. Tighten the bolts.

23   Install the jack support bracket and tighten the bolts securely.

24   Install the jack and spare tire and secure them.

---

21   Air conditioning condenser - removal and installation

Refer to illustrations 21.5 and 21.7
Warning: *The air conditioning system must be discharged before performing this procedure.*

Removal

1   Disconnect the cable from the negative terminal of the battery.

2   Remove the upper condenser mounting bolts.

3   Raise the vehicle and support it securely on jackstands.

4   Remove the grille.

5   Disconnect both condenser lines **(see illustration)**.

6   Remove the lower condenser mounting bolts.

7   Remove the condenser assembly **(see illustration)**.

Installation

8   Place the condenser in position and install the lower condenser mounting bolts.

9   Connect the condenser inlet and outlet lines and tighten the fittings to the specified torque. **Note:** *Be sure to install new O-rings at both condenser lines (lubricate with 525 viscosity refrigerant oil)*.

10   Install the grille and tighten the grille attaching screws securely.

11   Lower the vehicle.

12   Install the upper condenser mounting bolts.

---

22   Air conditioning evaporator - removal and installation

Refer to illustrations 22.3, 22.8 and 22.9
Warning: *The air conditioning system must be discharged before performing this procedure.*

Removal

1   Disconnect the cable from the negative terminal of the battery.

2   Disconnect the electrical connector and tube from the windshield washer reservoir and remove the reservoir.

3   Disconnect and set aside the air conditioning high blower relay and the power switching relay connectors **(see illustration)**.

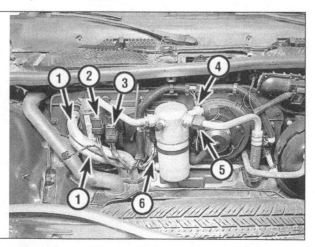

22.3 To remove the heater core/evaporator module front cover, remove the following items:

1   *Inlet and outlet heater hoses*
2   *High blower relay*
3   *Power switching relay*
4   *Cycling pressure switch*
5   *Accumulator outlet (low pressure) pipe fitting*
6   *Air conditioning resistor connector*

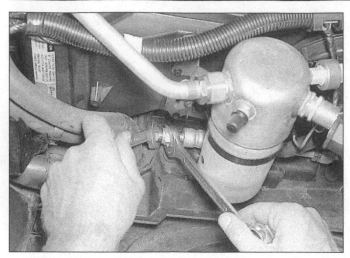

22.8 Disconnect the evaporator inlet (high pressure) pipe threaded fitting below the accumulator

22.9 Unplug the main wire harness junction box just to the right of the brake booster

4   Unplug the air conditioning resistor connector (immediately to the right of and behind the accumulator).

5   Loosen the hose clamps on the heater core inlet and outlet pipes. **Note:** *These two hoses fit tightly onto the inlet and outlet pipes. If you are unable to break them loose, slit the ends of both hoses with a razor or a sharp knife and slip them off.*

6   Disconnect the cycling pressure switch at the accumulator.

7   Disconnect the threaded fitting at the accumulator outlet.

8   Disconnect the threaded fitting on the evaporator inlet (high pressure) pipe below the accumulator **(see illustration)**.

9   Unplug the main wire harness junction box just to the right of the brake booster **(see illustration)**.

10   Disconnect the brake booster vacuum hose just to the right of the main wire harness

junction box.

11   Remove the bolts from the heater core/evaporator module front cover.

12   Remove the evaporator/heater core module front cover.

13   Disconnect the threaded fitting on the evaporator outlet pipe at the accumulator and the support bracket to the immediate right of the fitting.

14   Pull the evaporator from the module front cover.

Installation

15   Install the evaporator in the evaporator/heater core module front cover.

16   Install the evaporator/heater core module front cover and tighten the mounting bolts securely.

17   Reconnect the brake booster vacuum hose.

18   Plug in and bolt up the main wire har-

ness junction box.

19   Reconnect the threaded fitting on the evaporator inlet pipe and tighten it to the specified torque.

20   Reconnect the threaded fitting on the accumulator outlet pipe and tighten it to the specified torque.

21   Reconnect the cycling pressure switch on the accumulator.

22   Install the heater core hoses and tighten the hose clamps securely.

23   Plug in the air conditioning resistor connector.

24   Reattach the air conditioning high blower relay and power switching relay connectors.

25   Install the windshield washer reservoir and tighten the bolts securely. Reattach the windshield washer fluid tube and reconnect the electrical connector.

26   Reinstall the negative battery cable.

Chapter 4
Fuel and exhaust systems

Contents

Specifications

Torque specifications

Ft-lbs (unless otherwise indicated)

Throttle Body Injection (TBI) unit
Fuel meter cover screws	28 in-lbs
Idle Air Control (IAC) valve	156 in-lbs
TBI mounting bolts	120 to 180 in-lbs
TBI mounting stud	36 to 72 in-lbs
TBI mounting nut	120 to 180 in-lbs
Fuel feed line fitting at TBI	17
Fuel return line fitting at TBI	17

Multi-Port Fuel Injection (MPFI)
Throttle body mounting bolts	150 in-lbs
Idle Air Control (IAC) valve	156 in-lbs
Fuel rail bolts	18
Plenum chamber mounting bolts	18
Fuel pressure assembly-to-fuel rail	88 in-lbs
Fuel block attaching screw assemblies	44 in-lbs

2.2 Location of the fuel pump fuse (arrow)

2.5 The V6 engine has a Schrader valve (arrow) on the right end of the fuel rail for releasing fuel pressure

1   General information

Fuel system

The fuel system consists of a center tunnel mounted fuel tank, an electrically operated fuel pump, a fuel pump relay, an air cleaner assembly and either a Throttle Body Injection (TBI) system or a Multi-Port Fuel Injection (MPFI) system. The TBI system is used on four cylinder models and the multiport system on V6 engines.

The basic difference between the throttle body and multi-port systems is the number and location of the fuel injectors.

The throttle body system utilizes one injector, centrally mounted in a carburetor-like housing. The injector is an electrical solenoid, with fuel delivered to the injector at a constant pressure level. To maintain the pressure level, excess fuel is returned to the fuel tank. A signal from the ECM opens the solenoid, allowing fuel to spray through the injector into the throttle body. The amount of time the injector is held open by the ECM determines the fuel/air mixture ratio.

The multi-port system is equipped with the same type of injector, and the fuel/air ratio is controlled in the same manner. However, instead of one injector mounted in a centrally located throttle body, six injectors are used, one above each intake port. The throttle body serves only to control the amount of air passing into the system. Because individual injectors are used for each cylinder, mounted immediately adjacent to the intake valves, much better control of the fuel/air mixture ratio is possible.

Exhaust system

The exhaust system, which is similar for both four-cylinder and V6 powered vehicles, includes an exhaust manifold fitted with an exhaust oxygen sensor, a stainless steel exhaust pipe, a catalytic converter and a tri-flow design, transversely mounted muffler. Spring type hangers support the entire one-piece exhaust system.

The catalytic converter is an emission control device added to the exhaust system to reduce pollutants. A single-bed converter design is used in combination with a three-way (reduction) catalyst. The catalytic coating on the three-way (reduction) catalyst contains platinum and rhodium, which lowers levels of oxides of nitrogen (NOX) as well as hydrocarbons (HC) and carbon monoxide (CO).

2   Fuel pressure relief procedure

Refer to illustrations 2.2 and 2.5

1   **Warning:** *Gasoline is extremely flammable, so take extra precautions when you work on any part of the fuel system. Don't smoke or allow open flames or bare light bulbs near the work area, and don't work in a garage where a natural gas-type appliance (such as a water heater or a clothes dryer) with a pilot light is present. Since gasoline is carcinogenic, wear latex gloves when there's a possibility of being exposed to fuel, and, if you spill any fuel on your skin, rinse it off immediately with soap and water. Mop up any spills immediately and do not store fuel-soaked rags where they could ignite. The fuel system is under constant pressure, so, if any fuel lines are to be disconnected, the fuel pressure in the system must be relieved first. When you perform any kind of work on the fuel system, wear safety glasses and have a Class B type fire extinguisher on hand.*

Throttle Body Injection (TBI)

2   Remove the fuse marked Fuel Pump from the fuse block in the passenger compartment **(see illustration)**.

3   Crank the engine over. It will start and run until the fuel supply remaining in the fuel lines is depleted. When the engine stops, engage the starter again for another three seconds to ensure that any remaining pressure is dissipated.

4   With the ignition turned to Off, replace the fuel pump fuse. Unless this procedure is followed before servicing fuel lines or connections, fuel spray (and possible injury) may occur.

Multi-Port Fuel Injection (MPFI)

5   Connect a fuel pressure gauge equipped with a bleed hose (available at most auto parts stores) to the fuel rail valve **(see illustration)**. Wrap a shop rag around the fitting while connecting the gauge.

6   Route the bleed hose into an approved container, then open the valve to bleed off system pressure.

7   After servicing a fuel system component, cycle the ignition on and off several times (wait ten seconds between cycles) and check the system for leaks.

3   Fuel pump - testing

Refer to illustrations 3.4, 3.6 and 3.8

Warning: *Gasoline is extremely flammable, so take extra precautions when you work on any part of the fuel system. See the **Warning** in Section 2.*

1   The fuel pump for both four-cylinder and V6 powered models is located in the fuel tank, is sealed and is not repairable.

2   When the key is first turned on without the engine running, the ECM will turn the fuel pump relay on for two seconds. This builds up the fuel pressure quickly. If the engine is not started within two seconds, the ECM will shut the fuel pump off and wait until the engine starts. As soon as the engine is cranked, the ECM will turn the relay on and run the fuel pump.

3   As a backup system to the fuel pump relay, the fuel pump can also be turned on by the oil pressure switch. The oil pressure switch is a normally open switch which closes when the oil pressure reaches about 4 psi. If the fuel pump relay fails, the oil pressure switch will run the fuel pump.

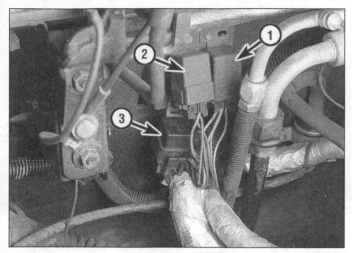

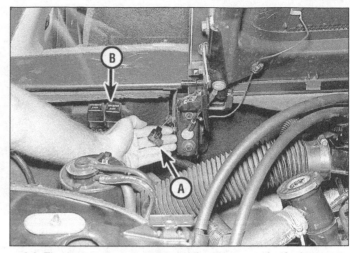

3.4 Relay locations

1	Fuel pump relay	3	Coolant fan relay
2	A/C compressor relay		

3.6 The fuel pump test terminal (A) will turn on the fuel pump when battery voltage is applied to either of its terminals; the fuel pump relay (B) will also turn on the pump when battery voltage is applied to terminal B

4   An inoperative fuel pump relay **(see illustration)** can result in long cranking times, particularly if the engine is cold. The oil pressure switch acts as a backup to the relay and will turn on the fuel pump as soon as the oil pressure reaches about 4 psi.

5   When the ignition is turned to On, the Electronic Control Module (ECM) will turn on the in-tank fuel pump. It will remain on as long as the engine is cranking or running and the ECM is receiving High Energy Ignition (HEI) distributor reference pulses.

6   The fuel pump test terminal is located in the left side of the engine compartment **(see illustration)**. When the engine is stopped, the pump can be activated by applying battery voltage to the test terminal or to fuel pump relay terminal B (the terminals are clearly marked).

7   Before beginning the following sequence of tests, make sure that the fuel tank has fuel in it.

8   Relieve fuel system pressure (refer to Section 2). Install a pressure gauge in the line between the fuel filter and the inlet fitting of the TBI **(see illustration)**. On V6 models, connect the fuel pressure gauge to the Schrader valve on the fuel rail. Use a rag to absorb any leakage that may occur when connecting the gauge.

9   Turn the ignition switch to the On position. The fuel pump should run for two seconds. The pump should pressurize the system at 9 to 13 psi on TBI systems and 40.5 to 47 psi on MPFI systems. Pressure may drop slightly when the pump stops.

10   If there is no fuel pressure, check the fuel pump fuse. If the fuse is blown, check the wiring associated with the fuel pump (see the wiring diagrams in Chapter 12).

11   If the fuse is good and there is no fuel pressure, listen for the fuel pump running while applying battery voltage to the fuel pump test connector. If it runs, check for a restriction in the fuel delivery line or a restricted fuel filter.

12   If there is fuel pressure but it is less than specified, pinch the fuel return line. Apply voltage to the fuel pump test terminal. If the pressure is still low, either the fuel pump, the fuel pump-to-fuel line coupling hose or the pulsator on the pump inlet is faulty.

13   If there is fuel pressure but it is above the specified pressure, disconnect the injector connector (TBI only), disconnect the primary coil connectors, then disconnect the fuel return line flexible hose and attach a 5/16-inch diameter flex hose to the throttle body side of the return line (TBI) or to the pressure regulator side of the return line (MPFI). Insert the other end into an approved fuel container. Note the fuel pressure within two seconds after the ignition is turned on. This test will determine if the high fuel pressure is due to a restricted fuel return line or a faulty fuel pressure regulator.

14   If the fuel pressure is now within the specified limits, locate and repair the restriction in the fuel return line.

15   If fuel pressure is still high, replace the fuel meter cover (TBI) or the fuel pressure regulator (MPFI).

16   If the specified fuel pressure is attained but bleeds down, pinch the return line and recheck the pressure gauge. If the pressure holds, replace the fuel meter cover (TBI) or the fuel pressure regulator (MPFI).

17   If the fuel pressure does not hold, check the pressure again, this time pinching the pressure hose after the specified pressure is obtained. If the pressure holds, check for a leaking pump coupling hose, a leaking pulsator or a faulty in-tank pump.

18   If the pressure does not hold with the pressure line pinched after pressure buildup, check for a leaking throttle body injector (TBI) or a leaking fuel injector or cold start valve (MPFI).

19   On TBI equipped vehicles, a leaking injector can be diagnosed by disconnecting the injector connector, turning on the ignition to allow fuel pressure to build up and looking

3.8 To test fuel pump operation on the 4-cylinder engine, install a pressure gauge in the line between the fuel filter and the inlet fitting of the TBI, then turn the ignition switch to On

into the throttle body to see if the injector drips. Inspect the injector and seals. If the seals do not appear to be damaged or leaking, replace the injector.

20   To diagnose a leaking fuel injector or cold start valve (MPFI), remove the spark plugs and look for a gasoline saturated plug. If there are no signs of fouling or saturation, remove the plenum, the cold start valve and the fuel rail bolts. Follow the fuel rail removal procedure in Section 10, but leave the fuel lines connected. Reconnect the cold start valve. Connect a hose to the cold start valve nozzle and insert it into a fuel container. Lift the fuel rail out just enough to leave the injector nozzles in the ports. With all of the injector electrical connectors disconnected, pressurize the fuel system. Lift each side of the fuel rail up and look for a leaking injector(s). Replace the leaking injector(s).

21   If there is no fuel pressure and the fuel pump fuse is good, check the fuel pump relay and connector for burned or loose wires.

4   Fuel lines and fittings - repair and replacement

Warning: *Gasoline is extremely flammable, so take extra precautions when you work on any part of the fuel system. See the* **Warning** *in Section 2.*

1   Always relieve the fuel pressure before servicing fuel lines or fittings (Section 2).

2   The fuel feed and return lines extend from the fuel gauge sending unit to the engine compartment. The lines are secured to the underbody with clip and screw assemblies. Both fuel feed lines must be properly routed and maintained and should be occasionally inspected for leaks, kinks or dents.

3   If evidence of dirt is found in the system or fuel filter during disassembly, the line should be disconnected and blown out. Check the fuel strainer on the fuel gauge sending unit (Section 7) for damage or deterioration.

4   If replacement of a fuel feed, fuel return or emission line is called for, use welded steel tubing meeting the manufacturer's specifications.

5   Do not use copper or aluminum tubing to replace steel tubing. These materials do not have satisfactory durability to withstand normal vehicle vibrations.

6   When replacing rubber fuel hose, use only hose designed for use in high-pressure fuel injection systems. Hose inside diameter must match pipe outside diameter.

7   Do not use rubber hose within four inches of any part of the exhaust system or within ten inches of the catalytic converter.

8   In repairable areas, cut a piece of fuel hose four inches longer than the portion of the line removed. If more than a six inch length of line is removed, use a combination of steel line and hose so that hose lengths will not be more than two inches. Follow the same routing as the original line.

9   Cut the ends of the line with a tube cutter. Using the first step of a double flaring tool, form a bead on the end of both line sections. If the line is too corroded to handle the flaring operation without damage, the line should be replaced.

10   Use a screw type hose clamp. Slide the clamp onto the line and push the hose on. Tighten the clamps on each side of the repair.

11   Secure the lines properly to the frame to prevent chafing.

5   Fuel tank - removal and installation

Refer to illustrations 5.4 and 5.7

Warning: *Gasoline is extremely flammable, so take extra precautions when you work on any part of the fuel system. See the* **Warning** *in Section 2.*

Removal

1   Relieve the fuel pressure (Section 2).

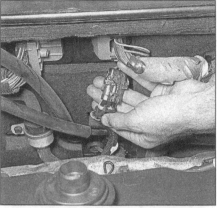

5.4 Before removing the fuel tank, disconnect the electrical connector for the in-tank fuel pump/sending unit assembly - the connector is in the front of the engine compartment

2   Remove the cable from the negative terminal of the battery.

3   Raise the vehicle and support it on jackstands.

4   Disconnect the electric fuel pump wire harness connector from the main wire harness in the engine compartment **(see illustration)**. **Note:** *The other end of this wire harness is an integral, permanent part of the fuel pump/sending unit assembly and cannot be disconnected from it.*

5   From underneath the vehicle, disconnect the fuel feed, fuel return, canister vent, filler and filler breather hoses. **Caution:** *Although it is impossible to confuse the fuel filler and breather hoses (one is much larger in diameter than the other), it is possible to mix up the three smaller hoses. To avoid confusion during reinstallation, note that the fuel feed (high pressure) line has two screw-type hose clamps and the return line only has one. The vent line to the charcoal vapor canister has a spring type clamp. It is also a good idea to mark the three lines with paint to avoid attaching the wrong metal line and hose.* **Warning:** *Any of these lines may spray residual droplets of fuel when disconnected. Additionally, if the tank is not at least half empty, it may slosh out through these lines. Therefore, it is a good idea to wear safety goggles to protect your eyes. If any fuel comes in contact with your skin, be sure to immediately rinse it off with water.*

6   Remove the catalytic converter heat shield.

7   While supporting the tank with either an assistant or a floor jack, remove the main support bracket and the two smaller support straps **(see illustration)**.

8   With the help of an assistant, carefully lower the tank. **Warning:** *Unless the tank is totally empty, it's a good idea to drop the front end first because the unplugged inlet and outlet pipes at the rear of the tank will drip gasoline onto you or your assistant.*

9   The fuel gauge sending unit/electric fuel pump assembly is a delicate mechanism. It

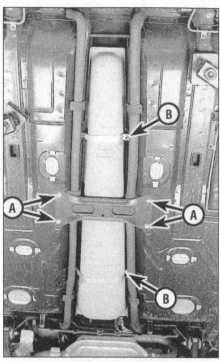

5.7 Remove the bolts (A) from the support bracket, then remove the retaining bolts on each support strap (B)

must be removed before the inside of the tank can be inspected or repaired (see Section 7).

10   Before reinstalling the tank make sure that all traces of dirt and corrosion are cleaned from it. A coat of rust preventative paint is recommended. If the tank is rusted internally, however, it should be replaced with a new one.

Installation

11   If any of the insulating pads above the fuel tank strap mounting brackets fell off the pan during removal of the fuel tank, glue them back into place.

12   With the help of an assistant, raise the tank into position. Make sure that the electrical leads for the sending unit/fuel pump assembly are facing to the rear.

13   Tighten the strap bolts securely.

14   Coat the four main support bracket bolts with RTV sealer and install the main bracket. Tighten the bolts securely.

15   Reconnect the fuel tank filler and breather hoses. Tighten the hose clamps securely.

16   Reconnect the fuel feed and return hoses and the vapor canister hose. Tighten the hose clamps securely.

17   Thread the electrical wiring for the fuel pump and sending unit up into the forward engine compartment and install the catalytic converter heat shield. Tighten the heat shield bolts securely.

18   Lower the vehicle.

19   Reattach the electrical leads for the fuel pump and sending unit to the main engine compartment wire harness.

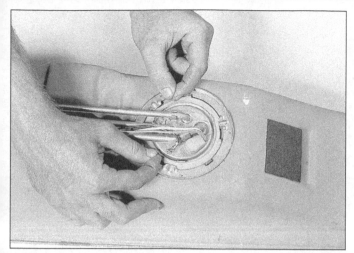

7.5 To unlock the lock ring mechanism, rotate the inner ring clockwise until the locking cams are free of the retaining tangs

7.6 Carefully lift the fuel pump/sending unit assembly out of the fuel tank - this assembly is very delicate so avoid banging it into the outer lock ring during removal

6   Fuel tank - cleaning and repair

1   Any repairs to the fuel tank or filler neck should be carried out by a professional who has experience in this critical and potentially dangerous work. Even after cleaning and flushing, explosive fumes can remain and ignite during repair of the tank.
2   If the fuel tank is removed from the vehicle, it should not be placed in an area where sparks or open flames could ignite the fumes coming out of the tank. Be especially careful inside garages where a natural gas-type appliance is located, because the pilot light could cause an explosion.

7   Fuel pump - removal and installation

Refer to illustrations 7.5 and 7.6
Warning: *Gasoline is extremely flammable, so take extra precautions when you work on any part of the fuel system. See the* **Warning** *in Section 2.*

Removal

1   Remove the cable from the negative battery terminal.
2   Relieve fuel pressure (Section 2).
3   Remove the fuel tank (Section 5).
4   The fuel pump/sending unit assembly is located inside the fuel tank. It is held in place by a lock ring mechanism consisting of an inner ring with three locking cams and an outer ring with three retaining tangs.
5   To unlock the fuel pump/sending unit assembly, turn the inner ring clockwise using a hammer and a brass punch until the locking cams are free of the retaining tangs **(see illustration)**. **Note:** *If the rings are locked together too tightly to release them by hand, gently knock them loose with a soft-face hammer.* **Warning:** *Do not use a steel punch to knock the lock rings loose. A spark could*

cause an explosion.
6   Carefully extract the fuel pump/sending unit assembly from the fuel tank **(see illustration)**. **Caution:** *The fuel level float and sending unit are delicate. Do not bump them into the lock ring during removal or the accuracy of the sending unit may be affected.*
7   Inspect the condition of the rubber gasket around the mouth of the lock ring mechanism. If it is dried, cracked or deteriorated, replace it.
8   Inspect the strainer on the lower end of the fuel pump. If it is dirty, remove it, clean it with a solvent and blow it out with compressed air. If it is too dirty to be cleaned, replace it.
9   If it is necessary to separate the fuel pump and sending unit, remove the pump from the sending unit by pulling the fuel pump assembly into the rubber connector and sliding the pump away from the bottom support. Care should be taken to prevent damage to the rubber insulator and fuel strainer during removal. After the pump assembly is clear of the bottom support, pull the pump assembly out of the rubber connector for removal.

Installation

10   Insert the fuel pump/sending unit assembly into the fuel tank.
11   Turn the inner lock ring counterclockwise until the locking cams are fully engaged by the retaining tangs. **Note:** *If you have installed a new O-ring type rubber gasket, it may be necessary to push down on the inner lock ring until the locking cams slide under the retaining tangs.*
12   Install the fuel tank (Section 5).

8   Fuel injection system - general information

Refer to illustration 8.5
Electronic fuel injection (EFI) provides optimum mixture ratios at all engine speeds

and loads and offers immediate throttle response characteristics. It also enables the engine to run at the leanest possible air/fuel mixture ratio, greatly reducing exhaust gas emissions.
On four-cylinder models, a throttle body injection (TBI) unit replaces a conventional carburetor atop the intake manifold. V6 powered vehicles are fitted with a multi-port fuel injection (MPFI) system. Both systems are controlled by an Electronic Control Module (ECM), which monitors engine performance and adjusts the air/fuel mixture accordingly (see Chapter 6 for a complete description of the fuel control system).

Throttle Body Injection (TBI)

The TBI unit is computer controlled by the Electronic Control Module (ECM) and supplies the correct amount of fuel during all engine operating conditions.
An electric fuel pump located in the fuel tank with the fuel gauge sending unit pumps fuel to the TBI through the fuel feed line and an inline fuel filter. The pump is designed to provide pressurized fuel at about 18 psi. A pressure regulator in the TBI keeps fuel available to the injector at a constant pressure between 9 and 13 psi. Fuel in excess of injector needs is returned to the fuel tank by a separate line. The injector, located in the TBI, is controlled by the ECM. It delivers fuel in one of several modes (see Chapter 6 for a complete description of ECM modes of operation).
The basic TBI unit is made up of two major casting assemblies: A throttle body with an Idle Air Control (IAC) valve controls air flow and a throttle position sensor monitors throttle angle. A fuel body consists of a fuel meter cover with a built-in pressure regulator and a fuel injector to supply fuel to the engine **(see illustration)**.
The throttle body portion of the TBI unit contains ports located at, above and below the throttle valve. These ports generate the

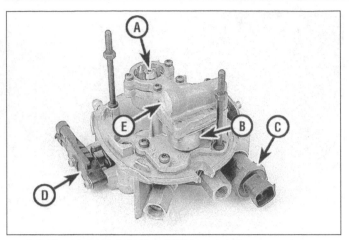

8.5 Throttle body injection (TBI) has very few components

A	Fuel injector
B	Fuel pressure regulator
C	Idle Air Control (IAC)
D	Throttle Position Sensor (TPS)
E	Fuel meter cover

9.7 The fuel pressure regulator is installed in the fuel meter cover and pre-adjusted by the factory - do not remove the four retaining screws (arrows) or you may damage the regulator

vacuum signals for the exhaust gas recirculation (EGR) valve, manifold absolute pressure (MAP) sensor and the canister purge system.

The fuel injector is a solenoid operated device controlled by the ECM. The ECM turns on the solenoid, which lifts a normally closed ball valve off its seat. The fuel, which is under pressure, is injected in a conical spray pattern at the walls of the throttle body bore above the throttle valve. The fuel which is not used by the injector passes through the pressure regulator before being returned to the fuel tank.

The pressure regulator is a diaphragm-operated relief valve with injector pressure on one side and air cleaner pressure on the other. The function of the regulator is to maintain a constant pressure at the injector at all times by controlling the flow in the return line.

The purpose of the idle air control valve is to control engine idle speed while preventing stalls due to changes in engine load. The IAC valve, mounted on the throttle body, controls bypass air around the throttle valve. By moving a conical valve in, to decrease air flow, or out, to increase air flow, a controlled amount of air can move around the throttle valve. If rpm is too low, more air is bypassed around the throttle valve to increase rpm. If rpm is too high, less air is bypassed around the throttle valve to decrease rpm.

During idle, the proper position of the IAC valve is calculated by the ECM based upon battery voltage, coolant temperature, engine load and engine rpm. If the rpm drops below a specified rpm, and the throttle valve is closed, the ECM senses a near stall condition. The ECM will then calculate a new valve position to prevent stalls based on barometric pressure.

Multi-Port Fuel Injection (MPFI)

An electric fuel pump, located in the fuel tank with the gauge sending unit, pumps fuel

to the fuel rail through an inline fuel filter. The pump is designed to provide fuel at a pressure above the pressure needed by the injectors. A pressure regulator in the fuel rail keeps fuel available to the injectors at a constant pressure. Unused fuel is returned to the fuel tank by a separate line.

The injectors are controlled by the ECM. They deliver fuel in one of several modes (see Chapter 6 for a complete description of the modes of operation).

The throttle body has a throttle valve to control the amount of air delivered to the engine. The throttle position sensor (TPS) and idle air control (IAC) valves are located on the throttle body. The throttle body contains vacuum ports located at, above and below the throttle valve. These ports generate the vacuum signals needed to operate various components.

The fuel rail is mounted to the top of the engine. It distributes fuel to the individual injectors. Fuel is delivered to the input end of the rail by the fuel lines, goes through the rail and then to the pressure regulator. The regulator keeps the pressure to the injectors at a constant level. The remaining fuel is returned to the fuel tank.

The fuel injectors are solenoid operated devices controlled by the ECM. The ECM turns on the solenoid, which opens a valve to allow fuel delivery. The fuel which is not used by the injectors passes through the pressure regulator before being returned to the fuel tank.

The pressure regulator is a diaphragm-operated relief valve with injector pressure on one side and manifold pressure on the other. The function of the regulator is to maintain a constant pressure differential across the injectors at all times, by controlling the flow in the return line. The pressure regulator is mounted to the fuel rail and is replaced as an assembly.

The purpose of the Idle Air Control (IAC) valve is to control engine idle speed, while

preventing stalls due to changes in engine load. The IAC valve, mounted in the throttle body, controls bypass air around the throttle valve. By moving a conical valve in, to decrease air flow, or out, to increase air flow, a controlled amount of air can move around the throttle plate. If rpm is too low, more air is bypassed around the throttle valve to increase rpm. If rpm is too high, less air is bypassed around the throttle valve to decrease rpm. During idle, the proper position of the IAC valve is calculated by the ECM based on battery voltage, coolant temperature and engine rpm.

9   Throttle Body Injection (TBI) - component removal and installation

Refer to illustrations 9.7, 9.8a, 9.8b, 9.8c, 9.9, 9.10, 9.11, 9.17, 9.18, 9.19, 9.25, 9.26 and 9.35.

Note: *Because of its relative simplicity, a throttle body assembly does not have to be removed from the intake manifold or completely disassembled during component replacement. However, for the sake of clarity, the following procedures are depicted in the accompanying photos on a TBI assembly removed from the vehicle.*

1   Relieve the fuel pressure (Section 2).
2   Disconnect the cable from the negative terminal of the battery.
3   Remove the air cleaner.

Fuel meter cover and fuel injector

Disassembly

4   Remove the injector electrical connector by squeezing the two tabs together and pulling straight up.
5   Unscrew the five fuel meter cover retaining screws and lockwashers securing the fuel meter cover to the fuel meter body. Note the

9.8a The best way to remove the fuel injector is to pry on it with a screwdriver, using a second screwdriver as a fulcrum

9.8b Note the position of the terminals on top and the dowel pin on bottom of the injector in relation to the fuel meter cover when you lift the injector out of the cover

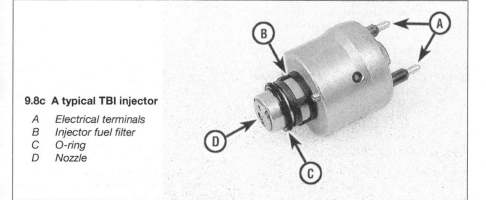

9.8c A typical TBI injector

A *Electrical terminals*
B *Injector fuel filter*
C *O-ring*
D *Nozzle*

location of the two short screws.

6   Remove the fuel meter cover. **Caution:** *Do not immerse the fuel meter cover in solvent. It might damage the pressure regulator diaphragm and gasket.*

7   The fuel meter cover contains the fuel pressure regulator. The regulator is pre-set and plugged at the factory. If a malfunction occurs, it cannot be serviced. It must be replaced as a complete assembly. **Warning:**

Do not remove the screws securing the pressure regulator to the fuel meter cover **(see illustration)**. *It has a large spring inside which is tightly compressed. If accidentally released, it could cause injury. Disassembly might also cause a fuel leak between the diaphragm and the regulator container.*

8   With the old fuel meter cover gasket in place to prevent damage to the casting, carefully pry the injector from the fuel meter body

with a screwdriver until it can be lifted free **(see illustrations)**. **Caution:** *Use care in removing the injector to prevent damage to the electrical connector terminals, the injector fuel filter, the O-ring and the nozzle* **(see illustration)**.

9   The fuel meter body should be removed from the throttle body if it needs to be cleaned. To remove it, remove the fuel feed and return line fittings **(see illustration)** and the Torx screws that attach the fuel meter body to the throttle body.

10   Remove the old gasket from the fuel meter cover and discard it. Remove the large O-ring and steel back-up washer from the upper counterbore of the fuel meter body injector cavity **(see illustration)**. Clean the fuel meter body thoroughly in solvent and blow dry.

11   Remove the small O-ring from the nozzle end of the injector. Carefully rotate the injector fuel filter back-and-forth and remove the filter from the base of the injector **(see illustration)**. Gently clean the filter in solvent and allow it to drip dry. It is too small and delicate to dry with compressed air. **Caution:** *The fuel injector itself is an electrical component. Do not immerse it in any type of cleaning solvent.*

12   The fuel injector is not serviceable. If it is malfunctioning, replace it as an assembly.

Reassembly

13   Install the clean fuel injector nozzle filter on the end of the fuel injector with the larger end of the filter facing the injector so that the filter covers the raised rib at the base of the injector. Use a twisting motion to position the filter against the base of the injector.

14   Lubricate a new small O-ring with automatic transmission fluid. Push the O-ring onto the nozzle end of the injector until it presses against the injector fuel filter.

15   Insert the steel back-up washer in the top counterbore of the fuel meter body injector cavity.

16   Lubricate a new large O-ring with automatic transmission fluid and install it directly

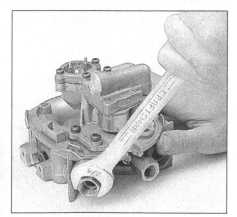

9.9 The fuel inlet and outlet fittings must be removed before the fuel meter body can be removed from the throttle body

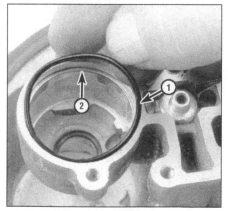

9.10 Remove the large O-ring and steel washer from the injector cavity of the fuel meter body

1   *O-ring*
2   *Steel back-up washer*

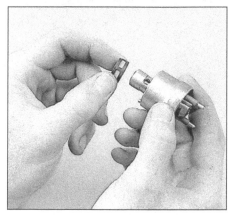

9.11 Gently rotate the fuel injector filter back-and-forth and carefully pull it off the nozzle

9.17 Push straight down with both thumbs to install the injector in the fuel meter body cavity

1   Fuel injector
2   Fuel meter body
3   Notch
4   Raised lug

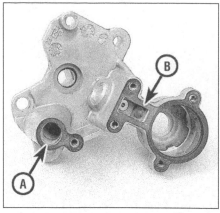

9.18 Position the fuel outlet passage gasket (A) and the fuel meter cover gasket (B) properly

over the back-up washer. Be sure that the O-ring is seated properly in the cavity and is flush with the top of the fuel meter body casting surface. **Caution:** *The back-up washer and large O-ring must be installed before the injector or improper seating of the large O-ring could cause fuel to leak.*

17   Install the injector in the cavity in the fuel meter body, aligning the raised lug on the injector base with the cast-in notch in the fuel meter body cavity. Push straight down on the injector with both thumbs **(see illustration)** until it is fully seated in the cavity. **Note:** *The electrical terminals of the injector should be approximately parallel to the throttle shaft.*

18   Install a new fuel outlet passage gasket on the fuel meter cover and a new fuel meter cover gasket on the fuel meter body **(see illustration)**.

19   Install a new dust seal into the recess on the fuel meter body **(see illustration)**.

20   Attach the fuel meter cover to the fuel meter body, making sure that the pressure regulator dust seal and cover gaskets are in place.

21   Apply a thread locking compound to the threads of the fuel meter cover screws. Install the screws (the two short screws go next to the injector) and tighten them to the specified

torque. **Note:** *Service repair kits include a small vial of thread compound with directions for use. If material is not available, use Loctite 262, GM part number 1052624, or equivalent. Do not use a higher strength locking compound than recommended, as this may prevent subsequent removal of the screws or cause breakage of the screwhead if removal becomes necessary.*

22   Plug in the electrical connector to the injector.

23   Install the air cleaner.

Idle Air Control (IAC) valve

Removal

24   Unplug the electrical connector at the Idle Air Control valve.

25   Remove the Idle Air Control valve with a wrench on the hex surface only **(see illustration)**.

Adjustment

26   Before installing a new Idle Air Control valve, measure the distance the valve is extended **(see illustration)**. The measurement should be made from the motor housing to the end of the cone. The distance should be no greater than 1-1/8 inch. If the cone is extended too far, damage may occur

to the valve when it is installed.

27   Identify the replacement IAC valve as either a Type I (with a collar at the electric terminal end) or a Type II (without a collar) (see illustration 9.26 above). If the measured dimension A is greater than 1-1/8 inch, the distance must be reduced as follows:

Type I - Exert firm pressure on the valve to retract it (a slight side-to-side movement may be helpful).

Type II - Compress the retaining spring of the valve while turning the valve in with a clockwise motion. Return the spring to its original position with the straight portion of the spring aligned with the flat surface of the valve.

Installation

28   Install the new Idle Air Control valve on the throttle body. Use the new gasket supplied with the assembly. Tighten the IAC valve to the specified torque.

29   Plug in the electrical connector.

30   Install the air cleaner.

31   Start the engine and allow it to reach operating temperature. The Electronic Control Module (ECM) will reset the idle speed when the vehicle is driven above 35 mph.

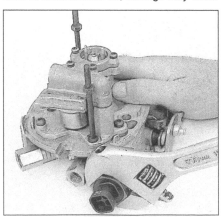

9.19 Install a new dust seal into the recess of the fuel meter body

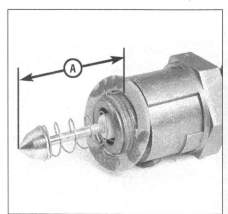

9.25 Remove the IAC valve with a large wrench, but be careful - it's a delicate device

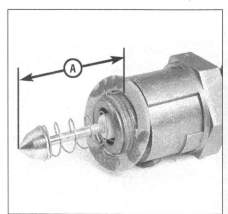

9.26 Distance A should be less than 1-1/8 inch for either type of Idle Air Control valve - if it isn't, determine what kind of IAC valve you have and adjust it accordingly

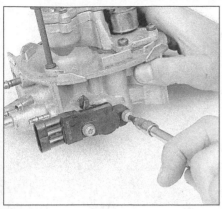

9.35 The Throttle Position Sensor (TPS) is mounted to the side of the TBI with two Torx screws

Throttle Position Sensor (TPS)

32   The Throttle Position Sensor (TPS) is connected to the throttle shaft on the TBI unit. As the throttle valve angle is changed (as the accelerator pedal is moved), the output of the TPS also changes. At a closed throttle position, the output of the TPS is below 1.25 volts. As the throttle valve opens, the output increases so that, at wide-open throttle, the output voltage should be approximately 5 volts.

33   A broken or loose TPS can cause intermittent bursts of fuel from the injector and an unstable idle, because the ECM thinks the throttle is moving. A problem in any of the TPS circuits will set either a Code 21 or 22 (see Trouble Codes, Chapter 6).

34   The TPS is not adjustable. The ECM uses the reading at idle for the zero reading. If the TPS malfunctions, it is replaced as a unit.

35   Unscrew the two Torx screws **(see illustration)** and remove the TPS.

36   Install the new TPS. **Note:** *Make sure that the tang on the lever is properly engaged with the stop on the TBI.*

---

10   Multi-Port Fuel Injection (MPFI) - component removal and installation

---

Warning 1: *Gasoline is extremely flammable, so take extra precautions when you work on any part of the fuel system. See the* **Warning** *in Section 2.*

Warning 2: *Before servicing an injector, fuel rail or pressure regulator, relieve the pressure in the fuel system to minimize the risk of fire and injury. After servicing the fuel system, cycle the ignition between On and Off several times (wait 10 seconds between cycles) and check the system for leaks.*

Throttle Body
Removal

1   Disconnect the cable from the negative terminal of the battery.
2   Unplug the IAC and TPS connectors.

3   Disconnect the coolant lines.
4   Disconnect the throttle linkages.
5   Disconnect the air inlet duct and cold start valve tube.
6   Remove the throttle body bolts and the throttle body.

Installation

7   Install the throttle body and gasket and tighten the bolts to the specified torque.
8   Reconnect the air inlet duct.
9   Reconnect the throttle linkage.
10   Reconnect the coolant lines and cold start valve tubes.
11   Plug in the TPS and IAC electrical connectors.
12   Reconnect the cable to the negative terminal of the battery.

Idle Air Control (IAC) valve
Removal

13   Unplug the electrical connector from the Idle Air Control (IAC) valve assembly.
14   Remove the IAC valve assembly from the idle air/vacuum signal housing assembly. **Caution:** *Do not remove any thread locking compound from the threads.*
15   Remove the IAC valve assembly gasket and discard it.
16   Clean the gasket mounting surface of the idle air/vacuum signal housing assembly to ensure a good seal. **Caution:** *The IAC valve assembly itself is an electrical component, and must not be soaked in any liquid cleaner or solvent, as damage may result.*

Installation

17   Before installing the IAC valve assembly, the position of the pintle must be checked. If the pintle is extended too far, damage to the assembly may occur.
18   Measure the distance from the gasket mounting surface of the IAC valve assembly to the tip of the pintle **(see illustration 9.26)**.
19   If the distance is greater than 1-1/8 inch, reduce it as follows:

a) *If the IAC valve assembly has a collar around its electrical connector end, use firm hand pressure on the pintle to retract it (a slight side-to-side motion may help).*
b) *If the IAC valve assembly has no collar, compress the pintle retaining spring toward the body of the IAC and try to turn the pintle clockwise. If the pintle will turn, continue turning until the 1-1/8 inch dimension is reached. Return the spring to its original position with the straight part of the spring end lined up with the flat surface under the pintle head. If the pintle will not turn, use firm hand pressure to retract it.*

20   Install the new gasket on the IAC valve assembly.
21   Install the IAC valve assembly in the idle air/vacuum signal housing assembly.
22   Tighten the IAC valve assembly to the specified torque.
23   Plug in the electrical connector at the

IAC valve assembly. **Note:** *No adjustment is made to the IAC assembly after reinstallation. IAC resetting is controlled by the ECM when the engine is running.*

Plenum
Removal

24   Remove the cable from the negative terminal of the battery.
25   Remove the vacuum lines and throttle cable bracket bolts.
26   Remove the EGR pipe-to-EGR valve base bolts.
27   Remove the throttle body.
28   Remove the plenum bolts.
29   Remove the plenum and gaskets.

Installation

30   Install the plenum and gaskets.
31   Install the plenum bolts and tighten them to the specified torque.
32   Install the throttle body. Tighten the throttle body bolts to the specified torque.
33   Install the EGR pipe.
34   Install the throttle cable bracket bolts.
35   Install the vacuum lines.
36   Install the negative battery cable.

Fuel rail and components
Fuel rail removal

37   Remove the negative battery cable.
38   Remove the plenum.
39   Remove the cold start valve line at the fuel rail.
40   Remove the fuel lines at the fuel rail.
41   Remove the vacuum line at the regulator.
42   Unplug the injector electrical connectors.
43   Remove the fuel rail retaining bolts.
44   Remove the fuel rail and injectors.
45   **Caution:** *Use care when handling the fuel rail assembly to avoid damaging the injectors.*

Cold start valve removal and installation

46   Disconnect the cold start valve electrical connector.
47   Remove the cold start valve retaining bolt.
48   Disconnect the valve from the tube and body assembly by bending the tab back to permit unscrewing of the valve.
49   Install a new cold start valve O-ring seal and body O-ring seal on the cold start valve. **Note:** *An eight digit identification number is stamped on the side of the fuel rail assembly .* Refer to this number if servicing or part replacement is required.
50   Install a new tube O-ring seal on the tube and body assembly.
51   Turn the valve completely into the body assembly.
52   Turn the valve back one full turn, until the electrical connector is at the top position.
53   Bend the tang of the body forward to limit rotation of the valve to less than one full turn.
54   Before installing into the engine, coat the O-ring seals with engine oil.

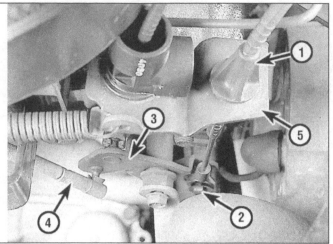

11.2 Details of the accelerator cable brackets and linkage for the V6 engine

1   Accelerator cable
2   Retainer clip
3   Lever assembly
4   Throttle linkage
5   Throttle cable/TV cable bracket

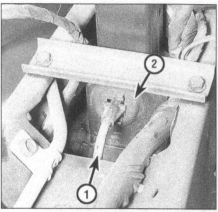

11.5a Details of the accelerator cable-to-center console area

1   Accelerator cable
2   Instrument panel support

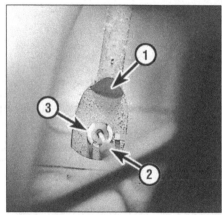

11.5b Details of the accelerator cable and linkage

1   Accelerator pedal assembly
2   Accelerator cable
3   Retainer clip

Fuel Injector removal and installation

Caution: *To prevent dirt from entering the engine, the area around the injectors should be cleaned before servicing.*

55   Rotate the injector retaining clip(s) to the release position.

56   Remove the port injector(s).

57   Inspect the injector O-ring seal(s). Replace if damaged.

58   Install the new O-ring seal(s), as required, onto the injector(s) and lubricate them with engine oil.

59   Install the injectors on the fuel rail.

60   Rotate the injector retainer clips to the lock position.

Pressure regulator removal and installation

61   The pressure regulator is factory adjusted and is not serviceable. *Do not attempt to remove the regulator from the fuel rail.*

Fuel rail installation

62   Lubricate all injector O-ring seals with engine oil.

63   Install the fuel rail and injector assembly.

64   Install the fuel rail retaining bolts and tighten them to the specified torque.

65   Plug in the injector electrical connectors.

66   Install the vacuum line at the regulator.

67   Install the fuel lines at the fuel rail.

68   Install the cold start valve line at the fuel rail.

69   Install the plenum.

70   Install the negative battery cable.

---

11   Throttle cable - removal, installation and adjustment

Refer to illustrations 11.2, 11.5a and 11.5b

Note: *The accelerator control system is a cable type. There are no linkage adjustments. Because there are no adjustments, the specific cable for each application must be used. Only the specific replacement part will work.*

When work has been performed on accelerator controls, always check to make sure that all components are installed as removed and that the cable is not rubbing or binding in any manner.

Removal

1   Disconnect the cable from the negative terminal of the battery.

2   Disconnect the cable from the EFI unit and related cable brackets **(see illustration)**. It is usually held in place by a clip or retainer at the throttle shaft.

3   Remove the shift knob and console covers and supports.

4   Disconnect the ECM electrical harness and remove the ECM unit.

5   Remove the accelerator cable at the instrument panel support and pedal assembly **(see illustrations)**.

6   Remove the accelerator cable through the rear body panel and out of the console.

Installation

7   Route the new cable through the rear body panel and out of the console. **Note:** *The conduit fitting at both ends of the cable must have locking tangs expanded and locked in attaching holes.*

8   Connect the cable at the instrument panel support and pedal assembly.

9   Install the ECM unit and connect the ECM electrical harness.

10   Install the console covers and shift knob.

11   Connect the cable to the EFI unit and related cable brackets. **Note:** *The retainer must be installed with the tangs secured over the head of the stud.*

12   Connect the cable to the negative terminal of the battery.

---

12   Exhaust system components - removal and installation

Refer to illustrations 12.5, 12.6, 12.7, 12.8 and 12.9

Caution: *The exhaust system components get very hot during engine operation. No part of the exhaust should be touched until the entire system has completely cooled. Be especially careful around the catalytic converter, where the highest temperatures are generated.*

Four-cylinder engine

Removal

1   Disconnect the cable from the negative terminal of the battery.

2   Raise the vehicle and support it securely on jackstands.

3   Remove the catalytic converter splash shield.

4   Disconnect the exhaust pipe from the exhaust manifold flange by removing the two flange bolts.

5   Remove the two springs from between the exhaust pipe bracket and the left front corner of the cradle **(see illustration)**.

6   Remove the springs from the exhaust pipe bracket and the right front corner of the cradle **(see illustration)**.

7   Disconnect the exhaust pipe U-bolt

12.5 Remove the hanger springs from the left front corner of the cradle

12.6 Remove the hanger springs from the right front corner of the cradle

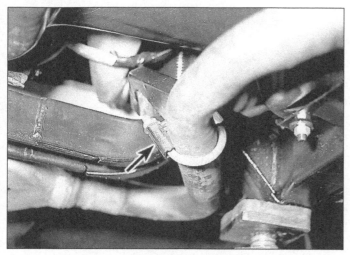

12.7 Remove the U-bolt clamp from the engine mount support bracket

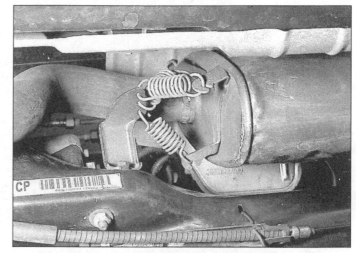

12.8 Remove the springs from both ends of the muffler

from the exhaust pipe support bracket (just behind the cradle front crossmember) **(see illustration)**.

8   Remove the springs from both ends of the muffler **(see illustration)**.

9   Disconnect the bracket bolt just forward of the tailpipe **(see illustration)**.

10   Remove the exhaust system.

11   If you are replacing the muffler or the catalytic converter, take the exhaust system to a dealer service department or muffler shop. The removal of either of these components requires a cutting torch and welding equipment.

Installation

12   Raise the exhaust system back into place.

13   Bolt up the exhaust pipe to the exhaust manifold. Remember to use a new "graph oil" type seal at the flange joint. Tighten the bolts securely.

14   Install all spring hangers.

15   Bolt up the support bracket in front of the tailpipe and tighten securely.

16   Attach the U-bolt clamp to the engine

support bracket.

17   Lower the vehicle.

18   Start the engine and check for exhaust leaks.

V6 engine

Removal

19   Disconnect the cable from the negative terminal of the battery.

20   Raise the vehicle and support it securely on jackstands.

21   Remove the splash shield retaining bolts, the U-clamp and the catalytic converter splash shield.

22   Remove the exhaust pipe U-bolt from the exhaust pipe support bracket (just behind the cradle front crossmember).

23   Remove the crossover pipe-to-exhaust manifold flange bolts.

24   Remove the exhaust pipe-to-crossover pipe flange bolts.

25   Remove the crossover pipe.

26   Remove the left and right side muffler-to-cradle spring hangers from the cradle rear crossmember exhaust pipe support brackets.

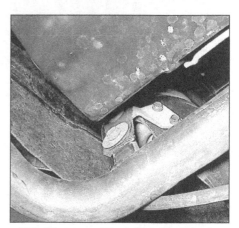

12.9 Disconnect the bracket bolt at the left rear corner of the cradle

27   Remove the left and right side exhaust pipe-to-cradle spring hangers from the front ends of the left and right cradle frame rails.

28   Remove the catalytic converter/exhaust pipe/muffler assembly as a unit.

Installation

29   Install the catalytic converter/exhaust pipe/muffler assembly as a unit.

30   Install all spring hangers. Note the direction in which the muffler retaining spring hooks face when they are attached to the muffler retaining bracket.

31   Attach the exhaust pipe U-bolt to the support bracket and tighten the nuts securely.

32   Install the crossover pipe-to-exhaust manifold bolts and tighten them securely.

33   Install a new "graph oil" type seal at the flange joint between the crossover pipe and the exhaust pipe, then install and tighten the exhaust pipe-to-crossover pipe flange bolts securely. Do not forget to install the springs with the wide end of the spring facing up, toward the exhaust pipe flange.

34   Attach the exhaust pipe U-bolt to the exhaust pipe support bracket. Tighten the U-bolt mounting nuts securely.

35   Install the catalytic converter splash shield and U-clamp, then tighten the splash shield bolts-to-U-clamp bolts finger tight. Install the splash shield-to-exhaust clamp bolt at the other end of the catalytic converter and tighten it securely. Tighten the two splash shield-to-U-clamp bolts securely.

36   Lower the vehicle.

37   Connect the cable to the negative terminal of the battery.

38   Start the engine and check for exhaust leaks.

Chapter 5
Engine electrical systems

Contents

Specifications

Cylinder numbers (right-to-left - transaxle is on left)

4-cylinder engine	1-2-3-4
V6 engine	
Rear bank (trunk)	1-3-5
Front bank (bulkhead)	2-4-6
Firing order	
4-cylinder engine	1-3-4-2
V6 engine	1-2-3-4-5-6
Distributor rotation	Clockwise

Torque specifications

	Ft-lbs
Alternator adjustment bolt	20 to 25
Alternator bracket-to-engine bolt	30 to 40
Alternator pivot bolt	20 to 30
Distributor clamp bolt	
4-cylinder engine	
1984 and 1985	22
1986 and 1987	15
V6	20 to 30
Starter motor bolts	26 to 37
Starter bracket bolts	34

2.5 Carefully lift the battery straight up out of the carrier - don't tilt it

4.4a Routing for the battery cables - four-cylinder engine

1   *Negative battery terminal*
2   *Negative cable-to-engine*
3   *Negative cable-to-body ground*
4   *Positive battery terminal*
5   *Retainer clamp*
6   *Battery*
7   *Battery heat shield*

1   Ignition system - general information and precautions

The ignition system consists of the ignition switch, the battery, the coil, the primary (low tension) and secondary (high tension) wiring circuits, the distributor and the spark plugs.

High Energy Ignition (HEI) distributor

A High Energy Ignition (HEI) distributor with Electronic Spark Timing (EST) is used on all engines. Some HEI distributors combine all the ignition components into one unit. The ignition coil is in the distributor cap and connects through a resistance brush to the rotor. On other HEI distributors, the coil is mounted separately.

The HEI distributor has an internal magnetic pick-up assembly which contains a permanent magnet, a pole piece with internal teeth and a pick-up coil.

All spark timing changes in the HEI/EST distributor are carried out electronically by the Electronic Control Module (ECM), which monitors data from various engine sensors, computes the desired spark timing and signals the distributor to change the timing accordingly. A backup spark advance system is incorporated to signal the ignition module in case of ECM failure. No vacuum or mechanical advance is used.

Electronic Spark Control (ESC)

Some engines are equipped with an Electronic Spark Control (ESC), which uses a knock sensor in connection with the ECM to control spark timing to allow the engine to have maximum spark advance without spark knock. This improves driveability and fuel economy.

Secondary (spark plug) wiring

The secondary (spark plug) wire used with the HEI system is a carbon impregnated cord conductor encased in an 8 mm (5/16-inch) diameter rubber jacket with an outer sil-

icone jacket. This type of wire will withstand very high temperatures and provides an excellent insulator for the HEI's high voltage. Silicone spark plug boots form a tight seal on the plugs. The boot should be twisted 1/2-turn before removing it (for more information on spark plug wiring refer to Chapter 1). **Warning:** *Because of the very high voltage generated by the HEI system, extreme care should be taken whenever an operation involving ignition components is performed. This not only includes the distributor, coil, control module and spark plug wires, but related items that are connected to the systems as well, such as the plug connections, tachometer and testing equipment.*

2   Battery - removal and installation

Refer to illustration 2.5

Note: *There are certain precautions to be taken when working on or near the battery: a) Never expose a battery to open flame or sparks which could ignite the hydrogen gas given off by the battery b) Wear protective clothing and eye protection to reduce the possibility of the corrosive sulfuric acid solution inside the battery harming you (if the fluid is splashed or spilled, flush the contacted area immediately with plenty of water); c) Remove all metal jewelry which could contact the positive terminal and another grounded metal source, thus causing a short circuit; d) Always keep batteries and battery acid out of the reach of children.*

1   The battery is located at the right front corner of the engine compartment, underneath the louvered cover panel.

Removal

2   Remove the two wing screws and detach the cover panel.
3   Disconnect both battery cables from the battery terminals. **Caution:** *To prevent arcing, disconnect the negative (-) cable first, then remove the positive (+) cable.*
4   Remove the hold-down clamp bolt and the clamp from the floor of the battery carrier.
5   Carefully lift the battery out of the carrier **(see illustration)**. **Warning:** *Always keep the*

battery in an upright position to reduce the likelihood of electrolyte spillage. If you spill electrolyte on your skin, rinse it off immediately with large amounts of water.

Installation

Note: *The battery carrier and hold-down clamp should be clean and free from corrosion before installing the battery. The carrier should be in sound condition, to hold the battery securely and to keep it level. Make certain that there are no parts in the carrier before installing the battery.*

6   Gently set the battery in position in the carrier. Don't tilt it.
7   Install the hold-down clamp and bolt. The bolt should be snug, but overtightening it may damage the battery case.
8   Install both battery cables, positive first, then negative. **Note:** *The battery posts and cable ends should be cleaned prior to connection* (see Chapter 1).
9   Install the cover panel and tighten both wing screws securely.

3   Battery - emergency jump starting

Refer to the *Booster battery (jump) starting* procedure at the front of this manual.

4   Battery cables - check and replacement

Refer to illustration 4.4

1   Periodically inspect the entire length of each battery cable for damage, cracked or burned insulation and corrosion. Poor battery cable connections can cause starting problems and decreased engine performance.
2   Check the cable-to-terminal connections at the ends of the cables for cracks, loose wire strands and corrosion. The presence of white, fluffy deposits under the insulation at the cable terminal connection is a sign the cable is corroded and should be replaced. Check the terminals for distortion, missing mounting bolts or nuts and corrosion.

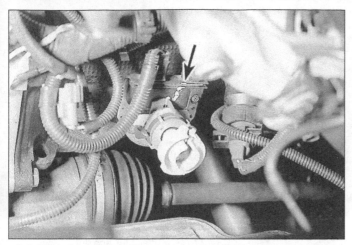

6.4 Mark the relationship of the rotor tip to the distributor base before removing the rotor

6.5 The distributor hold-down clamp and bolt must be removed before the distributor can be removed from the engine

3   If only the positive cable is to be replaced, be sure to disconnect the negative cable from the battery first.

4   Disconnect and remove the cable **(see illustrations)**. Make sure the replacement cable is the same length and diameter.

5   Clean the threads of the starter or ground connection with a wire brush to remove rust and corrosion. Apply a light coat of petroleum jelly to the threads to ease installation and prevent future corrosion.

6   Attach the cable to the starter or ground connection and tighten the mounting nut securely.

7   Before connecting the new cable to the battery, make sure it reaches the terminals without having to be stretched.

8   Connect the positive cable first, followed by the negative cable.

5   Ignition system check

Warning: *Because of the very high voltage generated by the High Energy Ignition (HEI) system, extreme care should be taken whenever an operation is performed involving ignition components. This not only includes the distributor, coil, control module and spark plug wires, but related items that are connected to the system as well, such as the plug connections, tachometer and any test equipment.*

1   If the engine turns over but will not start, remove the spark plug wire from a spark plug. Using an insulated tool, hold the wire about 1/4-inch from a good ground and have an assistant crank the engine.

2   If there is no spark, check another wire in the same manner. A few sparks followed by no spark is the same condition as no spark at all.

3   If there is good spark, check the spark plugs (refer to Chapter 1) and/or the fuel system (refer to Chapter 4).

4   If there is a weak spark or no spark, unplug the coil lead from the distributor, hold it about 1/4-inch from a good ground and

check for spark as described above.

5   If there is still no spark, have the system checked by a dealer service department or repair shop.

6   If there is a spark, check the distributor cap and/or rotor (refer to Chapter 1).

7   Further checks of the HEI ignition system must be done by a dealer service department or repair shop.

6   Distributor (four-cylinder engine) - removal and installation

Refer to illustrations 6.4 and 6.5

Removal

1   Disconnect the cable from the negative terminal of the battery.

2   Remove the coil wire from the distributor cap.

3   Remove the distributor cap.

4   Note the position of the rotor and the distributor-to-block alignment. Inscribe an alignment mark onto the distributor to indicate the position of the rotor **(see illustration)**.

5   Remove the distributor hold-down clamp bolt and clamp **(see illustration)**. Remove the distributor from the engine. **Caution:** *Do not turn the crankshaft while the distributor is removed from the engine. If the crankshaft is turned, the engine will have to be retimed.*

Installation (crankshaft not turned after distributor removal)

6   Insert the distributor into the engine in exactly the same relationship to the block in which it was removed. To mesh the gears, it may be necessary to turn the rotor slightly. At this point the distributor may not seat down against the block completely. This is due to the lower end of the distributor shaft not mating properly with the oil pump shaft. If this is the case, check again to make sure the distributor is aligned with the block in the same posi-

tion it was in before removal and that the rotor is correctly aligned with the distributor body. The gear on the distributor shaft is engaged with the gear on the camshaft and this relationship cannot change as long as the distributor is not lifted from the engine. Use a socket and breaker bar on the crankshaft bolt to turn the engine over in the normal direction of rotation. The rotor will turn, but the oil pump shaft will not because the two shafts are not engaged. When the proper alignment is reached, the distributor will drop down over the oil pump shaft and the distributor body will seat properly against the block.

7   Install the hold-down clamp and tighten the bolt to the specified torque.

8   Install the distributor cap and coil wire.

9   Connect the cable to the negative terminal of the battery.

Installation (crankshaft turned after distributor removal)

10   Remove the number one spark plug (refer to the Specifications).

11   Place your finger over the spark plug hole while turning the crankshaft with a wrench on the pulley bolt at the front of the engine.

12   When you feel compression, continue turning the crankshaft slowly until the timing mark on the crankshaft pulley is aligned with the 0 on the engine timing indicator.

13   Position the rotor to point between the number one and number three distributor terminals.

14   Insert the distributor into the engine in exactly the same relationship to the block in which it was removed. To mesh the gears, it may be necessary to turn the rotor slightly. If the distributor does not seat fully against the block it is because the oil pump shaft has not seated in the distributor shaft. Make sure the distributor drive gear is fully engaged with the camshaft gear, then use a socket on the crankshaft bolt to turn the engine over in the normal direction of rotation until the two shafts engage and the distributor seats against the block.

15   Install the hold-down clamp and tighten the bolt to the specified torque.
16   Install the distributor cap and coil wire.
17   Connect the cable to the negative terminal of the battery.

7   Distributor (V6 engine) - removal and installation

Removal

1   Disconnect the cable from the negative terminal of the battery.
2   Disconnect the ignition switch battery feed wire and the tachometer lead, if so equipped, from the distributor cap.
3   Disconnect the coil connector from the cap. Depress the locking tabs by hand. **Caution:** *Do not use a screwdriver or other tool to release the locking tabs.*
4   Turn the four distributor cap locking screws counterclockwise, remove the cap and position it out of the way.
5   Disconnect the four-terminal ECM wiring harness connector from the distributor.
6   If necessary for clearance on distributor caps with a secondary wiring harness attached to the cap, release the wiring harness latches and remove the wiring harness retainer. Note that the spark plug wire numbers are indicated on the retainer.
7   Make matching marks on the base of the distributor and the engine block to insure that you will be able to put the distributor back in the same position.
8   Remove the distributor hold-down clamp bolt and the clamp.
9   Make a mark on the distributor housing to show the direction the rotor is pointing. Lift the distributor up slowly. As you lift it the rotor will turn slightly. When it stops turning make another mark on the distributor body to show where it is pointing when the gear on the distributor is disengaged. This is the position the rotor should be in when you begin reinstallation.
10   Remove the distributor. **Caution:** *Avoid turning the crankshaft while the distributor is removed. Turning the crankshaft while the distributor is removed will require retiming the engine.*

Installation (crankshaft not turned after distributor removal)

11   Position the rotor in the exact location (second mark on the housing) it was in when the distributor was removed.
12   Lower the distributor into the engine. To mesh the gears at the bottom of the distributor it may be necessary to turn the rotor slightly. It is possible that the distributor may not seat down fully against the block because the lower part of the distributor shaft has not properly engaged the oil pump shaft. Make sure the distributor and rotor are properly aligned with the marks made earlier, then use a large socket and breaker bar on the

8.2 The ignition module is secured by two mounting screws (arrows)

crankshaft bolt to turn the engine in the normal direction of rotation until the two shafts engage and the distributor drops down against the block.
13   With the base of the distributor seated against the engine block turn the distributor housing to align the marks made on the distributor base and the engine block.
14   With the distributor properly seated, and the marks aligned, the rotor should point to the first mark made on the distributor housing.
15   Place the hold-down clamp in position and loosely install the bolt.
16   Reconnect the ignition wiring harness.
17   Install the distributor cap. If the secondary wiring harness was removed from the cap, reinstall it.
18   Reconnect the coil connector.
19   With the distributor in its original position, tighten the hold-down bolt and check the ignition timing (Chapter 1).

Installation (crankshaft turned after distributor removal)

20   Remove the number one spark plug (refer to the Specifications).
21   Place your finger over the spark plug hole while turning the crankshaft with a wrench on the pulley bolt at the front of the engine.
22   When you feel compression, continue turning the crankshaft slowly until the timing mark on the crankshaft pulley is aligned with the 0 on the engine timing indicator.
23   Position the rotor between the number one and six spark plug terminals on the cap.
24   Lower the distributor into the engine. To mesh the gears at the bottom of the distributor, it may be necessary to turn the rotor slightly. If the distributor does not drop down flush against the block it is because the distributor shaft has not mated to the oil pump shaft. Place a large socket and breaker bar on the crankshaft bolt and turn the engine over in the normal direction of rotation until the two shafts engage properly, allowing the distributor to seat flush against the block.
25   With the base of the distributor properly seated against the engine block, turn the distributor housing to align the marks made on the distributor base and the engine block.

26   With the distributor all the way down and the marks aligned, the rotor should point to the first mark made on the distributor housing.
27   Place the hold-down clamp in position and loosely install the bolt.
28   Reconnect the ignition wiring harness.
29   Install the distributor cap. If the secondary wiring harness was removed from the cap, reinstall it.
30   Reconnect the coil connector.
31   With the distributor in its original position, tighten the hold-down bolt and check the ignition timing.

8   Ignition module - replacement

Refer to illustrations 8.2 and 8.4
Note: *It is not necessary to remove the distributor from the engine to replace the ignition module.*

Removal

1   Remove the distributor cap and rotor (Section 6 for four-cylinder engine, Section 7 for V6 engine).
2   Remove both module mounting screws and detach the module from the distributor **(see illustration)**.
3   Disconnect both electrical leads from the module. Note that the leads cannot be interchanged.
4   Do not wipe the grease from the module or the distributor base if the same module is to be reinstalled. If a new module is to be installed, a package of silicone grease will be included with it. Wipe the distributor base and the new module clean, then apply the silicone grease to the face of the module and the distributor base where the module seats **(see illustration)**. This grease is necessary for heat dissipation. **Note:** *The module cannot be tested without a module tester. If you suspect that the module is malfunctioning, have it checked by a dealer service department.*

Installation

5   Connect both electrical leads to the new module.
6   Place the module in position on the distributor base and tighten the two mounting screws securely.
7   Install the distributor rotor and cap.

9   Ignition pick-up coil - testing and replacement

Refer to illustrations 9.5, 9.9, 9.20, 9.23, 9.24, 9.29 and 9.30

Four-cylinder engine

1   Remove the distributor cap (Section 6).
2   Remove the rotor (Section 6).
3   Remove the distributor from the engine (Section 6).
4   Remove the pick-up coil leads from the module.

8.4 Silicone lubricant applied to the distributor base in the area under the ignition module dissipates heat (arrows) - this is a distributor with a separately mounted coil

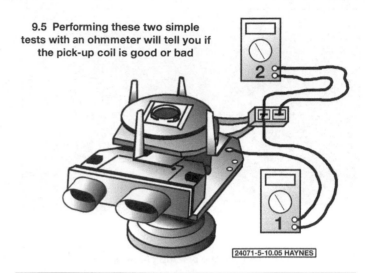

9.5 Performing these two simple tests with an ohmmeter will tell you if the pick-up coil is good or bad

Testing

5   Connect an ohmmeter to each terminal of the pick-up coil lead and ground as shown **(see illustration)**. Flex the leads by hand to check for intermittent opens. The ohmmeter should indicate infinite at all times. If it doesn't, the pick-up coil is defective.

6   Connect the ohmmeter to both terminals of the pick-up coil lead. Flex the leads by hand to check for intermittent opens. The ohmmeter should read one steady value between 500 and 1500 ohms as the leads are flexed by hand. If it doesn't, the pick-up coil is defective.

Replacement

7   Mark the distributor gear and shaft so that they can be reassembled in the same position.

8   Carefully mount the distributor in a soft-jawed vise and remove the roll pin from the distributor shaft and gear with a hammer and punch.

9   Remove the retainer **(see illustration)**.

10   Lift the pick-up assembly straight up

and remove it from the distributor.

11   Wipe the distributor base and module clean and apply silicone lubricant between the module and the base for heat dissipation.

12   Attach the module to the base.

13   Attach the pick-up coil connector to the module.

14   Assemble the pick-up pole piece and the retainer.

15   Assemble the shaft, the gear parts and the roll pin.

16   Spin the shaft to make sure that the teeth do not touch.

17   Install the rotor and cap (Section 6).

18   Install the distributor in the engine (Section 6).

V6 engine

19   Remove the distributor from the engine (Section 7).

20   Detach the wiring connector from the cap **(see illustration)**.

21   Turn the four latches and remove the cap and coil assembly from the lower housing.

22   Remove the rotor and pick-up coil leads from the ignition module.

Testing

23   Connect an ohmmeter between the pick-up coil lead connector and the distributor housing **(see illustration)**. The ohmmeter should indicate infinite resistance.

9.9 To remove the pick-up coil, pry off the retainer

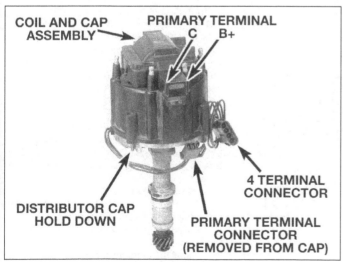

9.20 A typical coil-in-cap type distributor used on some V6 engines

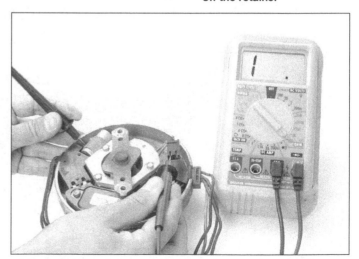

9.23 To test the pick-up coil on the coil-in-cap type distributor, perform these two tests with an ohmmeter - if the pick-up coil fails either test, replace it

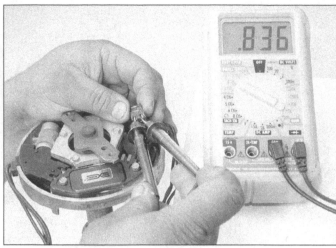

9.24 Checking the pick-up coil resistance

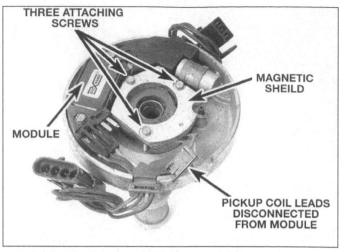

9.29 The anti-magnetic (aluminum) shield on the coil-in-cap type distributor is secured by three attaching screws (arrows) - remove the screws and the shield to get at the pick-up coil

24   Connect the ohmmeter between both terminals of the pick-up coil lead connector. The ohmmeter should indicate a steady value between 500 and 1500 ohms **(see illustration)**.
25   If the pick-up coil fails either test, replace it.

Replacement

26   Mark the distributor shaft and gear so that they can be reassembled in the same position.
27   Drive out the roll pin.
28   Remove the gear and pull the shaft from the distributor.
29   Remove the three attaching screws and detach the magnetic shield **(see illustration)**.
30   Remove the C-clip and detach the pick-up coil, magnet and pole piece **(see illustration)**.
31   Install the new pick-up coil, the magnet and the pole piece.
32   Install the thin washer.
33   Install the shaft, gear parts and roll pin.
34   Spin the shaft to ensure that the teeth do not touch.
35   Loosen, then retighten, the pick-up coil pole piece to eliminate contact.

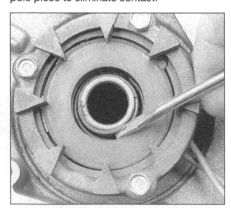

9.30 The pick-up coil and pole piece assembly can be removed after carefully prying out the C-clip

36   Install the rotor and cap (Section 7).
37   Install the distributor in the engine (Section 7).

---

10   Ignition coil - removal, testing and installation

Refer to illustrations 10.4, 10.11, 10.12 and 10.13
1   Disconnect the cable from the negative terminal of the battery.

Separate coil

2   Unplug the coil high tension wire and both electrical leads from the coil.
3   Remove both mounting nuts and remove the coil from the engine.
4   Check the coil for opens and grounds by performing the following three tests with an ohmmeter **(see illustration)**.
5   Check the coil primary resistance. Using

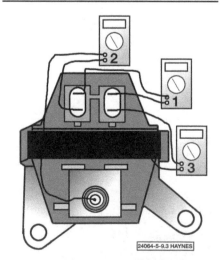

10.4 These three resistance checks will determine whether your separately mounted coil is good or bad

the low scale, hook up the leads as shown **(see test 1 in illustration 10.4)**. The ohmmeter should indicate a very low, or zero, resistance value. If it doesn't, replace the coil.
6   Check the coil secondary resistance. Using the ohmmeter's high scale, hook up the ohmmeter leads as shown **(see test 2 in illustration 10.4)**. The ohmmeter should indicate a very high, or infinite, resistance value. If it doesn't, replace the coil.
7   Check the ground circuit. Using the high scale, hook up the leads as shown **(see test 3 in illustration 10.4)**. The ohmmeter should not indicate an infinite resistance. If it does, replace the coil.
8   Installation of the new or old coil is the reverse of the removal procedure.

Coil-in-cap

9   Remove the distributor (Section 7).
10   Turn the four latches and remove the cap and coil assembly from the lower housing.
11   Connect an ohmmeter to both terminals of the cap coil assembly **(see illustration test 1)**. The reading should be zero, or nearly zero. If it isn't, replace the coil.
12   Connect the ohmmeter between each of the terminals of the coil assembly and the center contact of the cap as shown in test 2. Use the high scale on the ohmmeter **(see illustration)**. Replace the coil only if both readings are infinite.
13   Remove the coil cover attaching screws and lift off the cover **(see illustration)**.
14   Remove the ignition coil mounting screws and lift the coil, with the leads, from the cap.
15   Remove the ignition coil arc seal.
16   Clean with a soft cloth and inspect the cap for defects. Replace if necessary.
17   Assemble the new coil and cover in the cap.
18   Install the distributor cap.
19   Install the distributor in the engine (Section 7).

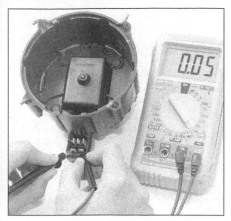

10.11 To test the HEI type coil-in-cap, attach the leads of an ohmmeter to the primary terminals and verify that the indicated resistance is zero or very near zero . . .

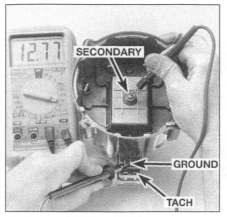

10.12 . . . then using the high scale, attach one lead to the high tension terminal and the other to GROUND terminal and then the TACH terminal. Verify that both readings are not infinite – if the indicated resistance is not as specified, replace the coil

10.13 Remove the two screws (arrows) and detach the coil cover

11 Hall effect switch - testing and replacement

Refer to illustration 11.2

1   Some four-cylinder models are equipped with a Hall effect switch which is located above the pick-up coil assembly. The Hall effect switch is used in place of the R terminal of the HEI distributor to send engine RPM information to the ECM.

2   Test the switch by connecting a 12-volt power supply and voltmeter as shown **(see illustration)**. Check the polarity markings carefully before making any connections.

3   When the feeler gauge is not inserted as shown, the voltmeter should read less than 0.5 volts. If the reading is more, the Hall effect switch is faulty and must be replaced by a new one.

4   With the feeler gauge inserted, the voltmeter should read within 0.5 volts of battery voltage. Replace the switch with a new one if the reading is more.

5   Remove the Hall effect switch by unplugging the connector and removing the

retaining screws.

6   Installation is the reverse of removal.

12 Charging system - general information and precautions

The charging system includes the alternator, voltage regulator and battery. These components work together to supply electrical power for the ignition system, lights, radio, etc. The alternator is driven by a drivebelt at the front of the engine.

The purpose of the voltage regulator is to limit the alternator's voltage to a preset value. This prevents power surges, circuit overloads, etc., during peak voltage output. On all models with which this manual is concerned, the voltage regulator is contained within the alternator housing.

The charging system does not ordinarily require periodic maintenance. The drivebelts, electrical wiring and connections should,

however, be inspected at the intervals suggested in Chapter 1.

Take extreme care when making circuit connections to a vehicle equipped with an alternator and note the following. When making connections to the alternator from a battery, always match correct polarity. Before using arc welding equipment to repair any part of the vehicle, disconnect the wires from the alternator and the battery terminal. Never start the engine with a battery charger connected. Always disconnect both battery leads before using a battery charger.

The charge indicator lights when the ignition switch is on, and goes out when the engine is running. If the charge indicator is on with the engine running, a charging system defect is indicated.

13 Charging system - checking

1   If a malfunction occurs in the charging circuit, do not immediately assume that the alternator is causing the problem. First check the following items:

   a) *The battery cables where they connect to the battery. Make sure the connections are clean and tight.*
   b) *The battery electrolyte specific gravity. If it is low, charge the battery.*
   c) *Check the external alternator wiring and connections. They must be in good condition.*
   d) *Check the drivebelt condition and tension (Chapter 1).*
   e) *Make sure the alternator mounting bolts are tight.*
   f) *Run the engine and check the alternator for abnormal noise.*

2   Using a voltmeter, check the battery voltage with the engine off. It should be approximately 12-volts.

3   Start the engine and check the battery voltage again. It should now be approximately 14-to-15 volts.

4   Locate the D-shaped test hole in the back of the alternator and ground the tab that is located inside the hole by inserting a screw-

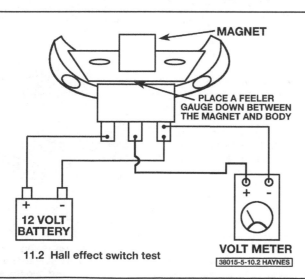

11.2 Hall effect switch test

38015-5-10.2 HAYNES

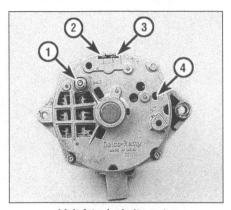

14.4 A typical alternator

1   *Battery BAT terminal*
2   *No. 1 terminal*
3   *No. 2 terminal*
4   *Test hole*

15.2 Mark the alternator end frame housings to ensure correct alignment during reassembly

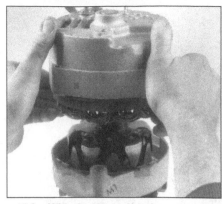

15.3a With the through-bolts removed, carefully separate the drive end frame and the rectifier end frame

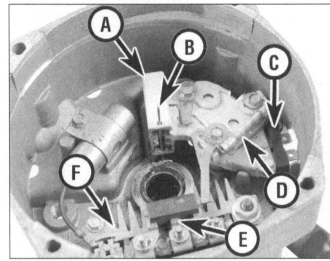

15.3b Inside a typical SI alternator

A   *Brush holder*
B   *Paper clip retaining brushes (for reassembly)*
C   *Regulator*
D   *Resistor (not all models)*
E   *Diode trio*
F   *Rectifier bridge*

driver blade into the hole and touching the tab and the case at the same time. **Caution:** *Do not run the engine with the tab grounded any longer than necessary to obtain a voltmeter reading. If the alternator is charging, it is running unregulated during the test. This condition may overload the electrical system and cause damage to the components.*

5   The reading on the voltmeter should be 15-volts or higher with the tab grounded in the test hole.
6   If the voltmeter indicates low battery voltage, the alternator is faulty and should be replaced with a new one (Section 15).
7   If the voltage reading is 15-volts or higher and a no charge condition is present, the regulator or field circuit is the problem. Remove the alternator (Section 15) and have it checked further by an auto electric shop.

14   Alternator - removal and installation

Refer to illustration 14.4

Removal

1   Disconnect the negative battery cable.
2   Loosen the adjusting bolt and remove the drivebelt.
3   Raise the rear of the vehicle and secure it on jackstands. Block the front wheels.
4   Unplug the two-terminal connector and remove the battery lead from the back of the alternator **(see illustration)**.
5   Remove the pivot through-bolt and loosen the alternator bracket bolt.
6   Remove the alternator by lowering it out the bottom of the engine compartment.

Installation

7   Place the alternator in position, plug in the two-terminal connector and attach the battery lead to the alternator terminal.
8   Install the through-bolt and nut and tighten it securely.
9   Tighten the alternator bracket bolt securely.

10   Install the adjusting bolt.
11   Install the drivebelt and adjust the tension (Chapter 1).
12   Connect the negative battery cable.

15   Alternator brushes - replacement

Refer to illustrations 15.2, 15.3a, 15.3b, 15.4, 15.5a, 15.5b, 15.6 and 15.9
Note: *This procedure applies only to the 10-S1 series alternator used from 1973 on.*
1   Remove the alternator from the vehicle (Section 13).
2   Scribe, punch or paint marks on the front and rear end frame housings of the alternator to facilitate reassembly **(see illustration)**.
3   Remove the four through-bolts holding the front and rear end frames together, then separate the drive end frame from the rectifier end frame **(see illustration)**.
4   Remove the nuts holding the stator leads to the rear end frame (rectifier bridge) and separate the stator from the end frame **(see illustration)**.
5   Remove the screws holding the regulator/brush assembly to the end frame and detach the brush holder **(see illustrations)**.
6   Remove the brushes from the holder by

slipping the brush retainer off **(see illustration)**.
7   Remove the springs from the brush holder.
8   Installation is the reverse of the removal procedure.
9   When installing the brushes in the brush holder, install the brush closest to the end frame first. Slip a paper clip through the rear of the endframe to hold the brush, then insert the second brush and push the paper clip in

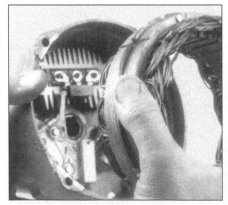

15.4 After removing the nuts holding the stator assembly to the rectifier bridge, remove the stator

15.5a Remove the screw attaching the diode trio to the end frame and remove the diode trio

15.5b After removing the screws that attach the brush holder and resistor (if equipped) to the end frame, remove the brush holder

to hold both brushes while reassembly is completed **(see illustration)**. The paper clip should not be removed until the front and rear end frames have been bolted together.

16 Starting system - general information

The function of the starting system is to crank the engine. The starting system is composed of a starting motor, solenoid and battery. The battery supplies the electrical energy to the solenoid, which then completes the circuit to the starting motor, which does the actual work of cranking the engine.

The solenoid and starting motor are mounted together at the lower front side of the engine. No periodic lubrication or maintenance is required.

The electrical circuitry of the vehicle is arranged so that the starter motor can only be operated when the clutch pedal is depressed (manual transmission) or the transmission selector lever is in Park or Neutral (automatic transmission).

Never operate the starter motor for more than 15 seconds at a time without pausing to allow it to cool for at least two minutes. Excessive cranking can cause overheating, which can seriously damage the starter.

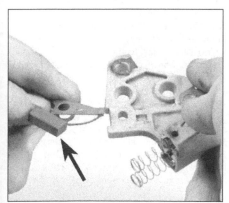

15.6 The brushes can be separated from the holder and the new ones installed (early models shown)

17 Starter motor - testing in vehicle

Note: *Before diagnosing starter problems, make sure that the battery is fully charged.*

1 If the starter motor does not turn at all when the switch is operated, make sure that the shift lever is in Neutral or Park (automatic transmission) or that the clutch pedal is depressed (manual transmission).

2 Make sure that the battery is charged and that all cables, both at the battery and starter solenoid terminals, are secure.

3 If the starter motor spins but the engine is not cranking, then the overrunning clutch in the starter motor is slipping and the starter must be removed from the engine and disassembled.

4 If, when the switch is actuated, the starter motor does not operate at all but the solenoid clicks, then the problem lies with either the battery, the main solenoid contacts or the starter motor itself.

5 If the solenoid plunger cannot be heard when the switch is actuated, the solenoid itself is defective or the solenoid circuit is open.

6 To check the solenoid, connect a jumper lead between the battery (+) and the S terminal on the solenoid. If the starter motor now operates, the solenoid is OK and the

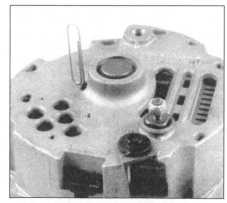

15.9 Insert a paper clip through the hole next to the bearing boss (arrow) to hold the brushes in the retracted position

problem is in the ignition switch, Neutral start switch or the wiring.

7 If the starter motor still does not operate, remove the starter/solenoid assembly for disassembly, testing and repair.

8 If the starter motor cranks the engine at an abnormally slow speed, first make sure that the battery is charged and that all terminal connections are tight. If the engine is partially seized, or has the wrong viscosity oil in it, it will crank slowly.

9 Run the engine until normal operating temperature is reached, then disconnect the coil wire from the distributor cap and ground it on the engine.

10 Connect a voltmeter positive lead to the starter motor terminal of the solenoid and then connect the negative lead to ground.

11 Actuate the ignition switch and take the voltmeter readings as soon as a steady figure is indicated. Do not allow the starter motor to turn for more than 15 seconds at a time. A reading of 9-volts or more, with the starter motor turning at normal cranking speed, is normal. If the reading is 9-volts or more but the cranking speed is slow, the motor is faulty. If the reading is less than 9-volts and the cranking speed is slow, the solenoid contacts are probably burned.

18 Starter motor - removal and installation

Refer to illustration 18.4

Removal

1 Disconnect the negative battery cable.

2 Raise the rear of the vehicle and support it securely on jackstands. Block the front wheels.

3 Remove the catalytic converter splash shield.

4 From under the vehicle, disconnect the solenoid wire and battery cable from the terminals on the rear of the solenoid **(see illustration)**.

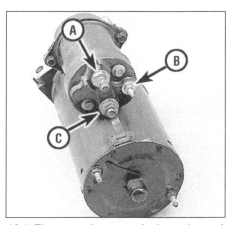

18.4 There are three terminals on the end of the typical starter solenoid

A   *Battery terminal*
B   *Switch terminal (S)*
C   *Motor terminal (M)*

5   Remove the support bracket, if equipped, and retaining bolts.
6   Remove the starter motor. Note the location of the spacer shim.

Installation

7   Place the starter motor in position.
8   Install the spacer shim.
9   Install the support bracket, if equipped, and tighten the starter motor mounting bolts.
10   Attach the solenoid wire and battery cable to the terminals on the rear of the solenoid.
11   Lower the vehicle.
12   Connect the negative battery cable.

---

19   Starter solenoid - removal and installation

Refer to illustration 19.5
1   Disconnect the cable from the negative terminal of the battery.
2   Remove the starter motor (Section 18).

Removal

3   Disconnect the strap from the solenoid to the starting motor (M) terminal.
4   Remove the two screws which secure the solenoid to the starter motor.
5   Twist the solenoid in a clockwise direction to disengage the flange from the starter body **(see illustration)**.

Installation

6   To install, first make sure the return spring is in position on the plunger, then insert the solenoid body into the starter housing and turn the solenoid counterclockwise to engage the flange.
7   Install the two solenoid screws and connect the motor strap.

---

20   Starter motor brushes - replacement

Refer to illustrations 20.6 and 20.7
1   Remove the starter and solenoid assembly from the vehicle (Section 18).
2   Remove the solenoid from the starter housing (Section 19).
3   Mark the relationship of the commutator end frame to the field frame housing to simplify reassembly.
4   Remove the starter motor through-bolts.
5   Remove the end frame from the field frame housing.
6   Mark the relationship of the field frame housing to the drive end housing **(see illustration)**.
7   Pull the field frame housing away from the drive end housing and over the armature **(see illustration)**.

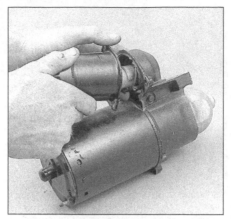

19.5 Remove the solenoid housing from the starter by twisting it clockwise

8   Unbolt the brushes and brush supports from the brush holders in the field frame housing.
9   To install new brushes, attach the brushes to the brush supports, making sure they are flush with the bottom of the supports, and bolt the brushes/supports to the brush holders.
10   Install the field frame over the armature, with the brushes resting on the first step of the armature collar at this point.
11   Make sure the field frame is properly aligned with the drive end housing, then push the brushes off the collar and into place on the armature.
12   The remaining installation steps are the reverse of removal.

---

21   Distributorless ignition - general information

Some engines in 1987 and later models feature distributorless ignitions, which use a "waste spark" method of spark distribution. Each cylinder is paired with its opposing cylinder in the firing order so that one cylinder on compression fires simultaneously with its opposing cylinder on exhaust. Since the cylinder on exhaust requires very little of the available voltage to fire its plug, most of the voltage is used to fire the cylinder on compression. The process reverses when the cylinders reverse roles. The "Direct Ignition System" (DIS) for the four-cylinder engine uses two coils; the V6 uses three coils (C3I).
The Direct Ignition System includes a coil pack, an ignition module, a crankshaft reluctor ring, a magnetic sensor and the ECM. The coil pack consists of two (or three) separate, interchangeable ignition coils. Two (three) coils are needed because each coil only fires two cylinders. The ignition module is located under the coil pack and is connected to the ECM by a six-pin connector.

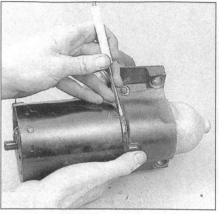

20.6 Be sure to scribe alignment marks on the starter motor housing before separating the field frame from the drive end housing

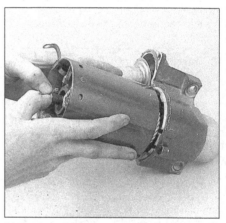

20.7 After removing the through-bolts and the commutator end frame, carefully separate the field frame from the drive end housing

The ignition module controls the primary circuit to the coils, turning them on and off, and controls spark timing below 400 rpm and if the ECM bypass circuit becomes open or grounded.
The magnetic pickup sensor inserts through the engine block, just above the oil pan rail, right next to the crankshaft reluctor ring. Notches in the crankshaft reluctor ring trigger the magnetic pickup sensor to provide timing information to the ECM. The magnetic pickup sensor provides a cam signal to identify the correct firing sequence and the crank signals trigger each coil at the proper sequence.
The system uses Electronic Spark Timing (EST) and control wires from the ECM, just like conventional distributor systems. The ECM controls timing using crankshaft position, engine rpm, engine temperature and manifold absolute pressure (MAP) sensing.

Chapter 6
Emissions control systems

Contents

Specifications

Torque specifications
Ft-lbs (unless otherwise indicated)

Oxygen sensor	30
EGR valve-to-manifold	
4-cylinder engine	
1984 and 1985	120 in-lbs
1986 and 1987	16
V6 engine	13 to 18
EGR solenoid bracket nut (V6 engine only)	17

1   General information

Refer to illustration 1.7

To prevent pollution of the atmosphere from incompletely burned and evaporating gases, and to maintain good driveability and fuel economy, a number of emission control devices are incorporated. They include the:

Fuel control system
Electronic Spark Timing (EST)
Exhaust Gas Recirculation (EGR) system
Evaporative Emission Control System (EECS)
Positive Crankcase Ventilation (PCV) system
Thermostatic Air Cleaner (Thermac)
Catalytic converter

All of these systems are linked, directly or indirectly, to the Computer Command Control (CCC or C3) system.

The Sections in this Chapter include general descriptions, checking procedures within the scope of the home mechanic and component replacement procedures (when possible) for each of the systems listed above.

Before assuming that an emissions control system is malfunctioning, check the fuel and ignition systems carefully. The diagnosis of some emission control devices requires specialized tools, equipment and training. If checking and servicing become too difficult or if a procedure is beyond the scope of your skills, consult your dealer service department.

This doesn't mean, however, that emission control systems are particularly difficult to maintain and repair. You can quickly and easily perform many checks and do most (if not all) of the regular maintenance at home with common tune-up and hand tools. **Note:**

The most frequent cause of emissions problems is simply a loose or broken vacuum hose or wiring connection. So always check the hose and wiring connections first.

Pay close attention to any special precautions outlined in this Chapter. It should be noted that the illustrations of the various systems may not exactly match the system installed on your vehicle because of changes made by the manufacturer during production or from year to year.

A Vehicle Emissions Control Information label is located in the engine compartment **(see illustration)**. This label contains important emissions specifications and setting procedures, as well as a vacuum hose schematic with emissions components identified. When servicing the engine or emissions systems, the VECI label in your particular vehicle should always be checked for up-to-date information.

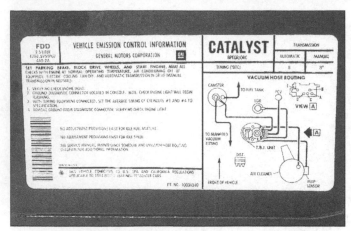

1.7 The Vehicle Emission Control Information (VECI) label provides information regarding engine size, exhaust emission system used, engine adjustment procedures and specifications and an emission component and vacuum schematic diagram

2.5 A typical Assembly Line Communications Link (ALCL) connector - the two terminals you will be concerned with are the A (ground) and B (diagnostic) terminals

2 Computer Command Control (CCC) system and trouble codes

Refer to illustration 2.5

The Computer Command Control (CCC) system consists of an Electronic Control Module (ECM) and nine information sensors which monitor various functions of the engine and send data back to the ECM.

The CCC system is analogous to the central nervous system in the human body: The sensors (nerve endings) constantly relay information to the ECM (brain), which processes the data and, if necessary, sends out a command to change the operating parameters of the engine (body).

Here's a specific example of how one portion of this system operates: An oxygen sensor, located in the exhaust manifold, constantly monitors the oxygen content of the exhaust gas. If the percentage of oxygen in the exhaust gas is incorrect, an electrical signal is sent to the ECM. The ECM takes this information, processes it and then sends a command to the fuel injection system, telling it to change the air/fuel mixture. This happens in a fraction of a second and it goes on continuously when the engine is running. The end result is an air/fuel mixture ratio which is constantly maintained at a predetermined ratio, regardless of driving conditions.

One might think that a system which uses an onboard computer and electrical sensors would be difficult to diagnose. This is not necessarily the case. The CCC system has a built-in diagnostic feature which indicates a problem by flashing a Service Engine Soon light on the instrument panel. When this light comes on during normal vehicle operation, a fault has been detected. More importantly, the source of each malfunction is determined and automatically stored in the ECM memory.

To retrieve this information from the ECM memory, you must use a short jumper wire to ground a diagnostic terminal. This terminal is part of a wiring connector known as the Assembly Line Communications Link (ALCL) **(see illustration)**. The ALCL is located just behind the rear console pad between the seats. To get at the ALCL, refer to Chapter 11 for rear pad console removal. With the connector exposed to view, push one end of the jumper wire into the diagnostic terminal (B) and the other end into the ground terminal (A).

When the diagnostic terminal is grounded with the ignition On and the engine stopped, the system will enter the Diagnostic Mode. In this mode the ECM will display a "Code 12" by flashing the Service Engine Soon light, indicating that the system is operating. A code 12 is simply one flash, followed

by a brief pause, then two flashes in quick succession. This code will be flashed three times. If no other codes are stored, Code 12 will continue to flash until the diagnostic terminal ground is removed.

After flashing Code 12 three times, the ECM will display any stored trouble codes. Each code will be flashed three times, then Code 12 will be flashed again, indicating that the display of any stored trouble codes has been completed.

When the ECM sets a trouble code, the Service Engine Soon light will come on and a trouble code will be stored in memory. If the problem is intermittent, the light will go out after 10 seconds, when the fault goes away. However, the trouble code will stay in the ECM memory until the battery voltage to the ECM is interrupted. Removing battery voltage for 10 seconds will clear all stored trouble codes. Trouble codes should always be cleared after repairs have been completed. **Caution:** *To prevent damage to the ECM, the key must be Off when disconnecting power to the ECM.*

Following is a list of the typical Trouble Codes which may be encountered while diagnosing the Computer Command Control System. Also included are simplified troubleshooting procedures. If the problem persists after these checks have been made, more detailed service procedures will have to be done by a dealer service department.

Trouble codes	Circuit or system	Probable cause
Code 12 (1 flash, pause, 2 flashes)	No reference pulses to ECM	This code will flash whenever the diagnostic terminal is grounded with the ignition turned On and the engine not running. If additional trouble codes are stored in the ECM they will appear after this code has flashed three times. If this code appears while the engine is running, no reference pulses from the distributor are reaching the ECM.
Code 13 (1 flash, pause, 3 flashes)	Oxygen sensor circuit	Check for a sticking or misadjusted throttle position sensor. Check the wiring and connectors from the oxygen sensor. Replace the oxygen sensor.
Code 14 (1 flash, pause, 4 flashes)	Coolant sensor circuit	If the engine is experiencing overheating problems the problem must be rectified before continuing. Then check all wiring and connectors associated with the coolant temperature sensor. Replace the coolant temperature sensor.*

Trouble codes	Circuit or system	Probable cause
Code 15 (1 flash, pause, 5 flashes)	Coolant sensor circuit	See Code 14, then check the wiring connections at the ECM.
Code 21 (2 flashes, pause, 1 flash)	Throttle position sensor	Check for a sticking or misadjusted TPS plunger.
Code 22 (2 flashes, pause, 2 flashes)	Throttle position sensor	Check the TPS adjustment (Chapter 4). Check the ECM connector. Replace the TPS (Chapter 4).*
Code 23 (V6 only) (2 flashes, pause, 3 flashes)	Manifold air temperature	Check the MAT sensor, wiring and connectors for an open sensor circuit. Replace the MAT sensor.*
Code 24 (2 flashes, pause, 4 flashes)	Vehicle speed sensor	A fault in this circuit should be indicated only when the vehicle is in motion. Disregard if it is set when the drive wheels are not turning. Check the connections at the ECM. Check the TPS setting.
Code 25 (V6 only) (2 flashes, pause, 5 flashes)	Manifold air temperature	Check the voltage signal from the MAT sensor to the ECM. It should be above 4 volts.
Code 32 (V6 only) (3 flashes, pause, 2 flashes)	EGR system	The EGR solenoid should not be energized and vacuum should not pass to the EGR valve. The diagnostic switch should close at about 2 inches of vacuum. With vacuum applied, the switch should close. Replace the EGR valve.*
Code 33 (3 flashes, pause, 3 flashes)	MAP sensor	Check the vacuum hoses from the MAP sensor. Check the electrical connections at the ECM. Replace the MAP sensor.*
Code 34 (3 flashes, pause, 4 flashes)	MAP sensor	Code 34 will set when the signal voltage from the MAP sensor is too low. Instead the ECM will substitute a fixed MAP value and use the TPS to control fuel delivery. Replace MAP sensor.
Code 35 (3 flashes, pause, 5 flashes)	Idle Air Control	Code 35 will set when the closed throttle speed is 50 rpm above or below the correct idle speed for 30 seconds. Replace the IAC.*
Code 42 (4 flashes, pause, 2 flashes)	Electronic Spark timing	If the vehicle will not start and run, check the wire leading to ECM terminal 12. **Note:** *A malfunctioning HEI module can cause this trouble code.* Check the EST wire (terminal 19 at the ECM) leading to the HEI module (E terminal). Check all distributor wires. Check the wire leading from EST terminal A to ECM terminal 12 and the wire from EST terminal A to ECM terminal 3. Replace the HEI module.*
Code 44 (4 flashes, pause, 4 flashes)	Lean exhaust	Check the ECM wiring connections, particularly terminals 15 and 8. Check for vacuum leakage at the TBI base gasket, vacuum hoses or the intake manifold gasket. Replace the oxygen sensor.*
Code 45 (4 flashes, pause, 5 flashes)	Rich exhaust	Check the evaporative charcoal canister and its components for the presence of fuel. Replace the oxygen sensor.*
Code 51 (5 flashes, pause, 1 flash)	PROM	Make sure that the PROM is properly installed in the ECM. Replace the PROM.*
Code 52 (V6 only) (5 flashes, pause, 2 flashes)	Fuel CALPAK	Check the CALPAK PROM to insure proper installation. Replace the PROM.*
Code 53 (V6 only) (5 flashes, pause, 3 flashes)	System over-voltage	Code 53 will set if the voltage at ECM terminal B2 is greaterthan 17.1 volts for 2 seconds. Check the charging system.
Code 55 (5 flashes, pause, 5 flashes)	ECM	Be sure that the ECM ground connections are tight. If they are, replace the ECM.*

Component replacement may not cure the problem in all cases. For this reason, you may want to seek professional advice before purchasing replacement parts.

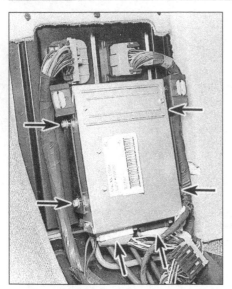

3.4 To remove the ECM, unplug the large electrical connectors at the bottom of the module (arrows) then remove the four mounting bolts (arrows)

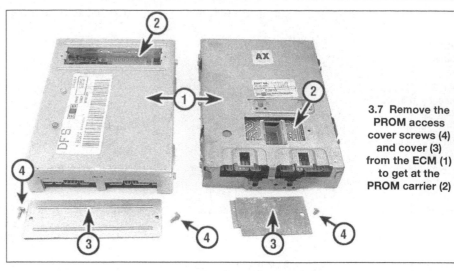

3.7 Remove the PROM access cover screws (4) and cover (3) from the ECM (1) to get at the PROM carrier (2)

3   Electronic Control Module/PROM removal and installation

Refer to illustrations 3.4 and 3.7

1   The Electronic Control Module (ECM) is located behind the padded console between the seats.

2   Disconnect the negative battery cable. **Caution:** *The ignition should always be off when installing or removing the ECM connectors.*

3   Remove the padded center console between the seats (Chapter 11).

4   Disconnect the electrical connectors and remove the retaining bolts from the ECM **(see illustration)**.

5   Carefully remove the ECM.

6   Remove the 7ROM from the old ECM and install it in the new one.

7   If you are installing a new PROM in the old ECM, remove the PROM access cover **(see illustration)** from the ECM. Use a rocker type PROM removal tool (available at your dealer). The PROM carrier with PROM in it should lift off the PROM socket easily. **Caution:** *The PROM carrier should only be removed with the special PROM rocker type removal tool. Removal without this tool or with any other type of tool may damage the PROM or the PROM socket.*

8   Check the part number of the new PROM to make sure that it is the same as the number of the old PROM. **Caution:** *Do not remove the PROM from the carrier to check the PROM number.*

9   Install the new PROM carrier in the PROM socket of the ECM. The small notch of the carrier should be aligned with the small notch in the socket. Press on the PROM carrier until it is firmly seated in the socket. **Caution:** *Do not press on the PROM - press only on the carrier.*

10   Install the access cover onto the ECM and tighten the two access cover screws.

11   Install the ECM back into its plastic support bracket between the seats, and plug in the electrical connectors to the ECM.

12   Install the padded console between the seats (see Chapter 11).

13   Turn the ignition On.

14   Enter the diagnostic mode (see Section 2). Code 12 should flash at least four times and no other codes should be present. This indicates that the PROM is installed properly.

15   If trouble code 51 occurs, or if the Service Engine Soon light comes on and remains constantly lit, the PROM is not fully seated, is installed backwards, has bent pins or is defective.

16   If it is not fully seated, pressing firmly firmly on the carrier should correct the problem.

17   It is possible to install the PROM backwards. If this occurs, and the ignition key is turned to ON, the PROM circuitry will be destroyed.

18   If the pins have been bent, remove the PROM, straighten the pins and reinstall it. If the bent pins break or crack when you attempt to straighten them, discard the PROM and replace it with a new one.

4   Information sensors

Refer to illustrations 4.3, 4.4 and 4.15

Engine coolant temperature sensor

1   A failure in the coolant sensor circuit should set either a Code 14 or a Code 15. These codes indicate a failure in the coolant temperature circuit, so the appropriate solution to the problem will be either repair of a wire problem or replacement of the sensor.

2   To remove the sensor, disengage the locking tab on the connector and unplug it from the sensor. Carefully unscrew the sensor itself.

3   **Caution:** *Handle the coolant sensor with care. Damage to this sensor will affect the proper operation of the entire fuel injection system. Wrap the threads of the new sensor with teflon sealing tape to prevent leakage and thread corrosion* **(see illustration)**. Install the new sensor and tighten it securely.

Manifold Air Temperature (MAT) sensor (2.8L V6 only)

4   The Manifold Air Temperature (MAT)

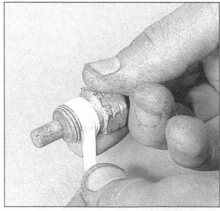

4.3 Wrap the threads of the coolant temperature sensor with Teflon tape to prevent leaks and thread corrosion.

4.4 The MAT sensor is located at the base of the air cleaner on the V6 engine (it's not used on the four)

sensor is mounted in the air cleaner housing **(see illustration)**.

5   A failure in the MAT circuit should set a Code 23 or Code 25.

6   To replace the sensor, disconnect the wire harness.

7   Unscrew the sensor from the air cleaner housing.

8   Installation is the reverse of removal.

Manifold Absolute Pressure (MAP) sensor

9   The Manifold Absolute Pressure (MAP) sensor measures the changes in the intake manifold pressure which result from engine load and speed changes.

10   The ECM uses the MAP sensor to control fuel delivery and ignition timing. A failure in the MAP sensor circuit should set a Code 33 or Code 34.

Oxygen sensor

11   The exhaust oxygen sensor is mounted in the exhaust system where it can monitor the oxygen content of the exhaust gas stream.

12   By monitoring the voltage output of the oxygen sensor, the ECM will know what fuel mixture command to give the injector.

13   The oxygen sensor, if open, should set a Code 13. A low voltage in the sensor circuit should set a Code 44. A high voltage in the circuit should set a Code 45. Codes 44 and 45 can also be set as a result of fuel system problems.

14   Refer to Section 5 for additional information on the oxygen sensor.

Throttle position sensor (TPS)

15   The throttle position sensor (TPS) is located on the end of the throttle shaft on the TBI unit for the four-cylinder engine and on the end of the throttle shaft on the throttle body for the MPFI system on the V6 engine **(see illustration)**.

16   By monitoring the output voltage from the TPS, the ECM can determine fuel delivery based on throttle valve angle (driver demand). A broken or loose TPS can cause intermittent bursts of fuel from the injector and an unstable idle because the ECM thinks the throttle is moving.

17   A problem in any of the TPS circuits will set either a Code 21 or 22. Once a trouble code is set, the ECM will use an artificial default value for TPS and some vehicle performance will return.

Park/Neutral switch (automatic transmission-equipped vehicles only)

18   The Park/Neutral (P/N) switch indicates to the ECM when the transaxle is in Park or Neutral. This information is used for the Transaxle Converter Clutch (TCC) and the Idle Air Control (IAC) valve operation. **Caution:** *This vehicle should not be driven with the Park/Neutral switch disconnected*

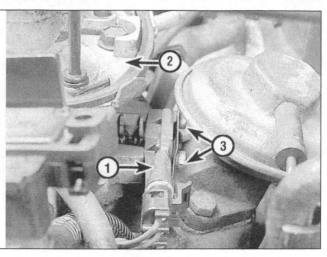

4.15 The throttle position sensor (TPS) (1) is attached to the front of the TBI unit (2) on the four-cylinder engine and to the rear of the throttle body on the V6 engine; it's secured by two screws (3)

because idle quality will be adversely affected and a false Code 24 (vehicle speed sensor) may be set. For more information regarding the P/N switch, which is part of the Neutral/start and backup light switch assembly, see Chapters 7 and 12.

Crank signal

19   The ECM is connected to the starter solenoid to determine when the engine is cranking.

A/C "On" signal

20   The A/C On signal tells the ECM that the A/C selector switch is turned to the On position and that the high side low pressure switch is closed. The ECM uses this information to turn on the A/C and adjust the idle speed when the air conditioning is working.

Vehicle speed sensor

21   The vehicle speed sensor (VSS) sends a pulsing voltage signal to the ECM, which the ECM converts to miles per hour. This sensor controls the operation of the TCC system.

Distributor reference signal

22   The distributor sends a signal to the ECM to tell it both engine rpm and crankshaft position.

---

5   Oxygen sensor

Refer to illustrations 5.5a and 5.5b

General description

1   The oxygen sensor is located in the exhaust manifold. The ECM monitors the sensor, which tells the ECM the ratio of oxygen in the exhaust gas. The ECM changes the air/fuel ratio to the engine by controlling the fuel injector. The most efficient mixture to minimize exhaust emissions is 14.7 to 1, which allows the catalytic converter to operate at maximum efficiency.

2   The oxygen sensor, if open, will set a Code 13. A low voltage in the sensor circuit,

or a shorted sensor circuit, will set a Code 44. A high voltage will set a Code 45. When any of these codes are set, the engine will run in the open loop mode.

3   The proper operation of the sensor depends on four conditions:

a)   *Electrical - The low voltages and low currents generated by the sensor depend upon good, clean connections which should be checked whenever a malfunction of the sensor is suspected or indicated.*

b)   *Outside air supply - The sensor is designed to allow air circulation to the internal portion of the sensor. Whenever the sensor is removed and installed or replaced, make sure the air passages are not restricted.*

c)   *Proper operating temperature - The ECM will not react to the sensor signal until the sensor reaches approximately 600ºF (315ºC). This factor must be taken into consideration when evaluating the performance of the sensor.*

d)   *Unleaded fuel - The use of unleaded fuel is essential for proper operation of the sensor. Make sure the fuel you are using is of this type.*

4   In addition to observing the above conditions, special care must be taken whenever the sensor is serviced.

a)   *The oxygen sensor has a permanently attached pigtail and connector which should not be removed from the sensor. Damage or removal of the pigtail or connector can adversely affect operation of the sensor.*

b)   *Grease, dirt and other contaminants should be kept away from the electrical connector and the louvered end of the sensor.*

c)   *Do not use cleaning solvents of any kind on the oxygen sensor.*

d)   *Do not drop or roughly handle the sensor.*

e)   *The silicone boot must be installed in the correct position to prevent the boot from being melted and to allow the sensor to operate smoothly.*

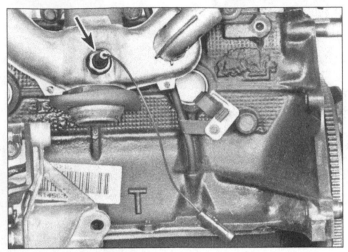

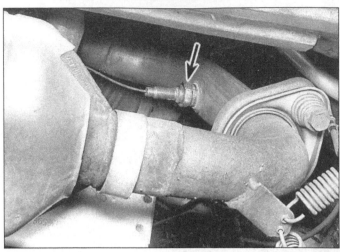

5.5a The oxygen sensor on the four-cylinder engine is located in the exhaust manifold

5.5b The oxygen sensor (arrow) on the V6 engine is located in the exhaust pipe just downstream from the front exhaust manifold

Replacement

5   The oxygen sensor is located in the exhaust manifold of the four cylinder engine **(see illustration)** and in the exhaust pipe (front left corner of the engine compartment) of the V6 engine **(see illustrations)**. The sensor may be difficult to remove if the engine is cold, so do not attempt to unscrew it until the engine has been warmed up to at least 120ºF. Be careful not to burn yourself.

6   Note the position of the silicone boot, if equipped, and carefully back out the sensor from the exhaust manifold. Excessive force may damage the threads.

7   Anti-seize compound must be used on the threads of the sensor to facilitate future removal. The threads of new sensors will already be coated with this compound, but if a sensor must be removed and reinstalled, the threads must be recoated with the compound.

8   Install the sensor and tighten it to the specified torque.

9   Connect the pigtail to the main engine wiring harness.

---

6   Electronic Spark Timing (EST)

General description

1   To provide improved engine performance, fuel economy and control of exhaust emissions, the Electronic Control Module (ECM) controls distributor spark advance (ignition timing) with the Electronic Spark Timing (EST) system.

2   The ECM receives a reference pulse from the distributor, which indicates both engine rpm and crankshaft position, determines the proper spark advance for the engine operating conditions and sends an EST pulse to the distributor.

Checking

3   The ECM will set timing at a specified value when the diagnostic test terminal in the

ALCL connector is grounded. To check for EST operation, the timing should be checked at 1500 rpm with the terminal ungrounded. Then ground the test terminal. If the timing changes at 1500 rpm, the EST is operating. A fault in the EST system will usually set a trouble code 42.

4   For further information regarding the testing and component replacement procedures for the EST distributor, refer to Chapter 5.

---

7   Exhaust Gas Recirculation (EGR) system

Refer to illustrations 7.14 and 7.31

General description

1   The Exhaust Gas Recirculation (EGR) system is used to lower NOX (oxides of nitrogen) emission levels caused by high combustion temperatures. It does this by decreasing combustion temperature. The main element of the system is the EGR valve, which feeds small amounts of exhaust gas back into the combustion chamber.

2   The EGR valve is opened by either ported manifold or full manifold vacuum to let exhaust gas flow into the intake manifold. The EGR valve is usually open during warm engine operation and anytime the engine is running above idle speed. The amount of gas recirculated is controlled by variations in vacuum and exhaust backpressure.

3   There are three types of EGR valves. Their names refer to the means by which they are controlled:

> *Positive backpressure*
> *Negative backpressure (2.5L four-cylinder engine only)*
> *Ported vacuum*

Positive backpressure EGR valve

4   The positive backpressure valve has an air bleed, located inside the EGR valve assembly, which acts as a vacuum regulator. This bleed valve controls the amount of vac-

uum in the vacuum chamber by bleeding vacuum to the atmosphere during the open phase of the cycle. When the bleed valve receives sufficient exhaust backpressure through the hollow shaft, it closes the bleed. At this point, maximum available vacuum is applied to the diaphragm and the EGR valve opens.

5   If there is little or no vacuum in the vacuum chamber, such as at idle or wide open throttle, or if there is little or no pressure in the exhaust manifold, the EGR valve will not open. This type of valve will not open if vacuum is applied to it with the engine stopped or idling.

Negative backpressure EGR valve

6   The negative backpressure EGR valve is similar to the positive backpressure EGR valve except that the bleed valve spring is moved from above the valve to below and the valve is normally closed. The negative backpressure valve varies the amount of exhaust gas flow into the manifold depending on manifold vacuum and variations in exhaust backpressure.

7   The diaphragm on this valve has an internal air bleed hole which is held closed by a small spring when there is no exhaust backpressure. Engine vacuum opens the EGR valve against the pressure of a large spring. When manifold vacuum combines with negative exhaust backpressure, the vacuum bleed hole opens and the EGR valve closes.

Ported vacuum EGR valve

8   The ported vacuum EGR valve uses ported vacuum connected directly to the EGR valve. The amount of exhaust gas recirculated is controlled by the throttle opening and the amount of manifold vacuum.

Port EGR valve (ECM controlled - 2.8L V6 only)

9   This valve is controlled by a flexible diaphragm which is spring loaded to hold the valve closed. Ported vacuum applied to the

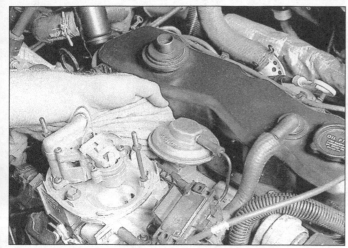

7.14 To check the EGR diaphragm for proper operation, warm-up the engine, remove the air cleaner assembly and, using a rag to protect your fingers, push up on the diaphragm - the engine should stumble and stall

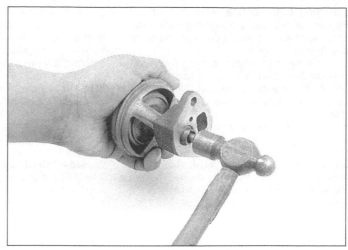

7.31 Deposits can be removed from the EGR pintle seating area by tapping the end of the pintle lightly with a plastic hammer

top side of the diaphragm overcomes the spring pressure and opens the valve in the exhaust gas port. This allows exhaust gas to be pulled into the intake manifold and enter the engine.

10 The EGR vacuum control has a vacuum solenoid that uses pulse width modulation. This simply means that the ECM turns the solenoid on and off many times a second and varies the amount of on time (the pulse width) to vary the amount of exhaust gas recirculated.

11 A diagnostic switch is part of the control and monitors vacuum to the EGR valve. This switch will trigger a Service Engine Soon light and set a Code 32 in the event of a vacuum circuit failure.

EGR valve identification

12 A series of numbers is stamped into the top of every EGR valve. This identification number indicates the assembly plant code, part number, date built and type of EGR valve.

a) *Positive backpressure EGR valves will have a P stamped on the top side of the valve after the part number.*

b) *Negative backpressure EGR valves will have an N stamped on the top side of the valve after the part number.*

c) *Port EGR valves have no identification stamped after the part number.*

Checking

Non-ECM-controlled EGR valves

13 Hold the top of the EGR valve and try to rotate it back and forth. If looseness is felt, replace the valve.

14 If no looseness is felt, place the transaxle in Neutral (manual) or Park (automatic), run the engine at idle until it warms up to at least 195ºF and push up on the underside of the EGR valve diaphragm **(see illustration)**. The rpm should drop. If there is no change in rpm, clean the EGR passages. If

there is still no change in rpm, replace the valve.

15 If the rpm drops, check for movement of the EGR valve diaphragm as the rpm is changed from approximately 2000 rpm to idle. If the diaphragm moves, there is no problem.

16 If the diaphragm does not move, check the vacuum signal at the EGR valve as the engine rpm is changed from approximately 2000 rpm to idle. If the vacuum is over six inches, replace the EGR valve. If it's under six inches, check the vacuum hoses for restrictions, leaks and poor connections.

ECM controlled port type EGR valves

17 Disconnect the EGR solenoid vacuum harness. Rotate the harness and reinstall only the EGR valve side. Install a vacuum pump with gauge on the manifold side of the EGR solenoid. Turn the ignition to On (engine stopped). Apply vacuum. Observe the EGR valve. The valve should not move.

18 If the valve moves, disconnect the EGR solenoid electrical connector and repeat the test. If the valve still moves, replace the solenoid.

19 If the valve does not move, ground the diagnostic terminal and repeat the test. If the valve still does not move, replace the EGR valve.

20 If the valve does move, start the engine. Lift up on the EGR valve and note the idle speed.

21 If there is no change in the idle, remove the EGR valve and check the passages for blockage. If the passages are not plugged, replace the EGR valve.

22 If the idle roughens, reconnect the EGR solenoid. Connect a vacuum gauge to the vacuum harness at the EGR valve. Warm up the engine to normal operating temperature. If your transaxle is an automatic, place it in Drive. Hold the brakes and accelerate

momentarily up to about 1800 rpm. Observe the gauge. It should indicate over two but less than ten inches of vacuum.

23 If it is zero or less than two inches of vacuum, check for any restrictions in the vacuum lines. If there are no restrictions in the vacuum lines, the Park/Neutral switch is probably faulty (see Chapter 7B).

24 If there is over ten inches of vacuum, replace the EGR filter.

25 If the vacuum is within two to ten inches, the EGR system is okay.

Component replacement

EGR valve

26 Disconnect the vacuum hose from the EGR valve.

27 Remove the nuts or bolts which secure the valve to the intake manifold or adapter.

28 Separate the EGR valve from the engine.

EGR valve cleaning

29 Inspect the valve pintle for deposits.

30 Depress the valve diaphragm and check for deposits around the valve seating area.

31 Hold the valve securely and tap lightly on the round pintle with a plastic hammer, using a light snapping action, to remove any deposits from the valve seat **(see illustration)**. Make sure to empty any loose particles from the valve. Depress the valve diaphragm again and inspect the valve seating area, repeating the cleaning operation as necessary.

32 Use a wire brush to carefully clean deposits from the pintle.

33 Remove any deposits from the valve outlet using a screwdriver.

34 If EGR passages in the intake manifold show an excessive buildup of deposits, the passages should be cleaned. Care should be taken to ensure that all loose particles are completely removed to prevent them from clogging the EGR valve or from being

ingested into the engine. **Note:** *It is a good idea to place a rag securely in the passage opening to keep debris from entering while cleaning the manifold.*

35   With a wire wheel, buff the exhaust deposits from the mounting surface.

36   Look for exhaust deposits in the valve outlet. Remove deposit buildup with a screwdriver.

37   Clean the mounting surfaces of the EGR valve. Remove all traces of old gasket material.

38   Place the new EGR valve, with a new lithium-base grease coated gasket, on the intake manifold or adapter and tighten the attaching nuts or bolts to the specified torque.

39   Connect the vacuum signal hose to the EGR valve.

EGR control solenoid

40   Disconnect the negative battery cable.

41   Disconnect the electrical connector at the solenoid.

42   Disconnect the vacuum hoses.

43   Remove the mounting nut and the solenoid.

44   Install the new solenoid and tighten the nut to the specified torque.

45   Install the vacuum hoses.

46   Connect the electrical connector.

47   Connect the negative battery cable.

8   Evaporative Emission Control System (EECS)

Refer to illustration 8.9

General description

1   This system is designed to trap and store fuel vapors that evaporate from the fuel tank, throttle body and intake manifold.

2   The Evaporative Emission Control System (EECS) consists of a charcoal-filled canister and the lines connecting the canister to the fuel tank, throttle body and intake manifold.

3   Fuel vapors are transferred from the fuel tank, throttle body and intake manifold to a storage canister to hold the vapors when the vehicle is not operating. When the engine is running, the fuel vapor is purged from the canister by intake air flow and consumed in the normal combustion process.

4   Poor idle, stalling and poor driveability can be caused by an inoperative purge valve, a damaged canister, split or cracked hoses or hoses connected to the wrong tubes.

5   Evidence of fuel loss or fuel odor can be caused by liquid fuel leaking from front fuel lines, fuel pump or the TBI, a cracked or damaged canister, an inoperative bowl vent valve, an inoperative purge valve, disconnected, misrouted, kinked, deteriorated or damaged vapor or control hoses or an improperly seated air cleaner or air cleaner gasket.

Checking

6   Inspect all hoses in and out of the canister for kinks, leaks and breaks along their entire lengths. Repair or replace as necessary.

7   Inspect the canister. It it's cracked or damaged, replace it.

8   Look for fuel leaking from the bottom of the canister. If fuel is leaking, replace the canister. Check the hoses and hose routing.

9   Check the filter at the bottom of the canister. If it's dirty, plugged or damaged, replace it **(see illustration)**.

10   Apply a short length of hose to the PCV tube of the purge valve assembly and attempt to blow through it. Little or no air should pass into the canister (A small amount of air will pass if the canister has a constant purge hole).

11   With a hand vacuum pump, apply vacuum through the control vacuum signal tube to the purge valve diaphragm. If the diaphragm does not hold vacuum for at least 20 seconds, the diaphragm is leaking and the canister must be replaced.

12   If the diaphragm holds vacuum, again try to blow through the hose connected to

8.9 If the filter on the bottom of the canister is dirty, replace it with a new one

the PCV tube while vacuum is still being applied. An increased flow of air should be noted. If it isn't, replace the canister.

9   Positive Crankcase Ventilation (PCV) system

Refer to illustrations 9.1 and 9.2

General description

1   The positive crankcase ventilation system reduces hydrocarbon emissions by scavenging crankcase vapors. It does this by circulating fresh air from the air cleaner through the crankcase, where it mixes with blow-by gases and is then rerouted through a PCV valve to the intake manifold **(see illustration)**.

2   The main components of this system are vacuum hoses and a PCV valve, which regulates the flow of gases in accordance with engine speed and manifold vacuum **(see illustration)**.

3   To maintain idle quality, the PCV valve restricts the flow when the intake manifold vacuum is high. If abnormal operating conditions arise, the system is designed to allow

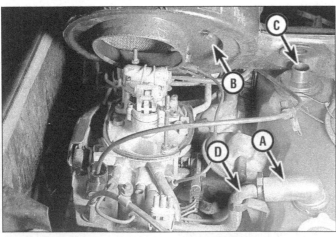

9.1 PCV system components

A PCV valve	C Crankcase vent
B Air cleaner	D Elbow

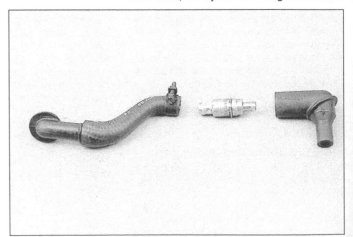

9.2 The components of a typical PCV system (this one is for the 2.5L four-cylinder engine) include an elbow fitting at each end (one for the rocker arm cover and one for the TBI unit), a short length of vacuum hose and the PCV valve

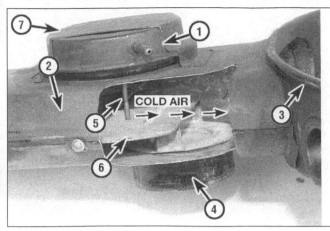

10.2 **Typical Thermac air cleaner**

1 *Vacuum diaphragm motor*
2 *Snorkel*
3 *Air cleaner housing*
4 *Heat stove duct*
5 *Linkage*
6 *Damper door*
7 *Retainer strap*

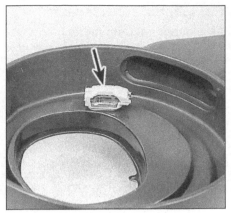

10.9 **The temperature sensor (arrow) for the Thermac system is inside the air cleaner housing**

excessive amounts of blow-by gases to backflow through the crankcase vent tube into the air cleaner to be consumed by normal combustion.

Checking

4   The PCV system can be checked quickly and easily for proper operation. It should be checked regularly because carbon and gunk deposited by blow-by gases will eventually clog the PCV valve and system hoses. The common symptoms of a plugged or clogged PCV valve include a rough idle, stalling or a slow idle speed, oil leaks, oil in the air cleaner or sludge in the engine .

5   To check for proper vacuum in the system, locate the PCV valve filter. It's located in the rocker arm cover directly under the air horn of the air cleaner housing on the four-cylinder engine and in the left end of the front rocker arm cover on the V6 engine.

6   Disconnect the hose leading to this filter. Be careful not to break the molded fitting on the filter.

7   With the engine idling, place your thumb lightly over the end of the hose. You should feel a slight vacuum. The suction may be heard as your thumb is released. This will indicate that air is being drawn all the way through the system. If a vacuum is felt, the system is functioning properly. Check that the filter inside the air cleaner housing is not clogged or dirty. If in doubt, replace it.

8   If there is very little or no vacuum at the end of the hose, the system is clogged and must be inspected further. A leaking valve or hose will cause a rough idle, stalling or a high idle speed.

9   Shut off the engine and locate the PCV valve. Carefully pull it from its rubber grommet. Shake it and listen for a clicking sound. If the valve does not click freely, replace it with a new one.

10   Start the engine and run it at idle speed with the PCV valve removed. Place your thumb over the end of the valve and feel for suction. There should be a relatively strong vacuum.

11   If little or no vacuum is felt at the PCV valve, turn off the engine and disconnect the vacuum hose from the other end of the valve. Run the engine at idle speed and check for

vacuum at the end of the hose just disconnected. No vacuum at this point indicates that the vacuum hose or inlet fitting at the engine is plugged. If it is the hose which is blocked, replace it with a new one or remove it from the engine and blow it out sufficiently with compressed air. A clogged passage at the manifold requires that the component be removed and thoroughly cleaned to remove carbon build-up. A strong vacuum felt going into the PCV valve, but little or no vacuum coming out of the valve, indicates a failure of the PCV valve itself. Replace it.

12   When purchasing a new PCV valve, make sure it is the correct one for your engine. An incorrect PCV valve may pull too little or too much vacuum, possibly leading to engine damage.

13   The component replacement procedures for the PCV valve and filter are in Chapter 1.

10   Thermostatic Air Cleaner (Thermac)

Refer to illustrations 10.2, 10.9 and 10.10

General description

1   The thermostatic air cleaner (Thermac) system improves engine efficiency and driveability under varying climatic conditions by controlling the temperature of the air coming into the air cleaner. A uniform incoming air temperature allows leaner air/fuel ratios during warm-up, which reduces hydrocarbon emissions.

2   The Thermac system uses a damper assembly, located in the snorkel of the air cleaner housing, to control the ratio of cold and warm air directed into the throttle body. This damper is controlled by a vacuum motor which is, in turn, modulated by a temperature sensor in the air cleaner **(see illustration)**. On some engines a check valve is used in the sensor, which delays the opening of the damper flap when the engine is cold and the vacuum signal is low.

3   When the engine is cold, the damper flap blocks off the air cleaner inlet snorkel, allowing only warm air from around the

exhaust manifold to enter the engine. As the engine warms up, the flap gradually opens the snorkel passage, increasing the amount of cold air allowed into the air cleaner. By the time that the engine has reached its normal operating temperature, the flap opens completely, allowing only cold, fresh air to enter.

4   Hesitation during warm-up can be caused by the heat stove tube becoming disconnected, the vacuum diaphragm motor becoming inoperative (open to the snorkel), loss of manifold vacuum, a sticking damper door, a missing or leaking seal between the TBI and the air cleaner, a missing air cleaner cover seal, a loose air cleaner cover or a loose air cleaner.

5   Lack of power, pinging or spongy throttle response can be caused by a damper door that does not open to the outside air or a temperature sensor that doesn't bleed off excess vacuum.

Checking

6   Refer to Chapter 1 for maintenance and checking procedures for the Thermac system. If any of the problems mentioned above are discovered while performing routine maintenance checks, refer to the following procedures.

7   If the damper door does not close off snorkel air when the cold engine is started, disconnect the vacuum hose at the snorkel vacuum motor and place your thumb over the hose end, checking for vacuum. If there is vacuum going to the motor, check that the damper door and link rod are not frozen or binding inside the air cleaner snorkel.

8   If a vacuum pump is available, disconnect the vacuum hose and apply vacuum to the motor to make sure the damper door moves. Replace the vacuum motor if the application of vacuum does not open the door.

9   If the above test indicates that there is no vacuum going to the motor, check the hoses for cracks, crimps and loose fitting connections. If the hoses are clear and in good condition, replace the temperature sensor **(see illustration)** inside the air cleaner housing.

Component replacement

Air cleaner vacuum motor

10   Remove the air cleaner assembly from the engine and disconnect the vacuum hose from the motor **(see illustration)**.

11   Drill out the two spot welds which secure the vacuum motor retaining strap to the snorkel tube.

12   Remove the motor attaching strap.

13   Lift up the motor, cocking it to one side to unhook the motor linkage at the control damper assembly.

14   To install, drill a 7/64-inch hole in the snorkel tube at the center of the retaining strap.

15   Insert the vacuum motor linkage into the control damper assembly.

16   Using the sheet metal screw supplied with the motor service kit, attach the motor and retaining strap to the snorkel. Make sure the sheet metal screw does not interfere with the operation of the damper door. Shorten the screw if necessary.

17   Connect the vacuum hose to the motor and install the air cleaner assembly.

Air cleaner temperature sensor

18   Remove the air cleaner from the engine and disconnect the vacuum hoses at the sensor.

19   Carefully note the position of the sensor. The new sensor must be installed in exactly the same position.

20   Pry up the tabs on the sensor retaining clip and remove the sensor and clip from the air cleaner.

21   Install the new sensor with a new gasket in the same position as the old one.

22   Press the retaining clip onto the sensor. Do not damage the control mechanism in the center of the sensor.

23   Connect the vacuum hoses and attach the air cleaner to the engine.

---

11   Catalytic converter

Refer to illustration 11.3

1   The catalytic converter is an emission control device added to the exhaust system to reduce pollutants from the exhaust gas

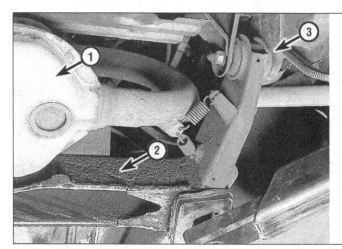

10.10 THERMAC motor and component locations for a typical TBI-equipped model

1   Air cleaner housing snorkel
2   Vacuum motor
3   Retaining strap
4   Spot welds
5   Vacuum hose

11.3 Location of the catalytic converter and surrounding components

1   Catalytic converter
2   Engine cradle
3   Left front engine cradle-to-body mount

stream. A single-bed converter design is used in combination with a three-way (reduction) catalyst. The catalytic coating on the three-way catalyst contains platinum and rhodium, which lower levels of oxide of nitrogen (NOx) as well as hydrocarbons (HC) and carbon monoxide (CO).

2   The test equipment for a catalytic converter is expensive and highly sophisticated. If you suspect that the converter on your vehicle is malfunctioning, take it to a dealer or authorized emissions inspection facility for diagnosis and repair.

3   The converter is located immediately below and in front of the engine **(see illustration)**. Therefore, whenever the vehicle is raised for servicing of underbody components, it is advisable to check the converter for leaks, corrosion and other damage. If damage is discovered, the converter cannot simply be unbolted from the exhaust system and replaced. It must be removed with a cutting torch because it is welded to the exhaust pipes. It must therefore be replaced by a dealer service department or muffler shop with the proper equipment.

Chapter 7 Part A
Manual transaxle

Contents

Specifications

Torque specifications

Ft-lbs (unless otherwise indicated)

Clutch slave cylinder retaining bolts	
1984 and 1985	20
1986 and later	14 to 20
Bellhousing-to-engine bolts	
1984 and 1985	15
1986 and later	120 in-lbs
Shifter assembly mounting nuts	18
Shift/select cable nut-to-shifter levers...................................	20
Transaxle ground cable stud nut..	30
Case-to-cover bolts	
1984 and 1985	18
1986 and later	16
Transaxle strut bolts	30
Transaxle-to-engine bolts...	55
Control box-to-case bolts	11 to 16
Rear cover bolts	11 to 16
Cradle bolts	
Front...	67
Rear..	76
Exhaust manifold-to-exhaust pipe bolts	22

Wheel lug nuts
    Steel 80
    Aluminum 100
Forward support bracket-to-transaxle case bolts
    1984 and 1985 48
    1986 and 1987 44
Forward mount-to-support bracket nut
    1984 and 1985 41
    1986 and 1987 35
Forward transaxle mount-to-crossmember nut
    1984 and 1985 41
    1986 and 1987 35
Rear mount-to-cradle nut
    1984 and 1985 47
    1986 and 1987 18
Rear mount-to-support bracket nut
    1984 and 1985 41
    1986 and 1987 35
Rear support bracket-to-transaxle case bolt
    1984 and 1985 47
    1986 and 1987 44

1   General information

The manual transaxle combines either a four or five-speed, constant mesh transmission with a differential unit into one compact assembly. Although the four and five-speed transaxles differ somewhat internally, they are identical in physical size and appearance.

All forward gears are in constant mesh. For ease of shifting and selection of the desired gear range, synchronizers with blocker rings, controlled by shift forks, are used. Reverse uses a sliding gear arrangement.

The basic components of both units include an aluminum transaxle case, an aluminum clutch cover, an input gear shaft, an output gear shaft and the differential assembly. The final output gear (an integral part of the output shaft) turns the ring gear and differential assembly, thereby turning the axleshafts which are attached to the rear wheels.

Because of the complexity of manual transaxles and because of the special tools and skills required to rebuild them, disassembly and reassembly procedures are not covered in this workshop manual. If your transaxle is malfunctioning, it is recommended that you take it to a dealer service department or automotive transmission shop for overhaul. The adjustment, removal and installation of a transaxle can, however, be accomplished at home, so those procedures have been included here.

Shifting is accomplished by a floor mounted shifter, which is connected to the transaxle shift linkage by cable assemblies. The shifter cables are called select and shift cables. The shifter cables attach to the shifter posts with pins. The shifter cables are adjusted at the rear where they are attached to the transaxle shift linkage.

Both four and five-speed units are attached to the engine and the cradle in a similar fashion, so the transaxle and cradle removal and installation procedures are the same for both. Both the four and five-speed transaxles are attached to the engine with six mounting bolts, five passing through the transaxle bellhousing and into the engine bosses while one attachment on the back side of the engine screws into a transaxle boss. Both four and five-speed transaxles also have front and rear mounting brackets that secure them to the cradle.

2   Transaxle mounts - check and replacement

Refer to illustrations 2.2 and 2.17

Check

1   Raise the vehicle and secure it on jackstands.
2   Watch the mount as an assistant pulls up and pushes down on the transaxle or place a pry bar between the transaxle and the cradle and lever it up and down. If the rubber separates from the metal plate of the mount or if the case moves up but not down (if the mount is bottomed out), replace it. If there is movement between a metal plate of the mount and its attaching point, tighten the bolts or nuts attaching the mount to the case or cradle crossmember **(see illustration)**.

Replacement

3   Disconnect the cable from the negative terminal of the battery.
4   Support the transaxle with a jack.

2.2 To check the forward transaxle mount for wear, place a crowbar or prybar between the front support bracket and the cradle crossmember, then lever it against the crossmember and note whether the transaxle moves

Forward transaxle mount

5   Remove the forward support bracket-to-transaxle bolts.
6   Remove the mount-to-support bracket nut.
7   Remove the mount-to-cradle nuts.
8   Remove the forward transaxle support bracket.
9   Remove the mount.
10   Place the new mount in position and install the mount-to-cradle nuts finger tight.
11   Place the transaxle support bracket in position and install the mount-to-support bracket nut and the bracket-to-transaxle bolts finger tight.
12   Make sure that all the mounting nuts and bolts are loose, then lower the jack supporting the transaxle until the transaxle centers the mount with its own weight.
13   Tighten the forward transaxle mounting hardware to the specified torque.
14   Connect the negative battery cable.

Rear transaxle mount

15   Disconnect the cable from the negative terminal of the battery.
16   Support the transaxle with a jack.
17   Remove the rear support bracket-to-transaxle bolts **(see illustration)**.
18   Remove the mount-to-support bracket nut.
19   Remove the mount-to-cradle bolt nuts.
20   Remove the support bracket.
21   Remove the mount.
22   Place the new mount in position and install the mount-to-cradle bolt nuts finger tight.
23   Place the rear transaxle support bracket in position and install the mount-to-support bracket nut and bracket-to-transaxle bolts finger tight.
24   Make sure that all the mounting nuts and bolts are loose, then lower the jack supporting the transaxle until the transaxle centers the mount with its own weight.
25   Tighten the rear transaxle mounting hardware to the specified torque.
26   Connect the negative battery cable.

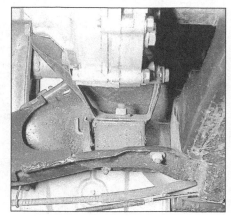

2.17 The rear transaxle support bracket and mount

3   Transaxle shift cables - removal and installation

Refer to illustrations 3.5, 3.7, 3.8a, 3.8b, 3.9 and 3.11

Removal

1   Disconnect the cable from the negative terminal of the battery.
2   Place the shifter in First gear.
3   Remove the front and rear console pad assemblies, the shift plate assembly and the console support assembly (see Chapter 11).
4   Remove the four bolts securing the Electronic Control Module (ECM) to the rear bulkhead of the passenger compartment and set the ECM aside.
5   Disconnect the ECM wire harness connector that is routed across the console **(see illustration)**.
6   Remove the four bolts from the ECM's plastic mounting bracket and detach the bracket.
7   Remove the clip on the end of the cable and disconnect the cable from the shifter pin **(see illustration)**.
8   Remove the large C-clip from the cable

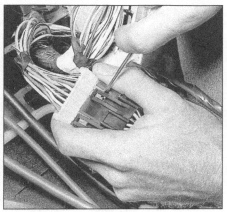

3.5 One of the ECM connectors lays on top of the console assembly - disconnect it and set it aside

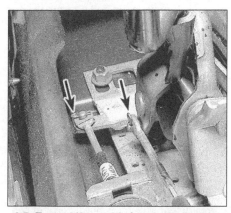

3.7 Each shifter cable is attached to the shifter assembly with a small hairpin clip (arrows) - remove the clip from the cable you are replacing

bracket just behind the shifter, then lift up on the large plastic cable ferrule until it's out of the bracket **(see illustrations)**.
9   In the engine compartment, disconnect the cable from the transaxle shift linkage **(see illustration)**.
10   Remove the shift cable bracket bolt

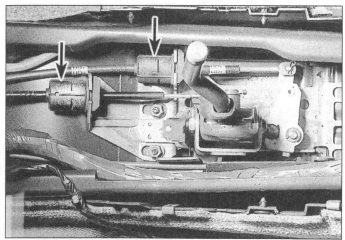

3.8a Each cable is also supported by a large C-clip and plastic ferrule (arrows) that attach it to the shifter assembly mounting bracket

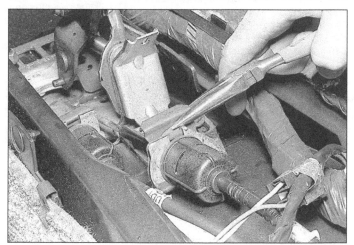

3.8b To remove the cable from the shifter assembly support bracket, grasp the large C-clip and pull it out, then lift the black plastic ferrule out of the slot in the bracket

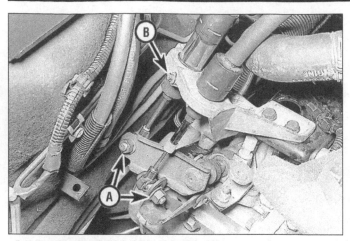

3.9 To disconnect the cable from the transaxle end, remove the nuts and washers (A) from the slotted linkage levers and remove the bolt (B) from the support bracket

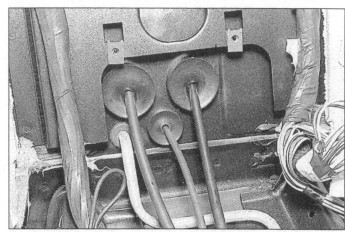

3.11 After removing the ECM module and its plastic support bracket, you will note several rubber grommets insulating various cables traveling between the engine compartment and the passenger compartment - the two big grommets are for the select and shift cables (pry out the grommet for the cable you intend to replace)

from the bracket, split the bracket halves and remove the cable.

11   In the passenger compartment, pry the rubber grommet from its hole in the bulkhead between the seats **(see illustration)**.

12   Pull the shift cable forward, through the bulkhead, from the passenger compartment side.

Installation

13   Thread the new shift cable through the hole in the bulkhead from the passenger compartment side (the rubber grommet cannot be installed into its hole from the engine compartment side of the bulkhead).

14   Push the rubber grommet into the hole until it is properly seated. **Note:** *The wider flange of the grommet should be facing forward, towards the passenger compartment.*

15   Place the new cable into position in the yoke of the shift cable bracket, install the other bracket half and install the bracket bolt. Tighten it securely.

16   Connect the cable to the transaxle shift linkage, but do not adjust it at this time. Leave the adjuster nut and bolt loose.

17   In the passenger compartment, install the large plastic ferrule into the bracket just behind the shifter assembly, then lock it into

place with the large C-clip.

18   Install the cable onto the shifter pin and lock it in place with the clip.

19   Install the plastic support bracket for the ECM. Tighten the four bolts securely.

20   Plug the large electrical connector that is routed across the console into the ECM.

21   Place the ECM in position in its support bracket and tighten the four ECM mounting bolts securely.

22   Install the console support assembly, front and rear console pads, radio and heater/air conditioner assemblies and trim plate and shifter trim plate (See Chapter 11).

23   Adjust and tighten the shift cable (Section 5).

24   Connect the cable to the negative terminal of the battery.

4   Shifter assembly - removal and installation

Refer to illustration 4.2

Removal

1   Refer to Chapter 11 for the procedure to

remove the shifter knob and trim plate.

2   Mark the location of the shifter assembly to ensure proper alignment upon reassembly **(see illustration)**.

3   Remove the four shifter assembly nuts.

Installation

4   Install the new shifter assembly. Tighten the assembly nuts to the specified torque.

5   Refer to Chapter 11 for the procedure to install the shifter knob and trim plate.

5   Transaxle shift cable linkage - adjustment

Refer to illustration 5.20

1   Disconnect the negative battery cable.

2   Place the transaxle in First gear.

3   Loosen the shift cable attaching nuts at both transaxle levers.

4   Remove the console and trim plates as required for access to the shifter (see Chapter 11).

5   With the shifter lever pulled to the left and held against the stop (First gear position), insert alignment pins into the alignment holes. **Note:** *You can use either 5/32-inch or No. 22 drill bits as alignment pins.*

6   Remove lash from the transaxle by rotating the vertical lever back while tightening the cable nut.

7   Next tighten the cable nut on the other lever.

8   Tighten both nuts to the specified torque.

9   Remove the alignment pins at the shifter assembly. **Note:** *To check the alignment, cycle the shifter from First to Second and from Second to First. The select cable (cable A) should not move. If it does move, repeat Steps 1 through 9.*

10   Lubricate the moving parts of the shift mechanism with white lithium base grease, using a stiff bristle brush.

11   Replace the console trim plate.

4.2 Scribe alignment marks from the front and left edges of the shifter assembly to ensure proper alignment when reinstalled, then remove the four mounting nuts (arrows)

SHIM PART NUMBER	DIMENSION C (mm)	COLOR and NUMBER of STRIPES
14008235	1.8	3 white
476709	2.1	1 orange
476710	2.4	2 orange
476711	2.7	3 orange
476712	3.0	1 blue
476713	3.3	2 blue
476714	3.6	3 blue
476715	3.9	1 white
476716	4.2	2 white

5.20 After you have determined the proper shim thickness for the shifter shaft, refer to this chart for the proper shim number to purchase

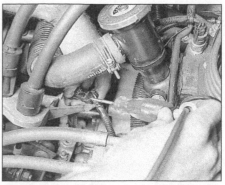

6.3 Insert a test light into the electrical connector just underneath the radiator hose and touch the dark blue wire terminal

12   Reconnect the negative battery cable. Road test the vehicle to check the shifting operation. If necessary, repeat the adjustment procedure to completely remove looseness and misalignment from the linkage.

13   If "hang-up" is encountered when shifting from First to Second, or vice versa, and the shift cables are properly adjusted, it may be necessary to check the shifter shaft shim. This shim helps position the shifter shaft for proper shifting characteristics. The following Steps will enable you to obtain the proper shim thickness.

14   Remove the reverse inhibitor spring and washer from the end of the housing.

15   Position the shifter shaft in Second gear.

16   Measure from the end of the housing to the shoulder just behind the end of the shaft. This is dimension A

17   Apply a load of 8 to 13 pounds on the opposite end of the shaft.

18   Measure from the end of the housing to the end of the shifter shaft major diameter. This is dimension B.

19   Subtract dimension B from dimension A to get dimension C.

20   Compare your result from the previous Step with the chart (see illustration) and choose the proper shim for use on reinstallation.

6   Backup light switch - check

Refer to illustration 6.3

1   With the ignition switch in the Run or Start position, voltage is applied through the turn signal/backup light fuse to the backup light switch. When the gear selector is shifted to the Reverse position, the backup switch closes, providing voltage to the backup lights.

2   If both backup lights don't work, check the turn/backup fuse by operating the turn signal lights. Check the backup switch adjustment by moving the gear selector lever to the Reverse position.

3   If both lights do not work, secure the vehicle so that it won't roll backwards. With the ignition switch in the Run position, place the gear selector lever in Reverse. Attach a test light to pin B (dark blue wire) of the backup switch and ground (see illustration).

Note: *On some models, the backup light switch is located on the transaxle housing instead of the shifter assembly. The procedure for checking it, however, is the same.*

4   If the test light does not glow, check the dark blue wire for an open.

5   If the test light does glow, attach it to pin A (light green wire) of the backup switch and ground. If the test light glows, replace the backup switch. If the test light does not glow, make sure that the bulbs are good.

7   Speedometer gear seal - removal and installation

Refer to illustrations 7.2 and 7.6

1   Disconnect the cable from the negative terminal of the battery.

2   Unplug the speedometer sensor harness from the speedometer fitting (see illustration).

3   Remove the retaining bolt and hold-down plate.

4   Wipe off any dirt or sludge around the speedometer fitting to prevent it from falling into the transaxle when the fitting is removed.

5   Carefully extract the speedometer fitting by pulling it straight up.

6   Peel off the old O-ring (see illustration).

7   Lubricate the new O-ring with engine oil or transaxle fluid and slide it onto the speedometer gear drive assembly body until it seats in the groove.

8   Reinstall the speedometer fitting, hold-down plate and retaining bolt. Be sure the fitting is fully seated. Tighten the bolt securely.

9   Plug the transmission sensor wiring harness into the top of the speedometer fitting.

10   Connect the cable to the negative terminal of the battery.

11   Drive the vehicle a short distance, then check for leaks.

8   Inner drive seal - removal and installation

Refer to illustrations 8.3 and 8.5

1   Refer to Chapter 8 for the driveaxle removal procedure.

2   Carefully note the exact position of the seal face. That is, how far it protrudes from

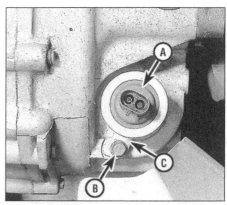

7.2 To get at the inner drive seal, disconnect the wire harness electrical connector from the plug atop the speedometer sensor (B) and remove the hold-down bolt (B) and plate (C)

the side of the transaxle case.

3   Gently pry the old inner driveaxle seal out with a screwdriver or tap it loose with a hammer and punch (see illustration).

4   Lubricate the walls of the new inner driveaxle seal with engine oil or gear oil and place it in position over the mouth of the driveaxle bore in the case.

5   Using a hammer and a block of wood or a large socket slightly smaller than the outside diameter of the seal, carefully tap the seal into the bore until it seats (see illustration). Do not drive it in too far or the seal may be damaged.

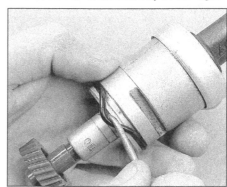

7.6 Remove the speedometer gear O-ring seal - be sure to lubricate the new O-ring with oil or transaxle fluid

6   Refer to Chapter 8 for the installation procedure for the driveaxle.

9   Transaxle - removal and installation

Note: *Because of the mid-engine layout and tight fit in the engine compartment, the following method of transaxle removal and installation cannot be performed without an engine support fixture shown in the accompanying illustrations, as well as a transaxle jack and a hydraulic hoist. An alternative method of removing and installing the transaxle, which can be found in Chapter 2, does not require the aforementioned special fixture but still requires the hoist. It also necessitates removal of the entire engine/transaxle/cradle assembly as a single unit.*

Removal

1   Disconnect the cable from the negative terminal of the battery and secure it out of the way.
2   Disconnect and remove the louvered engine compartment covers.
3   Disconnect and remove the engine compartment lid. Be sure to mark the hinge brackets with paint or scribe marks to facilitate proper alignment during reinstallation.
4   Disconnect and remove the air cleaner assembly.
5   Disconnect and remove the clutch slave cylinder and secure it out of the way.
6   Disconnect the shifter cable bracket (Section 3).
7   Disconnect the oxygen sensor wire.
8   Install an engine support fixture.
9   Slightly loosen, but do not remove, the rear wheel lug nuts.
10   Raise the vehicle on the hoist. **Note:** *You will need a large, sturdy pallet on which to rest the cradle.*
11   Remove the wheels.
12   Disconnect the parking brake cable from the brake calipers.
13   Disconnect the parking brake cable from the engine/transaxle cradle.
14   Disconnect the outer tie-rods from the knuckles.
15   Disconnect the lower A-arms from the knuckles.
16   Pry the inner CV joint housings from the transaxle (see Chapter 8).
17   Disconnect and remove the catalytic converter splash shield.
18   Disconnect the exhaust pipe from the exhaust manifold.
19   Loosen, but do not remove, the upper transaxle-to-engine bolts.
20   Disconnect the forward engine mount from the cradle.
21   If your vehicle is equipped with a V6, remove the crossover pipe heat shield and the crossover pipe.
22   Support the cradle with a sturdy pallet or an adjustable transmission stand.
23   Remove the front and rear transaxle bracket mounting bolts from the cradle.

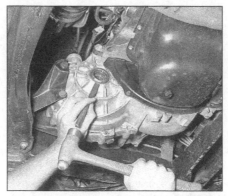

8.3 Remove the old inner drive seal from the transaxle case by either prying it out with a large screwdriver or knocking it out with a hammer and punch

24   Remove the upper transaxle to engine mounting bolts.
25   Loosen but do not remove the two front cradle mounting bolts and nuts.
26   Remove the two rear cradle mounting bolts.
27   Remove the two front cradle mounting bolts.
28   Raise the vehicle on the hoist just high enough so that the transaxle and the engine assembly clears the cradle.
29   Move the cradle out of the way.
30   Position a transaxle jack underneath the transaxle.
31   Remove the upper transaxle-to-engine mounting bolts.
32   Separate the transaxle from the engine by prying the two assemblies apart. **Note:** *Be sure to pry only at the right rear corner of the block.*
33   Lower the transaxle from the engine compartment on the adjustable stand.

Installation

34   Carefully raise the transaxle into position. Once the transaxle is aligned with the engine, install two four inch long bolts with the same threads as the mounting bolts in the top transaxle-to-engine bolt holes to use as guide pins when drawing the transaxle into place. Slide the transaxle toward the engine. If it does not move easily, have an assistant turn the engine over using a socket on the front pulley bolt as the transaxle is moved into position.
35   Install the transaxle-to-engine mounting bolts and tighten them to the specified torque.
36   Move the transaxle jack out of the way.
37   Place the cradle in position underneath the vehicle.
38   Lower the vehicle on the hydraulic hoist until it's close enough to the cradle that the front cradle bolts can be installed.
39   Install the front cradle through-bolts and nuts. Snug but do not tighten them.
40   Swing the rear of the cradle into position and install the rear cradle bolts. Tighten them to the specified torque.
41   Tighten the front cradle bolts to the

8.5 Install the new seal with a block of wood or a large socket slightly smaller in diameter than the outside diameter of the seal

specified torque.
42   Raise the vehicle on the hydraulic hoist and remove the pallet or adjustable stand supporting the cradle.
43   Install the front and rear transaxle mount-to-cradle bolts and tighten them to the specified torque.
44   Install the forward engine mount-to-cradle bolts and tighten them to the specified torque.
45   If your vehicle is equipped with a V6, install the crossover pipe heat shield and the crossover pipe.
46   Install the exhaust pipe-to-exhaust manifold bolts and tighten them to the specified torque.
47   Install the catalytic converter splash shield and tighten it securely.
48   Install the inner CV joint housings into the transaxle. **Note:** *If the splined driveaxle stub is difficult to push into the transaxle, rotate the driveaxle slightly and push at the same time.*
49   Connect the lower A-arms to the knuckles.
50   Connect the outer tie-rods to the knuckles.
51   Connect the parking brake cable halves together and attach the parking brake cable assembly to the cradle.
52   Connect the parking brake cable to the brake calipers.
53   Install the wheels and snug the wheel lug nuts.
54   Lower the vehicle until the wheels are touching the ground.
55   Tighten the wheel lug nuts to the specified torque.
56   Remove the engine support fixture.
57   Connect the oxygen sensor wire.
58   Install the shifter cable bracket.
59   Install the clutch slave cylinder, adjust it and tighten it to the specified torque.
60   Install the air cleaner assembly.
61   Install the engine compartment lid and tighten the hinge brackets securely.
62   Install the louvered engine compartment covers.
63   Connect the negative battery cable.

Chapter 7 Part B
Automatic transaxle

Contents

Specifications

Torque specifications

	Ft-lbs (unless otherwise indicated)
Shifter assembly mounting nuts	18
TV cable-to-transaxle mounting bolt	72 to 84 in-lbs
Transaxle control cable nut	15 to 25
Shift cable assembly pin nut	17
Backup light/Neutral start switch bolts	20
Shift cable-to-transaxle shaft nut	20
1984 and 1985 transaxle mounts	
Forward transaxle support bracket-to-transaxle mounting bolt	48
Forward transaxle support bracket-to-transaxle mount (insulator)	41
Forward transaxle insulator-to-crossmember	41
Rear transaxle support bracket-to-transaxle mounting bolt	47
Rear transaxle support bracket-to-transaxle mount (insulator)	41
Rear transaxle insulator-to-crossmember	47
1986 and later transaxle mounts	
Mount-to-bracket nuts	33
Upper case-to-transaxle case bolt	40
Bracket-to-transaxle case bolts	55
Transaxle-to-bracket nut	18

1   General information

Because of the complexity of the clutch mechanisms and the hydraulic control system, and because of the special tools and expertise needed to rebuild an automatic transaxle, this work is usually beyond the scope of the home mechanic. The procedures in this Chapter are therefore limited to general diagnosis, routine maintenance, adjustments and transaxle removal and installation.

Should the transaxle require major repair work, take it to a dealer service department or a transmission repair shop. You can, however, save the expense of removing and installing the transaxle by doing it yourself.

2   Diagnosis

Refer to illustrations 2.20 and 2.21
Note: *Automatic transmission malfunctions may be caused by four general conditions: poor engine performance, improper adjustments, hydraulic malfunctions and mechanical malfunctions. Diagnosis of these problems should always begin with a check of the easily repaired items: fluid level and condition (Chapter 1), shift linkage adjustment and throttle linkage adjustment. Next, perform a road test to determine if the problem has been corrected or if more diagnosis is necessary. If the problem persists after the preliminary tests and corrections are completed, additional diagnosis should be done by a dealer service department or transmission repair shop.*

Preliminary checks

1   Warm up the engine and transaxle to normal operating temperature.
2   Check the fluid level as described in Chapter 1.
  a) *If the fluid level is unusually low, add enough fluid to bring the level within the cross-hatched area of the dipstick, then check for external leaks and/or a defective vacuum modulator.*
  b) *If the fluid level is abnormally high, drain off the excess fluid, then inspect the excess fluid for contamination by coolant.*
  c) *If the fluid is foaming, drain it and refill the transaxle, then check for water in the fluid or a high fluid level.*
3   Check the engine idle speed. **Note:** *If the engine is not performing properly, do not attempt to proceed with the preliminary checks before first correcting any engine malfunctions.*
4   Check the throttle valve cable for freedom of movement and returnability at the cable activating lever. Check to be sure that the TV cable is adjusted to the proper length (Section 5). **Note:** *The TV cable may function properly when the engine is shut off and cold, but it may malfunction once the engine is hot. So check it both cold and at normal operating temperature.*

5   Inspect the shift control linkage (Section 3). Make sure that the linkage does not bind and is properly adjusted.

Fluid leak diagnosis

6   Most fluid leaks are easy to locate visually. Repair usually consists of replacing or repairing the leaking part. If a leak is difficult to find, the following procedure may help.
7   Identify the fluid. Make sure it is transaxle fluid and not engine oil or hydraulic brake or clutch fluid.
8   Try to pinpoint the source of the leak. Run the vehicle at normal operating temperature, then park it over a large sheet of paper. After a minute or two, you should be able to locate the leak by determining the source of the fluid dripping onto the paper.
9   Make a careful visual inspection around the suspected component. Pay particular attention to gasket mating surfaces. A mirror is often helpful for finding leaks in areas that are hard to see.
10   If the leak still cannot be found, clean the suspected area thoroughly with a degreaser, steam or spray solvent, then dry it.
11   Drive the vehicle for several miles at normal operating temperature and varying speeds. After driving the vehicle, visually inspect the suspected component again. If you still can't find the leak, try using the following method.
12   Clean the suspected area again.
13   Apply an aerosol-type powder (like foot powder) to the suspected area.
14   Drive the vehicle under normal operating conditions.
15   Visually inspect the suspected component. You should be able to trace the path of the leak back to its source over the white powder surface.
16   Once the leak has been located, the cause must be determined before it can be properly repaired. If a gasket is replaced but the sealing flange is bent, the new gasket will not repair the leak. The bent flange must also be repaired.
17   Before attempting to repair a leak, check to make sure that the following conditions are corrected or they may cause another leak. **Note:** *Some of the following conditions (a leaking torque converter, for instance) cannot be fixed without highly specialized tools and expertise. Such problems must be referred to a transmission specialist or a dealer service department.*

Gasket leaks

18   If the pan gasket is leaking, either the fluid level or the fluid pressure may be too high. The vent may be plugged. The pan bolts may be too tight. The pan flange or sealing surfaces may be warped. The sealing surface may be scratched, burred or damaged in some way. The gasket itself may be damaged or worn. The transaxle casting may be cracked or porous. If sealant instead of gasket material has been used to seal the joint between the pan and the transaxle housing, it may be the wrong sealant.

Seal leaks

19   If a transaxle seal is leaking, the fluid level or pressure may be too high. The vent may be plugged. The seal bore may be scratched, burred or nicked. The seal itself may be worn. Or it may have been improperly installed. There may be cracks in a component. The surface of the shaft protruding through the seal may be scratched, nicked or damaged. A loose or worn bearing may be causing excessive seal wear.

Transaxle pan or valve body cover leaks

20   Make sure that the attaching bolts are tightened correctly and that they're all there, that the gasket is properly installed and is not damaged and that the oil pan or valve body cover mounting face is flat **(see illustration)**.

Case leaks

21   Make sure that the filler pipe multi-lip seal is not damaged or missing. Make sure that the filler pipe is not dislocated. See if the TV cable multi-lip seal is missing, damaged or improperly installed. Make sure that the governor cover and O-rings are neither damaged nor missing. Check the speedometer driven gear/speed sensor seal for damage. Make sure that the manual valve bore plug is not loose. The oil cooler connector fittings should not be loose or damaged **(see illustration)**. The axle oil seals should not be worn or damaged. The parking pawl shaft cup plug, the governor pressure pickup plug or the line pressure pickup plug may be loose. The case-to-case cover gasket may be damaged. The casting may be porous.

Leak at torque converter end

22   The torque converter seal may be damaged. The seal lip could be cut, the bushing may have moved forward and become damaged or the garter spring may be missing from the seal. There may be a converter leak in the weld area. The casting may be porous. The turbine shaft oil seal may be worn or damaged.

Vent pipe or fill tube leaks

23   If this condition occurs, the transaxle is overfilled, there is water in the fluid, the case is porous, the dipstick is incorrect, the vent is plugged or the drain back holes are plugged.

3   Shift linkage - adjustment

Refer to illustration 3.3
Note: *The transaxle shift linkage must be adjusted so that the indicator quadrant and stops correspond with the transaxle detents. If the linkage is not adjusted properly, an internal leak could occur which could cause a clutch or band to slip.*
Warning: *If a manual linkage adjustment is made while the selector lever is in the Park position, the parking pawl should freely engage the reaction internal gear to prevent the vehicle from rolling. Transaxle or vehicle*

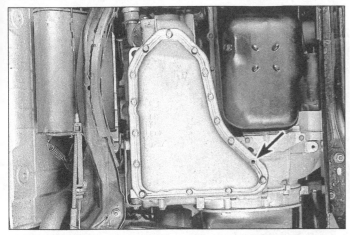

2.20 This missing transaxle oil pan flange bolt (arrow) can cause a leak as the transaxle fluid forces its way between the sealing flange of the pan and the sealing surface of the underside of the transaxle - check the pan bolts on your vehicle periodically for proper torque to prevent this from happening

2.21 The fittings for the transaxle oil cooler lines (arrows) are likely places for the transaxle to leak because over a period of time the clamps and fittings can loosen - check these fittings once in a while to make sure they're snug and in good condition

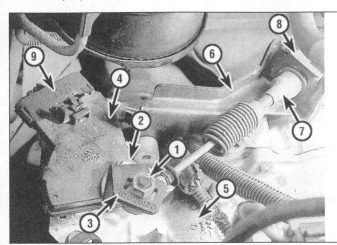

3.3 View of the transaxle shift linkage

1 Cable assembly threaded pin nut
2 Shift lever nut
3 Shift lever
4 Neutral start and back-up switch
5 Transaxle
6 Cable bracket
7 Shift cable
8 Retainer clip
9 Harness connector

4.12a The shift cables are attached to the shifter assembly and the shift linkage at the transaxle by small Teflon spherical bearings that pop on and off small pins

damage or personal injury may occur if not properly adjusted.

1   Place the shift lever in the Neutral (N) position.

2   Place the transaxle lever in the Neutral position. Find the Neutral position by rotating the transaxle lever clockwise from Park through R into N (Neutral).

3   Insert the threaded pin of the shift cable assembly up through the slotted hole in the shifting lever **(see illustration)** and hand start the cable assembly nut. The lever must be held out of Park when tightening the nut. Tighten the nut to the specified torque.

4   Cables - removal and installation

Refer to illustrations 4.12a, 4.12b, 4.13, 4.14, 4.18 and 4.38

Automatic shift cable
Removal

1   Disconnect the negative battery cable from the battery.

2   Remove the front trim plate and shift knob (Chapter 11).

3   Remove the shift trim plate, rear console pad assembly and front pad assembly (Chapter 11).

4   Disconnect the ECM electrical connectors (Chapter 6).

5   Remove the ECM (Chapter 6).

6   Remove the front carrier-to-instrument panel reinforcement frame (Chapter 11).

7   Remove the carrier reinforcement (Chapter 11).

8   Remove the carpet clips and rivets at the console assembly (Chapter 11).

9   Remove the heater control assembly (Chapter 3).

10   Remove the radio (Chapters 11 and 12).

11   Remove the carrier (Chapter 11).

12   Disconnect the shift cable from the shifter control assembly **(see illustration)**. **Note:** *If the cable has a plug, push on the plug to spread the tangs, then remove the cable end from the shifter and transaxle lever pins. If the cable end does not have a plug,*

4.12b Shift lever and cable mounting details

1 Shift lever handle
2 Shift cable snap fitting
3 Snap fitting release button
4 Shift cable
5 Cable retainer
6 Shifter assembly
7 Shifter assembly mounting nuts

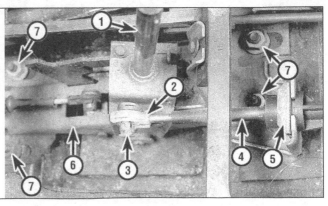

4.13 To disconnect the cable from the shifter base, pull straight up on the metal clip (A), then lift up on the cable (C) and remove it from the shifter base bracket (B)

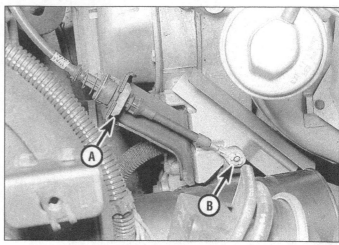

4.14 The cable assembly is attached to the transaxle by a bracket and retainer clip (A) - note that the cable end connector (B) on this vehicle does not have a plug, so it must be pried off with a screwdriver

use a screwdriver or flat tool **(see illustration)**. *Insert the tool between the lever and the nylon end at the pin center line. Rotate the tool and the cable end will snap off the pin.*

13   Remove the cable from the yoke clip securing it to the shifter base **(see illustration)**.

14   Disconnect the shift cable end from the shift lever at the transaxle and disconnect it from the yoke clip securing it to the transaxle mounting bracket **(see illustration)**.

15   Pull the cable through the body into the passenger compartment.

Installation

16   Guide the cable from the passenger side through the body into the engine compartment.

17   Clip the cable to the transaxle mounting bracket and snap the cable onto the shift lever.

18   Clip the cable in place at the fuel line

(see illustration).

19   Attach the cable to the yoke clip behind the shifter assembly.

20   Attach the cable end to the shift lever pin (see illustration for Step 12 above).

21   Install the console carrier assembly (Chapter 11).

22   Install the radio (Chapters 11 and 12).

23   Install the heater control assembly (Chapter 3).

24   Install the carpet clips and rivets at the console (Chapter 11).

25   Install the carrier reinforcements (Chapter 11).

26   Install the ECM (Chapter 6).

27   Plug in the ECM electrical connectors (Chapter 6).

28   Install the front pad assembly (Chapter 11).

29   Install the front pad trim plate (Chapter 11).

30   Install the rear console pad assembly (Chapter 11).

31   Install the shift trim plate and shift knob (Chapter 11).

32   Connect the cable to the negative terminal of the battery.

Park/lock control cable

Removal

33   Remove the console covers, the hush panel and the lower steering column as necessary for access to the park/lock cable.

34   Disconnect the cable from the negative terminal of the battery.

35   Place the shift lever in the Park position.

36   Turn the ignition key to the Run position.

37   Disconnect the cable from the inhibitor. To release the cable, insert a screwdriver into the inhibitor slot, depress the cable latch and pull the cable back from the inhibitor.

38   Disconnect the cable from the park lock lever pin **(see illustration)**.

39   Disconnect the cable from the shifter base.

40   Remove the cable.

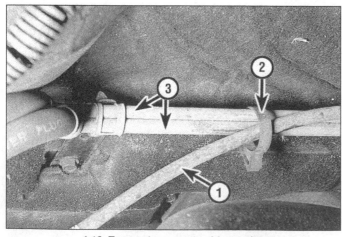

4.18 Transaxle control cable routing

1	Cable	3 Fuel lines
2	Clamp	

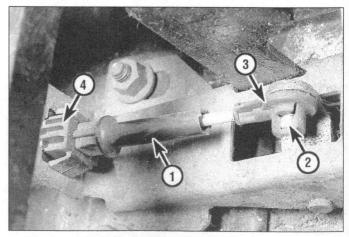

4.38 Park/lock cable routing

1 Park lock cable	3 Park lock cable snap
2 Park lock lever pin	fitting
	4 Adjuster button housing

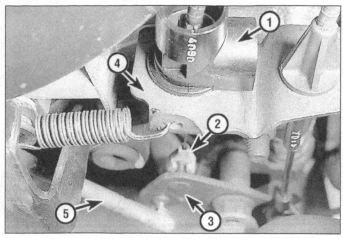

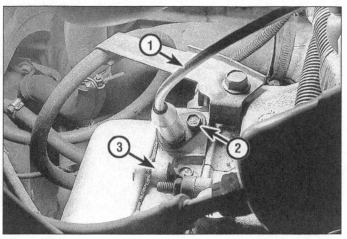

5.3 Throttle valve (TV) cable system

1   TV cable   4   Bracket
2   TV cable snap clip   5   TBI lever linkage
3   Throttle lever assembly

5.6 Typical TV cable-to-transaxle mounting details

1   TV cable   3   Transaxle
2   Retaining bolt

Installation

41   Place the shift lever in the Park position.
42   Snap the cable connector lock button to the up position.
43   Snap the cable connector to the base.
44   Turn the ignition key to the Off position.
45   Snap the cable into the inhibitor housing.
46   Snap the cable into the park lock lever pin.
47   Push the cable connector nose forward toward the connector to remove the slack.
48   With no load applied to the nose, snap the cable connector lock button down.
49   Place the shift lever in the Park position.
50   Turn the ignition key to the Lock position.
51   The shift lever should now be locked in position. The ignition key should be removable from the column.
52   Turn the ignition key to the Run position.
53   With the shift lever in Neutral, the ignition key should not be removable from the column.
54   If the above functional checks were met, adjustment is complete. If the key can be removed in Neutral, snap the connector lock button to the up position and repeat Steps 8

and 9. If the key cannot be removed in the Park position, snap the connector lock button to the up position and move the cable connector nose to the rear until the key can be removed from the ignition. Snap the connector lock button down.

---

5   Throttle Valve (TV) cable - removal, installation and adjustment

Refer to illustrations 5.3, 5.6 and 5.10

1   The throttle valve cable controls the transaxle line pressure and consequently the shift "feel" and timing, as well as the part throttle and detent downshifts.
2   The TV cable is attached to the link at the throttle lever and bracket on the transaxle and to the throttle lever on the TBI.

Removal

3   Disconnect the TV cable from the throttle body **(see illustration)**.
4   Pull up on the cable cover at the transaxle until the cable is visible. Remove the bolt securing the TV cable at the transaxle. Disconnect the cable from the transaxle rod.

5   Disconnect the clip securing the TV cable at the valve cover.

Installation

6   Install the new TV cable at the transaxle **(see illustration)**. Tighten the TV cable bolt to the specified torque.
7   Install the TV cable at the throttle body.
8   Install the clip securing the TV cable to the valve cover. **Caution:** *Any time the throttle valve cable is disconnected for any reason, it must be adjusted immediately upon reassembly prior to operation of the vehicle.*

Adjustment

Note: *Adjustment of the TV cable must be made by rotating the throttle lever at the throttle body. Do not use the accelerator pedal to rotate the throttle lever. The engine must be off during adjustment.*
9   The free movement of the TV cable can be checked by pulling the upper end of the cable. The cable should travel a short distance with light resistance due to the small coiled return spring. Pull the cable farther out to move the lever into contact with the plunger, thus compressing the heavier TV spring. When released, the cable should return to the closed position.
10   Depress and hold down the metal readjust tab at the engine end of the TV cable **(see illustration)**.
11   While holding the tab down, move the slider until it stops against the fitting.
12   Release the readjust tab.
13   Rotate the throttle lever to the maximum travel stop position. The cable will ratchet through the slider and automatically readjust itself.
14   Make sure that the cable moves freely. It may appear to function properly with the engine cold. Recheck it after the engine is hot.
15   Road test the vehicle. If delayed or only full-throttle shifts still occur, have the vehicle checked by a dealer service department.

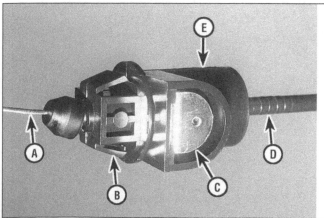

5.10 Depress the throttle valve (TV) cable release tab and pull the slider back until it rests on its stop – release the tab and open the throttle completely

A   TV cable
B   Locking lugs
C   Release tab
D   Cable casing
E   Slider

6.1 The Torque Converter Clutch (TCC) electrical connector is on the front of the automatic transaxle housing

8.2 The governor and speedometer sensor assembly is located at the right rear corner of the automatic transaxle housing

6   Torque Converter Clutch (TCC) - diagnosis

Refer to illustration 6.1

Description

1   The TCC is applied by fluid pressure which is regulated by an ECM controlled solenoid located inside the automatic transaxle assembly **(see illustration)**. When the vehicle reaches a specified speed, the ECM energizes the solenoid, enabling the torque converter to mechanically couple the engine to the transaxle. During mechanical lockup, emissions are reduced because engine rpm at any given speed is reduced.
2   Because the transaxle must also function in its normal, fluidcoupled mode during deceleration, passing, idle, etc., the ECM deenergizes the solenoid under these conditions and returns the transaxle to normal operation. The transaxle also returns to fluid

8.3 View of the governor and speedometer sensor assembly

1   Speed sensor
2   Retainer wire clip
3   Governor cover

coupling when the brake pedal is depressed.

Functional check

3   Install a tachometer.
4   Drive the vehicle until normal operating temperature is reached, then maintain a 50 to 55 mph speed.
5   Lightly touch the brake pedal and check it for a slight bumpy sensation, indicating that the TCC is releasing. A slight increase in rpm should also be noted.
6   Release the brake and check for reapplication of the converter clutch and a slight decrease in engine rpm.
7   If the TCC fails to perform satisfactorily during the above test, take your vehicle to a dealer for service.

7   Backup light/Neutral start switch - removal, installation and adjustment

Removal

1   Disconnect the electrical harness at the Neutral start switch. **Caution:** *The clip on the switch connection must be opened before attempting to detach the harness from the switch. Damage will occur to the switch assembly if the clip on the switch is closed when the harness is disconnected.*
2   Pry the cable from the pivot pin at the bottom of the shift lever (Section 4).
3   Remove the nut which attaches the lever to the transaxle shaft.
4   Remove the two bolts which attach the backup light/Neutral start switch to the transaxle.
5   Remove the backup light/Neutral start switch.

Installation

6   Make sure that the transaxle shaft is in the Neutral position when the switch is installed.

7   Attach the backup light/Neutral start switch to the transaxle and tighten the mounting bolts to the specified torque.
8   Install the nut attaching the lever to the transaxle shaft.
9   Attach the cable assembly to the pivot pin (Section 4).

Adjustment

10   The backup light and Neutral start switch has been adjusted at the factory and requires no further adjustment.

8   Speedometer gears - removal and installation

Refer to illustrations 8.2 and 8.3

Removal

1   Disconnect the cable from the negative terminal of the battery.
2   Disconnect the wire harness at the sensor **(see illustration)**.
3   Remove the sensor assembly retainer **(see illustration)**.
4   Remove the sensor assembly and gear.
5   Remove the governor cover screw.
6   Remove the governor cover.
7   Remove the O-ring.
8   Remove the speedometer thrust washer.
9   Remove the speedometer drive gear.

Installation

10   Install the speedometer drive gear.
11   Install the speedometer thrust washer.
12   Install a new O-ring.
13   Install the governor cover.
14   Install the sensor assembly and gear.
15   Install the sensor assembly retainer.
16   Attach the wire harness at the sensor.
17   Connect the cable at the negative terminal of the battery.

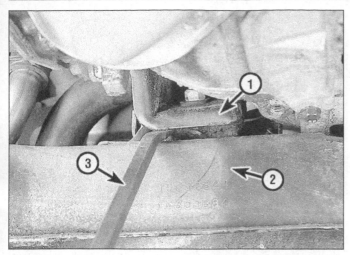

9.2a View of the forward automatic transaxle mounting bracket and mounting assembly (insulator)

| 1 | Bracket | 2 | Cradle | 3 | Pry bar |

9.2b To check the forward transaxle mounting assembly insulator for wear, insert a large screwdriver or prybar between the transaxle and the crossmember at the points indicated by the arrows and try to lever the transaxle housing - if the transaxle moves, the insulator rubber is worn out and must be replaced

9 Transaxle mounts - check and replacement

Refer to illustrations 9.2a, 9.2b. 9.2c and 9.2d

Check

1 Raise the vehicle and secure it on jackstands.

2 Watch the mount as an assistant pulls up and pushes down on the transaxle or place a pry bar between the transaxle and the cradle and lever it up and down. If the rubber separates from the metal plate of the mount or if the case moves up but not down (if the mount is bottomed out), replace it. If there is movement between a metal plate of the mount and its attaching point, tighten the bolts or nuts attaching the mount to the case or cradle crossmember **(see illustrations).**

Replacement

3 Disconnect the cable from the negative terminal of the battery.

4 Support the transaxle with a jack.

Forward transaxle mount

5 Remove the forward support bracket-to-transaxle bolts.

6 Remove the mount-to-support bracket nut.

7 Remove the mount-to-cradle nuts.

8 Remove the forward transaxle support bracket.

9 Remove the mount.

10 Place the new mount in position and install the mount-to-cradle nuts finger tight.

11 Place the transaxle support bracket in position and install the mount-to-support bracket nut and the bracket-to-transaxle bolts finger tight.

12 Make sure that all the mounting nuts and bolts are loose, then lower the jack supporting the transaxle until the transaxle centers the mount with its own weight.

13 Tighten the forward transaxle mounting hardware to the specified torque.

14 Connect the negative battery cable.

Rear transaxle mount

15 Disconnect the cable from the negative terminal of the battery.

16 Support the transaxle with a jack.

17 Remove the rear support bracket-to-transaxle bolts.

18 Remove the mount-to-support bracket nut.

19 Remove the mount-to-cradle bolt nuts.

20 Remove the support bracket.

21 Remove the mount.

22 Place the new mount in position and

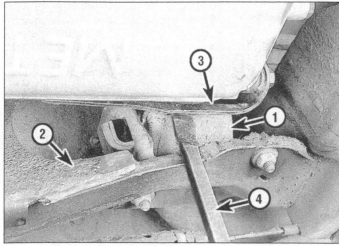

9.2c Exploded view of the rear automatic transaxle mounting bracket and mounting assembly (insulator)

| 1 | Mount | 2 | Cradle | 3 | Bracket | 4 | Pry bar |

9.2d To check the rear transaxle mounting assembly insulator for wear, insert a large screwdriver or prybar between the transaxle and the crossmember at the point indicated by the arrow and try to lever the transaxle housing - if the transaxle moves, the insulator rubber is worn out and must be replaced

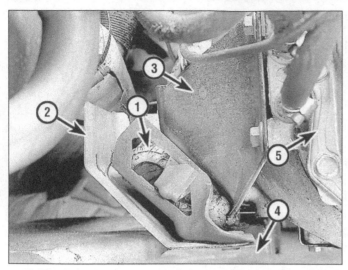

10.23a View of early model (thru mid-1986) transaxle mounts

1	*Rear mount*	4	*Cradle*
2	*Shield*	5	*Transaxle*
3	*Bracket*		

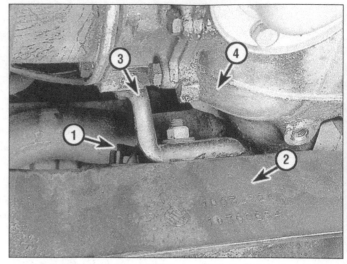

10.23b View of late model large one-piece transaxle mount

1	*Forward mount*	3	*Bracket*
2	*Cradle*	4	*Transaxle*

install the mount-to-cradle bolt nuts finger tight.

23   Place the rear transaxle support bracket in position and install the mount-to-support bracket nut and bracket-to-transaxle bolts finger tight.

24   Make sure that all the mounting nuts and bolts are loose, then lower the jack supporting the transaxle until the transaxle centers the mount with its own weight.

25   Tighten the rear transaxle mounting hardware to the specified torque.

26   Connect the negative battery cable.

10   Transaxle - removal and installation

Refer to illustrations 10.23a and 10.23b

1   **Note:** *Because of the mid-engine layout and tight fit in the engine compartment, the following method of transaxle removal and installation cannot be performed without an engine support fixture, as well as a transaxle jack and a hydraulic hoist. An alternative method of removing and installing the transaxle, which can be found in Chapter 2, does not require the aforementioned special fixture but still requires the hoist. It also necessitates removal of the entire engine/transaxle/cradle assembly as a single unit.*

Removal

2   Disconnect the cable from the negative terminal of the battery and secure it out of the way. Disconnect the oxygen sensor wire (see Chapter 6).

3   Unbolt and remove the engine compartment lid. Be sure to mark the hinge brackets with paint or scribe marks to facilitate proper alignment during reinstallation.

4   Disconnect and remove the air cleaner

assembly.

5   Disconnect the shift cable and Neutral safety/backup light switch.

6   Disconnect the speedometer sensor electrical connector and the TCC connector.

7   Disconnect the TV cable.

8   Disconnect the transaxle fluid cooler fittings.

9   Install the engine support fixture.

10   Raise the vehicle on the hoist. **Note:** *You will need a large, sturdy pallet on which to rest the cradle.*

11   Remove the wheels.

12   Disconnect the parking brake cable from the brake calipers.

13   Disconnect the parking brake cable from the engine/transaxle cradle.

14   Disconnect the outer tie-rods from the knuckles.

15   Disconnect the lower A-arms from the knuckles.

16   Pry the inner CV joint housings from the transaxle (see Chapter 8).

17   Disconnect and remove the catalytic converter splash shield.

18   Disconnect the exhaust pipe from the exhaust manifold.

19   Loosen, but do not remove, the upper transaxle-to-engine bolts.

20   Disconnect the forward engine mount from the cradle.

21   If your vehicle is equipped with a V6, remove the crossover pipe heat shield and the crossover pipe.

22   Support the cradle with a sturdy pallet or a transaxle jack.

23   Remove the front and rear transaxle bracket mounting bolts from the cradle **(see illustrations)**.

24   Remove the upper transaxle-to-engine mounting bolts.

25   Loosen but do not remove the two front cradle mounting bolts and nuts.

26   Remove the two rear cradle mounting

bolts.

27   Remove the two front cradle mounting bolts.

28   Raise the vehicle on the hoist just high enough so that the transaxle and the engine assembly clears the cradle.

29   Move the cradle out of the way.

30   Position a transaxle jack underneath the transaxle.

31   Remove the upper transaxle-to-engine mounting bolts.

32   Separate the transaxle from the engine by prying the two assemblies apart. **Note:** *Be sure to pry only at the right rear corner of the block.*

33   Lower the transaxle from the engine compartment on the adjustable stand.

Installation

34   Carefully raise the transaxle into position on the jack. Once the transaxle is aligned with the engine, install two four-inch long bolts with the same threads as the mounting bolts in the top transaxle-to-engine bolt holes to use as guide pins when drawing the transaxle into place. Slide the transaxle toward the engine. If it does not move easily, have an assistant turn the engine over using a socket on the front pulley bolt as the transaxle is moved into position.

35   Install the transaxle-to-engine mounting bolts and tighten them to the specified torque.

36   Move the transaxle jack out of the way.

37   Place the cradle in position underneath the vehicle.

38   Lower the vehicle on the hydraulic hoist until it's close enough to the cradle that the front cradle bolts can be installed.

39   Install the front cradle through-bolts and nuts. Snug but do not tighten them.

40   Swing the rear of the cradle into position and install the rear cradle bolts. Tighten them to the specified torque.

41   Tighten the front cradle bolts to the specified torque.

42   Raise the vehicle on the hoist and remove the pallet or adjustable stand supporting the cradle.

43   Install the front and rear transaxle mount-to-cradle bolts and tighten them to the specified torque.

44   Install the forward engine mount-to-cradle bolts and tighten them to the specified torque.

45   If your vehicle is equipped with a V6, install the crossover pipe heat shield and the crossover pipe.

46   Install the exhaust pipe-to-exhaust manifold bolts and tighten them to the specified torque.

47   Install the catalytic converter splash shield.

48   Install the inner CV joint housings into the transaxle. **Note:** *If the splined driveaxle stub is difficult to push into the transaxle, rotate the driveaxle slightly and push at the same time.*

49   Connect the lower A-arms to the knuckles.

50   Connect the outer tie-rods to the knuckles.

51   Connect the parking brake cable halves together and attach the parking brake cable assembly to the cradle.

52   Connect the parking brake cable to the brake calipers.

53   Install the wheels and snug the wheel lug nuts.

54   Lower the vehicle.

55   Tighten the wheel lug nuts to the specified torque.

56   Remove the engine support fixture.

57   Connect the oxygen sensor wire.

58   Install the shifter cable bracket.

59   Install the air cleaner assembly.

60   Install the engine compartment lid and tighten the hinge brackets securely.

61   Install the louvered engine compartment covers.

62   Connect the battery negative cable.

Notes

Chapter 8
Clutch and driveaxles

Contents

Specifications

Torque specifications

Ft-lbs (unless otherwise indicated)

Hub nut	192
Toe link rod	See Chapter 10
Control arm stud pinch bolt	See Chapter 10
Flywheel-to-crankshaft bolt	50
Clutch pressure plate-to-flywheel bolt	15
Clutch pedal assembly-to-cowl	156 in-lbs
Hydraulic line-to-clutch master cylinder	156 in-lbs
Clutch master cylinder-to-cowl nuts	156 in-lbs
Hydraulic line-to-clutch slave cylinder	144 to 156 in-lbs
Clutch release lever-to-clutch fork bolt/nut	20
Clutch slave cylinder-to-slave cylinder bracket nuts	20
Slave cylinder bracket-to-transaxle	32
Rear strut damper-to-knuckle nuts	See Chapter 10
Wheel lug nuts	See Chapter 1

1   Clutch - general information

All manual transaxle-equipped models use a single dry plate, diaphragm spring-type clutch. The clutch disc has a splined hub which allows it to slide along the splines of the input shaft. The clutch and pressure plate are held in contact by spring pressure exerted by a diaphragm spring in the pressure plate.

The clutch release system is operated by hydraulic pressure and consists of the clutch pedal, the clutch master cylinder, the line and hose assembly, the slave cylinder, the release lever and the shaft-and-fork assembly. The hydraulic clutch system determines clutch pedal height and provides automatic clutch adjustment. No adjustment of the clutch linkage or pedal height is required.

When pressure is applied to the clutch pedal to release the clutch, hydraulic pressure is exerted against the outer end of the clutch release lever. As the fork pivots on its shaft, its inner end pushes against the release bearing. The bearing pushes against the diaphragm spring levers of the pressure plate assembly, which releases the clutch/pressure plate from the flywheel.

2   Clutch operation - check

1   Several checks can be made to determine if there is a problem with the clutch itself.
2   With the engine running and the brake applied, hold the clutch pedal 1/2-inch from the floor and shift back-and-forth between First and Second several times. If the shifts are smooth, the clutch is releasing properly. If they aren't, the clutch is not releasing completely and the hydraulic system should be checked.

3   Clutch release lever - removal and installation

Removal

1   Remove the slave cylinder mounting nuts from the rear of the mounting bracket and move the slave cylinder aside. *Do not disconnect the hydraulic line from the slave cylinder.*
2   Scribe matching marks on the clutch release lever and shaft to ensure that the lever is positioned correctly upon reinstallation.
3   Remove the clutch release lever bolt/nut and remove the lever from the transaxle clutch fork shaft.

Installation

4   Install the clutch release lever on the transaxle clutch fork shaft. Tighten the clutch release lever bolt/nut to the specified torque.
5   Position the slave cylinder on the mounting bracket. Make sure that the slave cylinder

pushrod seats in the clutch release lever.
6   Tighten the clutch slave cylinder mounting nuts to the specified torque.

4   Clutch master cylinder - removal, overhaul and installation

Removal

1   Remove the wire clip and washer from the clutch pedal assembly pivot stud and detach the pushrod from the pedal.
2   Disconnect the hydraulic line from the master cylinder. **Note:** *Some fluid will drip out so place several shop rags underneath the master cylinder to protect the painted surface of the front compartment.*
3   Remove the nuts attaching the master cylinder to the vehicle cowl and remove the master cylinder.

Overhaul

Note: *To simplify reassembly, pay close attention to the relationship of the parts to each other and the direction in which each part faces during disassembly of the master cylinder.*
4   Unscrew the reservoir cap, drain the fluid and discard it. **Caution:** *Never reuse fluid bled or fluid drained from the hydraulic system.*
5   Pull back the dust cover.
6   Remove the circlip, retaining washer and pushrod.
7   Shake the cylinder to eject the plunger assembly.
8   Lift the leaf of the spring retainer and remove the spring assembly from the plunger.
9   Compress the spring to free the valve stem from the keyhole of the spring retainer to release spring tension.
10   Remove the spring, valve spacer and spring washer from the valve stem.
11   Remove the valve seal from the valve head.
12   Clean the parts thoroughly with denatured alcohol and place them on a clean sheet of paper or a clean rag.
13   Examine the bore of the master cylinder for visible scores and ridges and make sure that it feels smooth to the touch. If there is the slightest doubt regarding the condition of the master cylinder bore or plunger, the master cylinder must be replaced. **Note:** *A typical rebuild kit for the clutch master cylinder includes a new baffle, seal, center valve seal, circlip and dust cover. The rest of the components can be replaced individually if necessary.*
14   Install the plunger seal on the plunger.
15   Install the center valve seal (smallest diameter facing forward) on the valve head.
16   Position the spring washer on the valve stem so that it flares away from the valve stem shoulder.
17   Install the valve spacer (legs first) and the spring.
18   Attach the spring retainer to the spring and compress the spring until the valve stem

passes through the keyhole slot and engages in the center.
19   Join the spring to the plunger and press the leaf of the spring retainer to secure it.
20   Liberally lubricate the seal and the plunger bore with clean brake fluid.
21   Insert the plunger assembly, valve end first, into the cylinder body.
22   Position the pushrod and retaining washer and install the circlip to secure the assembly.
23   Coat the inside of the dust cover with silicone lubricant and install the dust cover.
24   Install the cap washer.
25   Screw the reservoir cap onto the master cylinder.

Installation

26   Attach the master cylinder to the cowl and tighten the mounting nuts to the specified torque. Hook up the hydraulic line and tighten the fitting.
27   Attach the pushrod and install the washer and clip.
28   Bleed the clutch hydraulic system (Section 6).

5   Clutch slave cylinder - removal, overhaul and installation

Removal

1   Disconnect the hydraulic line at the slave cylinder. **Caution:** *Place several rags underneath the slave cylinder before removing the hydraulic line.*
2   Remove the slave cylinder mounting nuts and detach the slave cylinder.

Overhaul

3   Pull back the dust cover and remove the retaining ring and pushrod.
4   Shake the cylinder to remove the piston and seal.
5   Clean the parts thoroughly with denatured alcohol then place them on a clean sheet of paper or a clean rag.
6   Examine the bore of the cylinder for visible scores and ridges and check to make sure that it is smooth to the touch. If there is any doubt about the condition of the bore or the plunger, replace the slave cylinder assembly.
7   Liberally lubricate the seal and piston bore with new brake fluid.
8   Insert the seal and piston.
9   Insert the pushrod and secure it with the retaining ring.
10   Coat the inside of the dust cover with silicone lubricant and install the dust cover.

Installation

11   Install the slave cylinder on the mounting bracket and tighten the mounting nuts to the specified torque.
12   Reconnect the hydraulic line to the slave cylinder.
13   Bleed the hydraulic clutch system (Section 6).

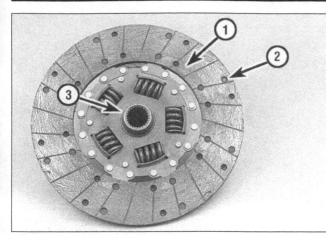

7.8a The clutch plate

1 **Lining** – this will wear down in use
2 **Rivets** – these secure the lining and will damage the flywheel or pressure plate if allowed to contact the surfaces
3 **Markings** – "Flywheel side" or something similar

6 Hydraulic clutch system - bleeding

Note: *Extreme cleanliness must be maintained throughout the entire bleeding operation. Use lint-free rags and make sure that dirt and grit do not enter the system, especially at the reservoir.*

1 Whenever any part of the hydraulic clutch system - master or slave cylinder or a hydraulic line - is disconnected for servicing, or any time that the level of fluid in the reservoir has been allowed to fall low enough for air to be drawn into the master cylinder, the air must be bled from the system.

2 When seals are worn, it is possible for air to enter the cylinders without any visible evidence of leaking fluid and cause a spongy pedal (the usual symptom of air bubbles in the system).

3 Fill the reservoir with brake fluid. **Caution:** *Do not fill the reservoir with old fluid which has been bled from a hydraulic system. It could be contaminated or aerated or it might have moisture in it.*

4 Make sure that the reservoir is not allowed to run dry during the following bleeding operation. Should the fluid be allowed to sink to a level that admits air into the system, you will have to start over.

5 Unscrew the bleeder screw at the slave cylinder enough to allow fluid to be pumped out (half a turn is normally sufficient). Attach a clear plastic hose to the bleeder screw and position the loose end in a small container.

6 Push the pedal down through its full stroke.

7 While the pedal is held all the way down, close the bleeder screw.

8 Allow the pedal to return quickly to its stop by removing your foot from the clutch pedal.

9 Repeat this procedure until the air is completely dispelled at the bleeder screw.

10 Close the bleeder screw immediately after the last downward stroke of the pedal when the air bubbles no longer appear in the plastic hose.

7 Clutch - removal, inspection and installation

Refer to illustrations 7.8a, 7.8b, 7.9a and 7.9b
Warning: *Dust produced by clutch wear and deposited on clutch components may contain asbestos, which is hazardous to your health. DO NOT blow it out with compressed air and DO NOT inhale it. DO NOT use gasoline or petroleum-based solvents to remove the dust. Brake system cleaner should be used to flush the dust into a drain pan. After the clutch components are wiped clean with* a rag, dispose of the contaminated rags and cleaner in a covered, marked container.

Note: *Because a special engine support beam, a telescoping transaxle stand, a hydraulic hoist and an engine hoist are needed to remove the transaxle, gaining access to the clutch by using the procedure outlined in Chapter 7 may be impossible for most home mechanics. The alternative, outlined in Chapter 2, involves removal of the engine, transaxle and cradle as an assembly, but doesn't require either the special engine support beam or the telescoping transaxle stand. The hydraulic hoist and engine hoist, however, must still be available. Once the engine/transaxle/cradle assembly is removed from the vehicle, you can disconnect the transaxle from the engine to gain access to the clutch.*

Removal

1 Disconnect and remove the transaxle assembly from the engine (Chapters 2 and 7). Remove the release bearing (Section 8).

2 Mark the relationship of the pressure plate assembly to the flywheel to assure that they are reassembled in the same position.

3 In order to avoid warping the cover, loosen the pressure plate retaining bolts evenly, one turn at a time, in a criss-cross pattern, until spring pressure is relieved.

4 Remove the pressure plate and clutch disc.

5 Handle the disc carefully, taking care not to touch the lining surface, and set it aside.

Inspection

6 Inspect the clutch release bearing for damage or wear (Section 8).

7 Clean the dust out of the clutch housing, using a vacuum cleaner or clean cloth. It's a good idea to wear a paper mask during cleaning. **Warning:** *Do not use compressed air to clean the housing or the clutch/pressure plate assembly. Asbestos dust can endanger your health if inhaled.*

8 Inspect the friction surfaces of the clutch disc **(see illustration)** and flywheel for signs of uneven contact, indicating improper mounting or damaged pressure plate springs **(see illustration)**. Check the surfaces for burned areas, grooves, cracks and other signs of wear. It may be necessary to remove a badly grooved flywheel and have it machined to restore the surface. Light glazing of the flywheel surface can be removed with fine sandpaper. Inspect the clutch lining for contamination by oil, grease or any other substance and replace the disc with a new one if any is present. Slide the disc onto the input shaft temporarily to make sure the fit is snug and the splines are not burred or worn.

Installation

9 Place the clutch disc in position on the flywheel, centering it with an alignment tool **(see illustrations)**. The disc spring offset and the stamped Flywheel Side letters must face the flywheel.

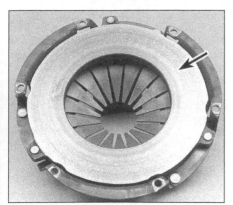

7.8b The machined face of the pressure plate must be inspected for score marks and other damage

7.9a A clutch alignment tool can be purchased at most auto parts stores and eliminates all guesswork when centering the clutch in the pressure plate

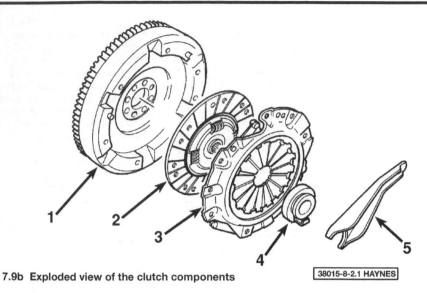

7.9b Exploded view of the clutch components

1   *Flywheel*
2   *Clutch disc*
3   *Clutch cover (pressure plate)*

4   *Clutch release bearing (throwout bearing)*
5   *Clutch release lever*

38015-8-2.1 HAYNES

8.5 Lightly lubricate the clutch fork pads (arrows) before installing the release bearing

10   With the disc held in place by the alignment tool, place the pressure plate assembly in position and align it with the marks made prior to removal.
11   Install the bolts and tighten them in a criss-cross pattern, 1/2-turn at a time, until they are tightened to the specified torque.
12   Install the clutch release bearing (Section 8).
13   Remove the alignment tool and install the transaxle (Chapter 7).

8   Clutch release bearing - removal and installation

Refer to illustration 8.5
Note: *Because a special engine support beam, a telescoping transaxle stand, a hydraulic hoist and an engine hoist are needed to remove the transaxle, gaining access to the clutch release bearing by using the procedure outlined in Chapter 7 may be impossible for most home mechanics. The alternative, outlined in Chapter 2, involves removal of the engine, transaxle and cradle as an assembly, but doesn't require either the special engine support beam or the telescoping transaxle stand. The hydraulic hoist and engine hoist, however, must still be available. Once the engine/transaxle/cradle assembly is removed from the vehicle, you can disconnect the transaxle from the engine and gain access to the clutch release bearing.*

Removal

1   Remove the transaxle (refer to Chapters 2 and 7).
2   Remove the clutch release bearing from the clutch fork shaft assembly.
3   Hold the center of the bearing and spin the outer portion. If the bearing doesn't turn smoothly or if it is noisy, replace it with a new

one.
4   Wipe the bearing with a clean rag and inspect it for damage, wear and cracks. Clean the release bearing. **Caution:** *Do not place the bearing in solvent or damage to the seals may result.*

Installation

5   Lubricate the clutch fork pads (where they contact the bearing) with white lithium base grease **(see illustration)**.
6   Pack the recesses in the bearing with the same grease.
7   Install the release bearing on the retainer so that both of the fork tangs fit into the outer diameter of the bearing groove.
8   Install the transaxle (Chapters 2 and 7). **Caution:** *The clutch lever must not be moved toward the flywheel until the transaxle is bolted to the engine or damage to the transaxle could occur.*

9   Flywheel - removal and installation

Note: *Because a special engine support beam, a telescoping transaxle stand, a hydraulic hoist and an engine hoist are needed to remove the transaxle, gaining access to the flywheel by using the procedure outlined in Chapter 7 may be impossible for most home mechanics. The alternative, outlined in Chapter 2, involves removal of the engine, transaxle and cradle as an assembly, but doesn't require either the special engine support beam or the telescoping transaxle stand. The hydraulic hoist and engine hoist, however, must still be available. Once the engine/transaxle/cradle assembly is removed from the vehicle, you can disconnect the transaxle from the engine and gain access to the flywheel.*

Removal

1   Remove the transaxle assembly from the engine (Chapters 2 and 7).
2   Remove the pressure plate and clutch disc (Section 7).
3   Remove the flywheel mounting bolts and detach the flywheel assembly.

Installation

4   Install the flywheel and bolts. Tighten the bolts to the specified torque following a criss-cross pattern.
5   Install the pressure plate and clutch disc (Section 7).
6   Attach the transaxle assembly to the engine (Chapters 2 and 7).

10   Driveaxles - general information

Power is transmitted from the transaxle to the rear wheels through driveaxles. The driveaxles are splined solid axles with constant velocity (CV) joints at each end. The inner CV joint is completely flexible and has the capability of in-and-out movement. On some models it consists of a spider bearing assembly and a tri-pot housing to allow angular movement. On other models it is similar to the outer joint (but can move in-and-out). The outer joint, which uses ball bearings running between an inner race and an outer cage, is also flexible, but cannot move in-and-out.

All driveaxles except the one used on the left side with an automatic transaxle incorporate a male spline and interlock with the transaxle gears through the use of barrel-type snap-rings. The left inboard shaft attachment on the automatic transaxle utilizes a female spline which installs over a stub shaft protruding from the transaxle.

The shaft end mating with the knuckle and hub assembly incorporates a helical spline to assure a tight, press-type fit. This is to assure a no end play condition between the hub bearing and the driveaxle assembly.

11.11 A large screwdriver will usually remove the inner CV joint assembly

12.3 Carefully tap around the circumference of the seal retainer to remove it from the housing

The boots should be inspected periodically (Chapter 1) for damage, leaking lubricant and cuts. Damaged CV joint boots must be replaced immediately or the joints can be damaged. Boot replacement involves removing the driveaxles (Section 11). The outer boots can be replaced with the driveaxles in place, using an aftermarket boot kit featuring split boots (Section 14). The most common symptom of worn or damaged CV joints, besides lubricant leaks, is a clicking noise in turns, a clunk when accelerating from a coasting condition or vibration at highway speeds.

11 Driveaxles - removal and installation

Refer to illustration 11.11

Removal

1   Loosen the wheel lug nuts and the hub nut.
2   Raise the rear of the vehicle and support it securely on jackstands.
3   Remove the rear wheel.
4   Disconnect the parking brake cable (Chapter 9).
5   Remove the brake caliper (Chapter 9) and hang it out of the way with a piece of wire.
6   Remove the driveaxle hub nut.
7   Disconnect the toe link rod from the knuckle (Chapter 10). **Note:** *The knuckle and the outer end of the driveaxle must be separated. Before they can be separated, however, the knuckle must be disconnected from the control arm.*
8   To separate the knuckle from the control arm, remove the pinch bolt from the clamp at the bottom of the knuckle and lift the knuckle off the balljoint stud (Chapter 10).
9   Using a puller, separate the driveaxle from the hub and bearing assembly.
10   Support the driveaxle with a piece of wire to prevent damage to the CV joints and boots.
11   Remove the driveaxle from the transaxle by carefully prying between the transaxle

case and the inner CV joint housing with a large screwdriver **(see illustration)**.
12   Refer to Section 13 for driveaxle overhaul procedures.

Installation

13   Insert the inner splined end of the driveaxle into the transaxle case. **Note:** *If the shaft won't fully seat, rotate it slowly while gently pushing on it. If more force is required, a groove in the tri-pot housing has been provided against which the tip of a large screwdriver can be inserted. Gently tap the screwdriver with a large hammer until the driveaxle is seated.*
14   Insert the outer splined end of the driveaxle through the knuckle and into the hub and bearing assembly. **Note:** *If you are replacing the driveaxle itself, use a new knuckle seal (Chapter 10).*
15   Lower the clamp on the underside of the knuckle onto the balljoint stud protruding from the lower control arm, install the pinch bolt and tighten it to the specified torque if you separated the knuckle from the lower control arm.
16   Attach the toe link rod to the knuckle assembly and tighten it to the specified torque.
17   Install the hub nut and tighten it to the

specified torque while immobilizing the hub flange with a large screwdriver.
18   Install the caliper and attach the parking brake cable (Chapter 9).
19   Install the wheel. Tighten the wheel lug nuts finger tight.
20   Lower the vehicle and tighten the wheel lug nuts to the specified torque.

12 Driveaxle boots - replacement (driveaxle removed)

Refer to illustrations 12.3, 12.4, 12.5, 12.16a and 12.16b
1   Remove the driveaxle (Section 11).
2   Place the driveaxle in a vise.

Double-offset CV joint (inner and outer)

3   Tap lightly around the outer circumference of the seal retainer with a hammer and drift to remove it **(see illustration)**. Take care not to deform the retainer.
4   Cut off the band retaining the boot to the shaft **(see illustration)**.
5   Remove the snap-ring and slide the joint assembly off **(see illustration)**.
6   Slide the old boot off the driveaxle.
7   Clean the old grease from the joint.
8   Repack the CV joint with half the grease supplied with the new boot and put the remaining half in the boot.
9   Slide the retainer and boot into position on the driveaxle.
10   Install the joint and snap-ring.
11   Seat the inner end of the boot in the seal groove and install the retaining clamp.
12   Install the seal retainer securely in place by tapping evenly around the outer circumference with a hammer and punch.

Inner CV joint (Tri-pot type)

Note: *See Section 13 for illustrations pertaining to the disassembly and reassembly procedures (particularly illustration 13.20a).*
13   Cut off the boot clamps and detach the boot from the housing. Some boots are held in place with a retainer which can be tapped with a brass drift and hammer to detach it

12.4 Wire cutters can be used to cut the inner boot clamp

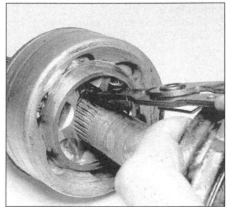

12.5 Use snap-ring pliers to remove the inner snap-ring

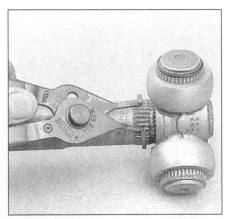

12.16a Spread the ends of the snap-ring apart and lift it out of the groove in the axleshaft, then slide it toward the center of the shaft so the spider can be moved in

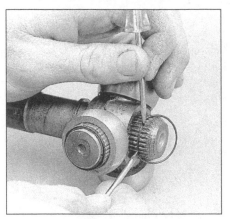

12.16b Once the spider is clear of the outer retaining ring, remove it with a pair of small screwdrivers

13.4 Cut the boot clamp with diagonal cutters and tap the boot retainer loose with a brass drift and a small hammer

from the housing. Mark the housing, axle and spider/bearing assembly to ensure that all the parts are mated correctly during reassembly.

14 Remove the joint housing from the axle by pulling straight out on it (support the three balls attached to the spider as the axle is removed from the housing).

15 If the balls are held in place with retainers and snap-rings, they cannot fall off the spider. If they aren't, wrap tape around the spider/bearing assembly to retain the bearings as the spider assembly is removed from the shaft.

16 Remove the inner spider retaining ring, then slide the spider back to expose the outer ring. After the outer ring is removed, the spider/bearing assembly can be removed from the axle **(see illustrations)**.

17 Slide the boot off the axle.

18 Clean all old grease from the housing and spider assembly.

19 Pack the housing with half of the grease furnished with the new boot and place the remainder in the boot.

20 Slide the boot onto the axle.

21 Align the marks and install the spider spider/bearing assembly with the recess in the counterbore facing away from the end of the driveaxle (see Section 13). Make sure the retaining rings are seated in the grooves. Remove the tape (if used).

22 Install the housing (make sure the marks are aligned).

23 Seat the boot in the housing and axle grooves and install the clamps.

13 Driveaxles - overhaul

Note: *The following procedure applies to both the outer and inner CV joints on all 1984 and most 1985 Fieros, which utilize a double-offset design where the inner and outer CV joints are essentially identical. In 1985, some driveaxles were equipped with tri-pot design inner CV joints. On 1986 vehicles, all driveaxles are equipped with tri-pot inner CV joints.*

Double-offset CV joint

Refer to illustrations 13.4, 13.5, 13.6, 13.7, 13.8, 13.12a, 13.12b, 13.12c, 13.13, 13.15a and 13.15b

Disassembly

1 Remove the driveaxle (Section 11). **Note:** *Refer to the accompanying exploded view of the driveaxle assembly for help in disassembling and reassembling the inner and outer CV joints.*

2 If the CV joint is equipped with a deflector ring, remove it from the groove in the end of the housing. **Note:** *Some deflector rings are rubber and some are steel. A rubber deflector ring can be stretched and pushed out of the groove. A steel ring must be tapped out of the groove with a brass drift.*

3 Cut, remove and discard the boot retaining clamps. These clamps are not reusable. Slide the boot back until it clears the CV joint outer race housing.

4 Lightly tap the boot retainer all the way around until it comes loose **(see illustration)**.

5 Spread the large snap-ring in the CV joint housing and pull out the axleshaft **(see**

13.5 Remove the snap-ring, slide the joint off the shaft and remove the old boot

illustration)**. Mark the housing, cage and inner race to ensure that all parts will be mated correctly during reassembly.

6 Gently tap on the bearing cage with a brass drift until it is tilted enough to remove the first ball bearing. Remove the other balls in a similar manner **(see illustration)**.

7 Pivot the cage and inner race at 90 degrees to the centerline of the outer race with the cage windows aligned with the lands of the outer race, then lift out the cage and the inner race **(see illustration)**.

8 Rotate the inner race until it is at a 90 degree angle to the cage **(see illustration)**, then lift it out of the large end of the cage.

9 Clean all the parts thoroughly in solvent and blow them dry with compressed air or allow them to air dry.

10 Inspect the parts for signs of scoring, scuffing or galling. If there is evidence of damage, replace the CV joint.

Reassembly

11 Put a light coat of CV joint grease on the ball grooves of the inner race and housing. Install the inner race into the cage. The inner race lobes should be centered in the win-

13.6 Gently tap on the cage with a brass drift until it is tilted enough to remove the first ball bearing - repeat this procedure until all the balls are removed

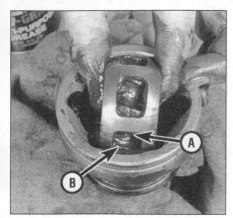

13.7 Pivot the cage and inner race at a 90 degree angle to the centerline of the outer race with the cage windows (A) aligned with the lands of the outer race (B), then lift out the cage and inner race

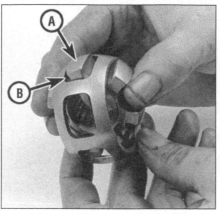

13.8 Align the inner race lands (A) with the cage windows (B) and rotate the inner race out of the cage

13.12a With the race and cage tilted at 90-degrees, lower the assembly into the housing

dows of the cage. Also check to make sure that the retaining ring on the inner race faces towards the small end of the cage before installing any balls.

12   Assemble the parts as shown **(see illustrations)**. **Note:** *Be sure that the retaining ring side of the inner race faces the axleshaft.*

13   Pack the joint with CV joint grease **(see illustration)**.

14   Slide the boot onto the axleshaft.

15   Push the outer CV joint housing onto the axleshaft until the retaining ring is seated in its groove, then install the boot retainer with an arbor press **(see illustration)** or a brass punch and hammer **(see illustration)**.

16   Slide the boot into place and install the boot clamps. **Note:** *The OEM boot kit includes a pair of clamps that require a special tool. If you do not have access to this special tool, use regular screw type hose clamps and tighten them securely.*

17   If the CV joint is equipped with a deflector ring, install it onto the housing. A rubber deflector ring can be stretched and pushed into the groove by hand. A steel ring should

13.12b Rotate the assembly by gently tapping with a hammer and brass punch, then . . .

be installed with a short section of 2 1/2-inch pipe, a small square piece of sheet steel and a 20 mm nut, or a similar setup.

18   If you are overhauling an axle with an inner tri-pot CV joint, refer to Steps 19 through 31.

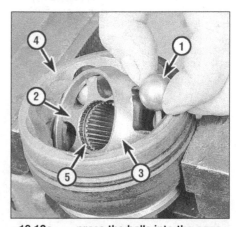

13.12c . . . press the balls into the cage windows, repeating until all of the balls are installed

1   Ball
2   Inner race
3   Cage
4   Housing
5   Retainer ring groove

13.13 Apply grease through the splined hole, then insert a wood dowel through the splined hole and push down – the dowel will force the grease into the joint

13.15a Position the CV joint assembly on the driveaxle, aligning the splines, then use a soft-face hammer to drive the joint onto the driveaxle until the snap-ring is seated in the groove

13.15b On models with a retaining ring, carefully tap around the circumference of the retaining ring to install it on the housing

Tri-pot type CV joint

Refer to illustrations 13.20, 13.21a, 13.21b, 13,22a, 13.22b, 13.23a, 13.23b, 13.23c, 13.23d, 13.23e, 13.25a, 13.25b, 13.25c, 13.26a, 13.26b, 13.28, 13.30a and 13.30b

Disassembly

19   If you have not already done so, remove the driveaxle (Section 12).

20   Cut the boot retaining clamps and tap the boot retainer loose with a small brass drift **(see illustration)**.

21   Refer to Section 12, Paragraphs 14 through 17, and remove the spider/bearing assembly and boot from the driveaxle **(see illustrations)**.

22   If the tri-pot joint balls are held in place by retainers and snap-rings, they must be removed to detach the balls and gain access to the needle bearings **(see illustrations)**.

23   Disassemble the driveshaft components **(see illustrations)**. Clean all the parts thoroughly with cleaning solvent. When you disassemble the needles and the tri-pot joint balls, be very careful not to lose any needles.

13.20 Cut off the boot seal retaining clamps, using wire cutters or a chisel and hammer

It's a good idea to keep the needles in a wire mesh tray, like a tea strainer, to prevent loss.

24   Inspect the parts for signs of galling, scoring or scuffing. If any damage is evident, replace the tri-pot CV joint assembly.

13.21a Slide the housing off the spider assembly

Reassembly

25   Carefully assemble the spider, needle rollers and tri-pot joint balls, using CV joint grease **(see illustration)**. If snap-rings are used, be sure they are seated in the grooves

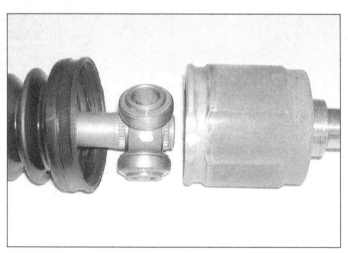

13.21b Be sure to mark the driveaxle and the housing with paint to insure correct alignment for reassembly

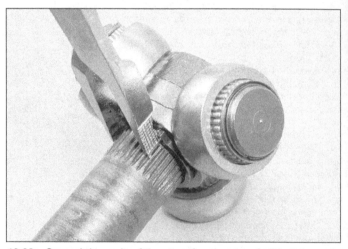

13.22a Spread the ends of the stop ring apart and slide it towards the center of the shaft

13.22b Slide the spider assembly back to expose the retaining ring and pry off the ring

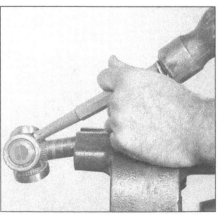

13.23a Carefully tap the spider off the axleshaft with a brass punch

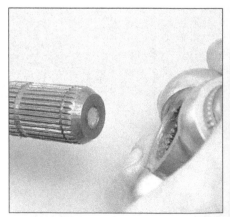

13.23b When you slide the spider off the driveaxle, hold the bearings in place with your hand; even better, use tape or a cloth wrapped around the spider bearing assembly to retain them

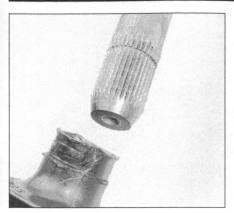

13.23c Slide the stop ring and the boot off the axleshaft

13.23d Clean all of the old grease out of the housing and spider assembly, then remove each bearing, one at time

13.23e Carefully disassemble each section of the spider assembly, clean the needle bearings with solvent and inspect the rollers, spider cross, bearings and housing for scoring, pitting and other signs of abnormal wear

(see illustrations).

26   Slide the spider onto the splined tip of the axleshaft far enough to install the outer retaining ring into the groove on the end of the axleshaft **(see illustrations).**

27   Slide the spider toward the end of the axleshaft until it covers the retaining ring. The

groove for the inner ring should now be exposed. Slide the inner ring into position in this groove.

28   Pack the tri-pot housing with approximately half the CV joint grease supplied with

the boot kit **(see illustration).**

29   Slide the axleshaft and spider assembly into the tri-pot housing.

13.25a Apply a coat of CV joint grease to the inner bearing surfaces to hold the needle bearings in place and slide the bearing over them

13.25b Wrap the axleshaft splines with tape to avoid damaging the boot, then slide the small clamp and boot onto the axleshaft

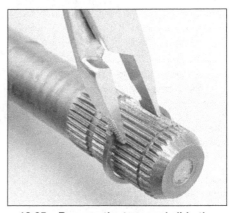

13.25c Remove the tape and slide the stop ring onto the axleshaft, past the groove in which it seats

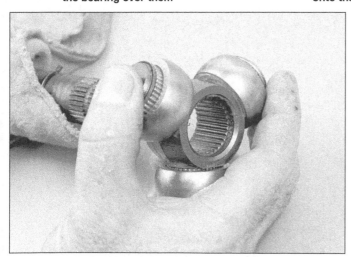

13.26a Install the spider assembly with the recess in the counterbore facing the end of the driveaxle

13.26b Use a screwdriver to install the retaining ring, then slide the spider (or ball-and cage) assembly against it and install the stop ring in its groove

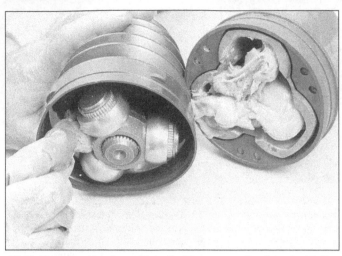

13.28 Pack the housing with half of the grease furnished with the new boot and place the remainder in the boot

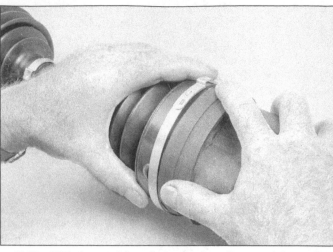

13.30a With the retaining clamps in place (but not tightened), install the joint housing

30   Slide the boot into place and install the boot clamps. **Note:** *The OEM boot kit includes a pair of clamps that require a special tool. If you do not have access to this special tool, use regular screw type hose clamps and tighten them securely* **(see illustrations).**

31   Install the driveaxle (Section 12).

---

14   Outer driveaxle boot - replacement (driveaxle installed)

---

1   The outer boot can be replaced with the axle installed, using an aftermarket boot replacement kit. These boots are split so they can be installed with the driveaxle in place.

2   Raise the vehicle, support it securely on jackstands and remove the rear wheel. Raise the lower control arm with a jack so that the driveaxle is level.

3   Remove the brake caliper and rotor and wire the caliper out of the way (Chapter 9).

4   Using a chisel and hammer, cut the boot clamp off the joint housing.

5   Remove the remaining boot clamp.

6   Cut the old boot and remove it.

7   Inspect the CV joint to determine if the damaged boot has allowed the grease to become contaminated with dirt or water. If it has, wipe the old grease off and apply new grease from the replacement boot kit, working it in with your fingers. **Note:** *The following steps describe a typical installation. Follow the instructions included with the replacement boot kit.*

8   Place the new boot in position over the axle. Typically these kits use a special fluid, which when applied to the sealing grooves, "welds" the boot into one piece. The retaining straps included are installed to hold the boot in place until the glue has set.

9   Install the boot in the sealing grooves on the driveaxle shaft and joint and securely install the clamps, following the included instructions.

13.30b Seat the boot in the housing and axle seal grooves - a small screwdriver can make the job easier (make sure the boot isn't dimpled, stretched or out of shape)

Chapter 9   Brakes

Contents

Specifications

Pedal travel..	2.52 in
Master cylinder piston diameter..............................	1.00 in
Front rotors	
Minimum refinish thickness	
1984 through 1987................................	0.44 in
1988	0.70 in
Discard thickness*	
1984 through 1987................................	0.39 in
1988	0.68 in
Runout	
1984 through 1987................................	0.004 in maximum
1988	0.003 in maximum
Rear rotors	
Minimum refinish thickness	
1984 and 1985..................................	0.44 in
1986 and 1987..................................	0.50 in
1988	0.70 in
Discard thickness*	
1984 and 1985..................................	0.39 in
1986 and 1987..................................	0.45 in
1988	0.68 in
Runout	
1984 through 1987................................	0.004 in maximum
1988	0.003 in maximum

* Refer to the dimension stamped into the disc (it supersedes information printed here).

Torque specifications

	Ft-lbs (unless otherwise indicated)
Brake pedal-to-power brake booster nut	19
Brake pedal bracket-to-dash nut	21
Parking brake control handle assembly bolt	25
Master cylinder-to-power brake booster nuts	28
Brake lines-to-master cylinder	
1984 and 1985	19
1986 through 1988	15
Brake line-to-combination valve	19
Brake line-to-brake hose	12
Power brake booster-to-front mounting nuts	
1984 and 1985	15
1986 through 1988	27
Caliper mounting bolts	
1984 through 1987	40
1988	74
Caliper bridge bolts (1988 models)	74
Parking brake lever-to-rear caliper nut	35
Caliper mounting knuckle assembly bolt	35
Caliper bleeder valve	
1984 through 1987	132 in-lbs maximum
1988	116 in-lbs maximum
Front brake hose-to-caliper bolt	
1984 and 1985	39
1986 through 1988	30
Rear brake hose-to-caliper bolt	
1984 and 1985	39
1986 through 1988	15
Front brake hose bracket-to-wheel well	34 in-lbs
Front brake hose-to-front brake line fitting	17
Wheel lug nuts	See Chapter 1

1   General information

The Fiero is equipped with hydraulically operated disc brakes at all four wheels. When the brake pedal is depressed, two pistons - one for the front and one for the rear brakes - in the master cylinder move forward, increasing the fluid pressure inside the piston bore of each caliper. This pressure is exerted on the caliper piston. The pressure applied to each piston is transmitted to a brake pad, forcing it against the inner rotor surface. The caliper body itself is free to slide in and out on its mounting bolts, so the pressure buildup inside the caliper bore also forces the caliper inward (toward the vehicle). Since the caliper is one piece, its movement toward the vehicle brings the outer section of the caliper body closer to the outer surface of the rotor, applying force against the pad material between it and the outer rotor surface.

The master cylinder has an aluminum body and a translucent nylon reservoir with minimum fill indicators (two windows in the side of the reservoir).

The combination valve releases outlet pressure to the rear brakes after a predetermined rear input pressure has been reached. This feature prevents early rear wheel lockup during heavy braking. The combination valve is designed to have a "bypass" feature which assures full system pressure to the rear brakes in the event of a front brake system failure. Similarly, full front pressure is retained in the event of a rear brake pressure failure.

The hydraulic system consists of separate front and rear circuits. The master cylinder has twin reservoirs - one for each circuit. In the event of a leak or failure in one hydraulic circuit, the other remains operative. A warning light is mounted in the dash which indicates circuit failure. Air in the system or other pressure differential conditions in the brake system is signaled by a switch in the combination valve.

The valve and switch are designed so that the switch will remain in the warning position once a failure has occurred. The only means by which the light can be turned off is

repair of the failure and application of a pedal force equal to approximately 450 psi line pressure.

The power brake booster, located between the master cylinder and the rear bulkhead of the front compartment, uses engine manifold vacuum and atmospheric pressure to provide assistance to the hydraulic disc brake system. It may have a single or dual function vacuum switch to activate the brake warning light in case of low booster vacuum or vacuum pump malfunction.

The parking brake mechanically operates the rear brake calipers through a cable system. It is activated by a pull-handle between the driver's seat and the left door. The parking brake lamp switch is bolted to the hand brake mounting bracket and is actuated by the parking brake hand lever.

After completing any service procedure involving disassembly of a brake system component, always test drive the vehicle to check for proper braking performance before resuming normal driving. Test the brakes

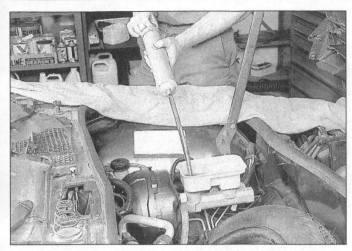

2.1 Always siphon the fluid out of the master cylinder reservoir before depressing the caliper piston back into its bore

2.5 Use a C-clamp to push the piston back into its bore - tighten the clamp until the piston bottoms - as the piston is depressed to the bottom of the caliper bore, the fluid in the master cylinder will be pushed back into the reservoir (make sure it doesn't overflow)

while driving on a clean, dry, flat surface. Conditions other than these can lead to inaccurate test results. Tires, vehicle load and front end alignment are also factors which affect braking performance. Test the brakes at various speeds with both light and heavy pedal pressure. The vehicle should stop evenly without pulling to one side or the other. Avoid locking the brakes because this slides the tires and diminishes braking efficiency and control.

2   Front disc brake caliper - removal and installation

Refer to illustrations 2.1, 2.5, 2.7, 2.8, 2.9 and 2.10

Warning: *Whenever you are working on the brake system, be aware that asbestos dust may be present. Asbestos is extremely harmful to your health, so avoid inhaling it. Wash the brake and surrounding area with brake system cleaner before beginning work.*
Warning: *All brake related fasteners are critical*

components that can affect the performance of the most important system on your vehicle. Always replace brake parts with identical components of the same quality. Do not substitute with replacement parts of lesser or unknown quality. Observe specified torque values to assure proper retention of brake parts.

Removal

1   Remove the cover from the brake fluid reservoir, siphon about 2/3 of the fluid from the reservoir **(see illustration)** and discard it.
2   Loosen but do not remove the front wheel lug nuts.
3   Raise the front of the vehicle and support it securely on jackstands.
4   Remove the front wheel. **Note:** *Work on only one side at a time so that you can use the other brake assembly as a guide, if necessary.*
5   Push the piston back into its bore with a C-clamp **(see illustration)**. Tighten the C-clamp until the piston bottoms in its bore. **Note:** *If a C-clamp is not available, you can also depress the piston with a screwdriver after the caliper is removed* (see Step 10). Once the

piston is bottomed, remove the C-clamp.
6   As the piston is depressed to the bottom of the caliper bore, the fluid in the master cylinder will be pushed back up into the reservoir. Make sure that it does not overflow. If necessary, siphon off more of the fluid.
7   If you are going to overhaul or replace the caliper, remove the bolt holding the brake caliper inlet hose fitting **(see illustration)**. If you are simply replacing the pads, do not disconnect this line.
8   Remove the two caliper mounting bolts, the sleeves and the bushings from the caliper **(see illustration)**. Keep the mounting bolts, but discard the sleeves and the bushings. It is recommended that you replace these items with new ones during reassembly. **Note:** *On 1988 models, the caliper mounting bolts require a 55 Torx socket for removal.*
9   If you are replacing the pads, hang the caliper from the upper A-arm with a piece of wire **(see illustration)** to protect the brake hose from damage during pad replacement. **Caution:** *Do not allow brake components to*

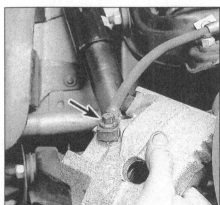

2.7 Don't disconnect the brake hose inlet fitting unless you intend to rebuild or replace the caliper - if you loosen it, you will have to bleed the brakes

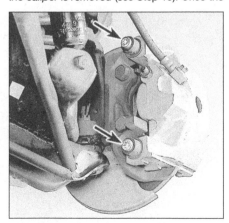

2.8 Don't confuse these Torx bolts with Allen bolts or you will strip the heads out

2.9 Always hang the caliper from the A-arm or a suspension component - letting it dangle by the brake hose can damage the hose

2.10 If you don't have a C-clamp to depress the caliper piston, this method will work - but we don't recommend it

3.2 The easiest way to remove the outboard pads is with a pair of water pump pliers

hang from the flexible hoses as damage to the hoses may occur. Some brake hoses have protective rings or covers to prevent direct contact with other chassis parts. Excessive tension could cause the rings to move out of their proper locations or cause structural damage.

10   If you did not have a C-clamp to depress the piston in Step 5, use a pair of screwdrivers or a pry bar to do so at this time.

Installation

Caution: *Inspect the mounting bolts for corrosion. If corrosion is evident, use new bolts when installing the caliper. Do not attempt to polish away corrosion.*

11   Lubricate the new sleeves and bushings with silicone grease.

12   Install the caliper and mounting bolts. Make sure that the clearance between the caliper housing and the caliper bracket stops is within between 0.005 to 0.012 inches to allow free movement of the caliper housing when the brakes are applied.

13   Tighten the caliper mounting bolts to the specified torque.

14   If the brake inlet hose has been discon-

nected, install the brake inlet hose mounting bolt through the banjo fitting and tighten it to the specified torque. **Caution:** *Use new copper washers when installing the bolt.*

15   Bleed the brake system (see Section 13).

16   Install the wheels, then lower the vehicle and tighten the lug nuts to the specified torque.

3   Front disc brake pads - replacement

Warning: *Whenever you are working on the brake system, be aware that asbestos dust may be present. Asbestos is extremely harmful to your health, so avoid inhaling it. Wash the brake and surrounding area with brake system cleaner before beginning work.*
Note: *Disc brake pads should be replaced on both wheels at the same time.*

1984 through 1987 models

Refer to illustrations 3.2, 3.3, 3.7 and 3.8
Removal
1   Remove the front calipers (Section 2).

2   Extract the outer pad with a pair of water pump pliers **(see illustration)**.

3   Remove the inner pad by popping the retainer clip loose from the piston **(see illustration)**.

4   Inspect the caliper bolts for scoring and the contact surfaces of the caliper mounting bolt bosses for corrosion.

5   Carefully peel back the edge of the piston dust seal and inspect the exposed portion of the piston for corrosion and leaking fluid. If there is evidence of serious contamination or leakage of hydraulic brake fluid, the caliper should be rebuilt (see Section 4).

Installation
6   Lubricate the new bushings and sleeves with silicone grease, then install them in their grooves in the caliper mounting bolt bosses.

7   Install the retainer clip on the inboard pad **(see illustration)**.

8   Install the inboard pad with the wear sensor at the leading edge of the pad **(see illustration)**.

9   Install the outboard pad by popping the spring clip into the detents.

10   Install the caliper (see Section 2).

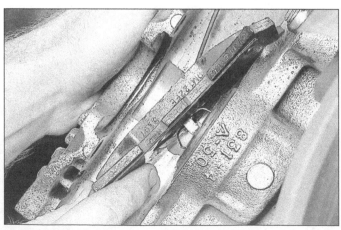

3.3 The inboard pads have a retaining clip that can be popped out of the face of the caliper piston

3.7 The retaining clip on the backside of the inboard pads (arrow) comes in several shapes, but it doesn't always come with the new pads, so remove it from the old ones and install it on the new ones

3.8 Install the inboard pad by pressing the spring clip into the piston - make sure that the wear sensor is at the leading edge

1988 models

Removal

Note: *Two special tools are required for this procedure - a slide hammer and a spring pin remover - for removing the front and rear disc calipers. However, if these tools are not available, try using a hammer and brass drift punch to knock out the spring pins.*

11   Drain 2/3 of the brake fluid from the master cylinder assembly.

12   Loosen the wheel lug nuts, raise the vehicle and place it securely on jackstands.

13   Mark the relationship of the wheel to the axle flange.

14   Remove the wheel.

15   Install two inverted lug nuts to retain the disc rotor.

16   Position a pair of large adjustable pliers over the caliper housing and flange to the inboard brake pad, then squeeze the pliers to compress the piston back into the caliper bore. You must bottom the piston into the caliper bore to provide enough clearance for the new brake pads.

17   Remove the threaded tip from the rod on the slide hammer, insert the rod completely through the spring pin and install the threaded tip as far as it will go. Slam the weight outward, against the tool handle, to pull out the pin. **Note:** *If you don't have the special tools, remove the pins with a brass punch and a hammer. Make sure the outside diameter of the punch is slightly smaller than the outside diameter of the pins. If you use a punch that's too small in diameter, it may jam into the end of the hollow pins and flare the ends (making them impossible to remove from the caliper).*

18   Remove the springs from the inboard and outboard pad flanges. **Warning:** *Wear safety goggles during his procedure - the springs can fly off forcefully when pried loose.*

19   Lift the outer pad through the caliper opening and remove it.

20   Lift the inner pad through the caliper opening and remove it. If necessary, push on the bridge and move the caliper housing

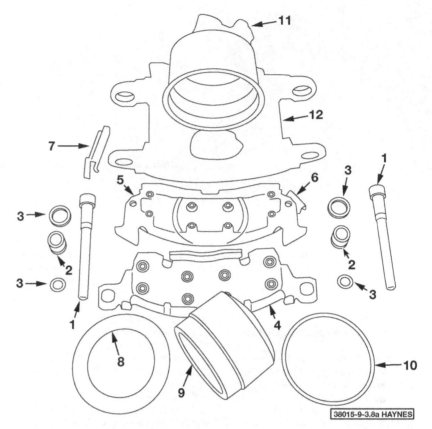

4.3a Exploded view of a typical caliper assembly on 1984 through 1987 models

1	Mounting bolt	5	Inner pad	9	Piston
2	Sleeve	6	Wear sensor	10	Piston seal
3	Bushing	7	Pad retainer	11	Bleeder screw
4	Outer pad	8	Dust boot	12	Caliper housing

toward the center of the vehicle to provide enough clearance for removal.

Installation

21   Make sure the piston is bottomed in the caliper housing bore. If it's not, bottom it with a large pair of adjustable pliers and a block of wood (to protect the piston).

22   Install the inner pad with the wear sensor at the bottom of the pad.

23   Install the outer pad.

24   Using a small mallet and a soft brass punch (not a steel one), tap in one spring pin until it is through both pads an slightly into the inner section of the caliper housing.

25   Tap in the other pin the same way. Stop when the pin is just through the *outer* section of the caliper housing.

26   Install the springs one at a time. Hook the end of the spring under the pin you installed with the midsection of the springs over the pad flange.

27   Press down on the other end of the spring with a screwdriver as you slide the remaining pin in.

28   Complete pin installation by tapping in both pins until the end of each pin just protrudes from the inner face of the caliper housing.

29   Make sure the springs are centered on the pad flanges with each spring end projecting under the pins an equal amount.

30   Remove the wheel lug nuts securing the rotor to the hub.

31   Using the alignment marks you made during disassembly, install the wheels. Tighten the lug nuts securely.

32   Lower the vehicle and tighten the wheel nuts to the specified torque.

33   Fill the master cylinder to the proper level with clean brake fluid.

34   Depress the brake pedal three or four times to seat the linings.

4   Front disc brake caliper - overhaul

Refer to illustrations 4.3a, 4.3b, 4.5, 4.7, 4.11, 4.15 and 4.16

1   Remove the caliper from the vehicle (Section 2).

2   Remove the pads from the caliper (Section 3).

3   Place a block of wood next to the piston **(see illustrations)** and have an assistant pump the brake pedal until the piston is nearly out of its bore in the caliper (this sim-

4.3b If you don't have compressed air, have an assistant pump the brake pedal until the piston is pushed out of the caliper bore - be sure to protect the piston with a small block of wood

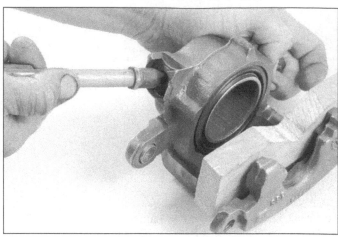

4.5 If you're removing the piston from the caliper with compressed air, disconnect and remove the caliper from the vehicle and lay it securely on a workbench, then use an air nozzle to force the piston from its bore - be sure to cushion the piston with a block of wood to prevent damage

plifies piston removal).

4   Disconnect the inlet hose fitting from the caliper.

5   Remove the caliper piston by blowing compressed air into the caliper inlet hole **(see illustration)**. **Warning:** *Do not place fingers in front of the piston in an attempt to catch or protect it when applying compressed air.*

6   Inspect the piston for scoring, nicks, corrosion and worn or damaged chrome plating. If any damage is evident, replace the piston with a new one.

7   Remove the boot by prying it out of the caliper with a screwdriver **(see illustration)**. Be careful not to scratch the caliper housing bore.

8   Carefully pry the piston seal from the groove inside the caliper bore. A wood or plastic tool is ideal for this task because there is little chance of scratching the piston bore.

9   Inspect the caliper bore for scoring, nicks, corrosion and wear. If only slight corrosion is noted, clean out the bore with crocus cloth. If the bore will not clean out with crocus cloth, replace the caliper housing.

10   Remove the bleeder valve and examine it for corrosion. If it is corroded, replace it.

11   Inspect the mounting bolts for corrosion. If they are even slightly corroded, replace them. Remove the four rubber bushings (two per hole) from the caliper mounting bolt holes **(see illustration)**.

12   Wash all brake parts in clean, denatured alcohol and blow dry with compressed air. Be sure to blow out all passages in the caliper housing, including the bleeder valve.

13   Install the bleeder valve in the caliper housing and tighten it to the specified torque.

14   Lubricate the new seal and the caliper housing bore with clean brake fluid. Install the new seal in the groove inside the caliper housing bore. Make sure that the seal is properly seated. It must not be twisted.

15   Install the boot on the end of the piston and install the piston in the bore. Push the piston all the way to the bottom of the bore. **Note:** *If the piston is difficult to depress into the bore, it can be levered down with a couple of screwdrivers* **(see illustration)**.

16   Press the boot into the counterbore in

4.7 Pry the dust boot from its counterbore with a screwdriver - do not scratch the piston bore

the caliper housing by carefully pushing on the outer edge with the tip of a screwdriver **(see illustration)**. **Caution:** *Do not apply force to the boot itself or it may be damaged.*

17   Install the brake pads (Section 3).

18   Install the caliper (Section 2).

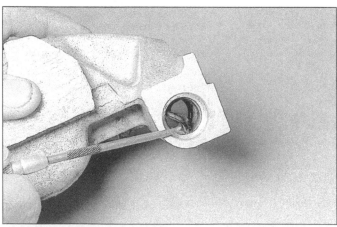

4.11 The bushings can be pried out of the caliper mounting bolt bosses - be careful not to scratch the inner walls of the hole

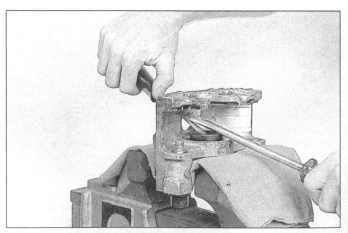

4.15 If you don't have a C-clamp, a couple of screwdrivers can be used, with caution, to depress the piston into the caliper bore

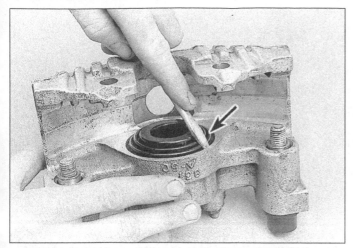

4.16 If you don't have a seal driver or a large socket to press the new dust boot into its counterbore, press it into place by hand and tamp it down carefully with a screwdriver - don't press on the new boot anywhere except along its outer hard edge

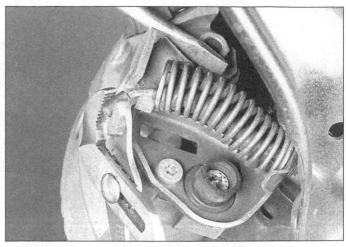

5.5 Hold the parking brake lever with a screwdriver and disconnect the parking brake cable from the lever with a pair of water pump pliers

5   Rear disc brake caliper - removal, inspection and installation

Refer to illustrations 5.5, 5.6, 5.7, 5.8 and 5.12

Removal

1   **Warning:** *When you are working on the brake system, be aware that asbestos dust is present. Asbestos must not be inhaled because it is harmful to your health.*

2   Remove the cover from the brake fluid reservoir, siphon off two-thirds of the fluid into a container and discard it.

3   Loosen the wheel lug nuts, apply the parking brake, raise the rear of the vehicle and place it securely on jackstands, then release the parking brake.

4   Remove the wheel, then reinstall one wheel lug, flat side toward the rotor, to hold the rotor in place. **Note:** *Work on one brake assembly at a time, using the assembled brake for reference, if necessary.*

5   Disconnect the parking brake cable from the lever on the caliper **(see illustration).**

6   Once the cable is freed from the lever, depress the spring by levering it with a screwdriver **(see illustration)** and slide it off the parking brake cable.

7   Holding the parking brake lever with a screwdriver, remove the nut, the lever, the lever seal and the anti-friction washer from the caliper actuator screw **(see illustration).** Note the relationship of these parts to one another before taking them apart. **Note:** *While caliper removal does not require the removal of these parts, it's a good idea to inspect the lever seal and anti-friction washer even if you are simply replacing the pads. If either the seal or the washer is damaged or worn, replace it.*

8   Position a C-clamp over the inboard surface of the caliper housing and the out-board surface of the mounting bracket **(see illustration).** Make sure the C-clamp does not contact the actuator screw. If you do not have a C-clamp, use a large water pump pli-

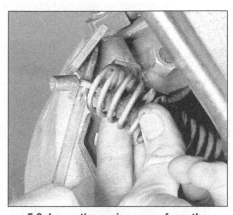

5.6 Lever the spring away from the parking brake lever and slide the spring off the cable

ers. **Note:** *1988 models require a special tool to rotate the piston to the bottom.*

9   Tighten the C-clamp until the piston bottoms in the cylinder bore (to provide enough clearance between the linings and

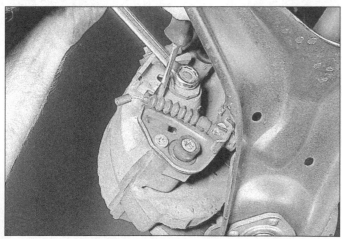

5.7 Hold the parking brake lever with a screwdriver and break the lever nut loose

5.8 When depressing the piston by squeezing the caliper into the disc with a C-clamp, do not allow the C-clamp to contact the actuator screw

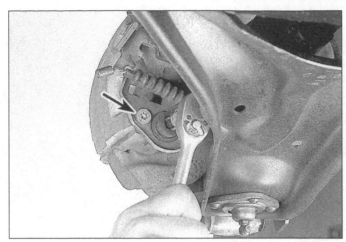

5.12 If you are replacing the pads, the caliper can be removed from its mounting bracket by removing a pair of Torx #50 mounting bolts - if you are going to rebuild or replace the caliper, you should first disconnect the smaller Torx bolt (arrow) and remove the parking brake lever cable bracket

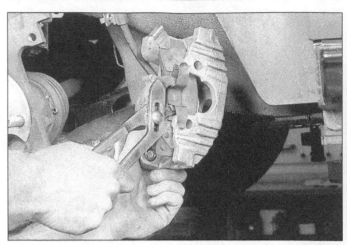

6.2 Grasp the retaining clip on the outboard pad with a pair of water pump pliers and remove the pad

the rotor), then remove the clamp.

10   If you are simply inspecting or replacing the pads or the rotor, go to Step 14. **Note:** *Do not disconnect the hydraulic hose or you will have to bleed the hydraulic system (Section 13).*

11   If you are replacing or overhauling the caliper, remove the brake hose inlet fitting bolt, then disconnect the brake hose from the caliper. To prevent fluid loss and contamination, plug the openings in the caliper and brake hose.

12   Remove the caliper mounting bolts. **Note:** *If you are simply replacing the pads or the rotor, it is not necessary to remove the parking brake cable bracket because the caliper can be lifted free with sufficient room to remove and install the pads or replace the rotor even with the bracket installed. If you are rebuilding or replacing the caliper, however, the cable bracket mounting bolt (the small Torx bolt adjacent to the lower caliper mounting bolt) must be removed and the bracket detached before you remove the caliper mounting bolts* **(see illustration)**.

13   Remove the caliper from the vehicle.

Inspection

14   Inspect the lever seal and anti-friction washer for wear. Replace any damaged parts.

15   Inspect the mounting bolts and sleeves for corrosion. If any damage is discovered, use new bushings, bolts and/or sleeves when the caliper is installed. **Caution:** *Do not attempt to polish away corrosion.*

Installation

16   Liberally fill both cavities in the mounting bolt bore between the bushings with silicone grease.

17   Install the sleeves, bolt boots and bushings in the caliper mounting bolt bores.

18   Install the caliper over the rotor and onto the mounting bracket.

19   Install the caliper mounting bolts and

tighten them to the specified torque.

20   If the brake hose was disconnected, reconnect and tighten it to the specified torque. **Caution:** *Use two new copper crush washers. Do not re-use the old washers as they will produce an incorrect torque reading.*

21   If any contamination in the area of the lever seal is evident, clean it thoroughly. Use a new lever seal and anti-friction washer. Lubricate both with silicone brake lube before installation. Also use a new antifriction washer.

22   Install the parking brake lever on the actuator screw hex with the lever pointing down, then rotate the lever toward the front of the vehicle and hold it while installing the retaining nut. Tighten the nut to the specified torque, then rotate the lever back against the stop on the caliper. **Caution:** *Make sure that the lever stays properly installed on the actuator screw hex as the nut is tightened.*

23   Install the spring and the parking brake

6.3 The inboard pad is removed by prying it free from the four clips (A), two on the top and two on the bottom, of the steel ring that is locked onto the face of the piston (when removing the pad, note the relationship between the two pins (B) on the backing plate and the D-shaped notches in the face of the piston)

cable.

24   After connecting the parking brake cable, tighten it at the equalizer until the lever starts to move off the stop on the caliper, then loosen the adjustment until the lever is just against the stop.

25   Remove the lug nut securing the rotor to the hub, install the wheel and snug the wheel lug nuts.

26   Lower the vehicle and tighten the wheel lug nuts to the specified torque.

27   If you have disconnected the hydraulic hose for any reason, fill the master cylinder and bleed the hydraulic system (Section 13).

6   Rear disc brake pads - replacement

Warning: *Whenever you are working on the brake system, be aware that asbestos dust may be present. Asbestos is extremely harmful to your health, so avoid inhaling it. Wash the brake and surrounding area with brake system cleaner before beginning work.*

1984 through 1987 models

Refer to illustrations 6.2 and 6.3

Note: *Disc brake pads should be replaced on both wheels at the same time. Do not replace the pads on only one wheel.*

1   Remove the brake caliper from the rotor (Section 5). Don't forget to compress the piston into the caliper with a C-clamp to allow room between the piston and the rotor for the new pads.

2   Use water pump pliers to grasp the retaining clip on the backside of the outboard pad and remove it from the caliper **(see illustration)**.

3   The inboard pad is secured to the piston via four small tangs on the edge of a circular spring clip which is mounted on the face of the piston. The pad is easily removed by hand from the piston **(see illustration)**. As

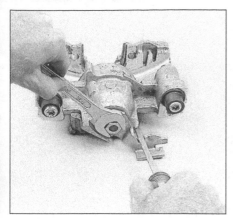

7.4a To remove the parking brake cable lever from the caliper on the bench, lock the lever in place with a screwdriver and break the nut loose

7.4b Remove the old lever seal from the actuator screw - always use a new lever seal when reassembling the caliper

7.4c Remove the anti-friction washer and replace it with a new washer when reassembling the caliper

you remove the pad, note the two small pins on the backing plate that fit into the two small D-shaped notches in the face of the piston. These pins prevent the actuator screw from rotating the piston when the parking brake is applied. The pins on the pad backing plate must line up with these notches.

4   Remove the bolt boots, bushings and sleeves from the caliper mounting bolt holes.

5   Bottom the piston in the cylinder bore before installing new pads.

6   Check behind the piston dust boot for leakage around the edge of the piston/caliper interface. If leakage is present, overhaul the caliper (Section 7).

7   Lubricate the new bushings, bolt boots and sleeves with silicone grease and install them.

8   Position the inboard pad assembly in the caliper. Make sure that the D-shaped tab on the shoe engages the D-shaped notch in the piston. If the tab and notch do not line up, turn the piston until they do.

9   With the tab and notch properly aligned, install the inboard pad assembly in the caliper with the wear sensor positioned so that it is at the leading edge. Slide the edge of the pad backing plate under the ends of the dampening spring and snap the assembly into place against the piston, making sure that the plate lies flat against the piston. **Note:** *If the assembly fails to lie flat, recheck the alignment of the D-shaped notch and tab.*

10   Install the outboard pad assembly in the caliper.

11   Reinstall the caliper (refer to Section 5).

12   Seat the linings by applying the brake pedal firmly at least three times.

1988 models

Removal

13   Remove the brake caliper from the rotor (Section 5). Don't forget to compress the piston into the caliper with a C-clamp to allow room between the piston and the rotor for the new pads.

14   After you have removed the old pads, use a small screwdriver to remove the two-

way check valve from the end of the piston assembly.

15   If you note any leakage from the piston hole after you've removed the check valve, remove and overhaul the caliper.

Installation

Note: *A special tool is required to bottom out the pistons prior to installing new pads. If you don't have this special tool, you'll have to unbolt the caliper housing assembly and slide it off the rotor to bottom out the piston.*

16   The piston rotator wrench has pins which engage the holes in the face of the piston assembly. Insert the pins into these holes and rotate the piston assembly until it bottoms into the caliper bore. **Note:** *The pistons are not threaded the same way: To bottom the pistons, turn the left caliper piston assembly counterclockwise and turn the right caliper piston assembly clockwise.*

17   If you don't have a rotator wrench, remove the rear caliper assembly (see Section 5), stick the tips of a pair of needle-nose pliers into the holes in the face of the piston and rotate the piston until it bottoms in the bore.

18   After you have bottomed the piston in the caliper bore, lubricate a new two-way check valve and install it in the piston face.

19   Install the inner pad with the wear sensor at the top of the pad. Make sure the pad pins on the back of the inner pad engage the holes on the front of the piston. (If you have a piston rotator wrench, you can use it to turn the piston until the holes align with the pins. If you don't have a rotator wrench, turn the piston with a pair of needle-nose pliers until the holes are aligned with the pins on the back of the pad).

20   Install the outer pads.

21   The procedure for installing the spring pins in the rear calipers is substantially the same as the one used for installing the pins in the front calipers (see Section 4).

22   If you removed the caliper assembly to bottom the piston, install the caliper (see Section 5).

23   The remainder of installation is the reverse of removal.

7   Rear disc brake caliper - overhaul

Refer to illustrations 7.4a, 7.4b, 7.4c, 7.5, 7.6, 7.8a, 7.8b, 7.9a, 7.9b, 7.10, 7.11a, 7.11b, 7.24a, 7.24b, 7.26 and 7.28

Disassembly

Note: *If you are going to rebuild one rear brake caliper, it is a good idea to do the other as well. But work on one caliper at a time so that you can use the other one as a guide, if necessary, for proper positioning of the pads, parking brake lever assembly, etc.*

1   Disconnect and remove the caliper from the vehicle (Section 5).

2   Remove the pad assemblies from the caliper (Section 6).

3   Place the caliper on a clean workbench. If you have a bench vise, secure the caliper in it. Be sure to cushion the jaws with wood or shop towels. The cast aluminum caliper housing can be cracked by too much direct pressure.

4   If you did not remove the parking brake lever nut, lever, seal and anti-friction washer while the caliper was installed on the vehicle, do so now **(see illustrations)**. If the parking

7.5 Press the actuator screw loose from the inside of the caliper with your finger - the only thing holding it is the varnish residue buildup between the seal and the caliper body

7.6 Removing the piston with a C-clamp - use a socket as a driver to push the actuator screw through the hole in the caliper housing

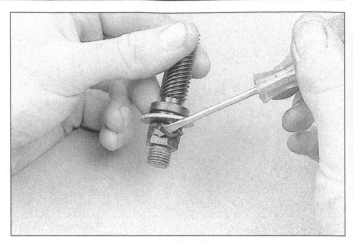

7.8a Removing the seal from the actuator screw

7.8b Removing the thrust washer from the actuator screw - note that the bearing surface (the copper colored side) faces toward the threads (toward the piston) and the grayish surface faces toward the caliper

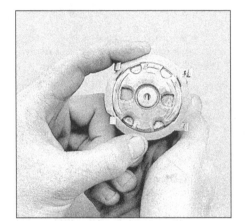

7.9a The lock ring type pad retaining clip is held on the piston face by four small locking tabs

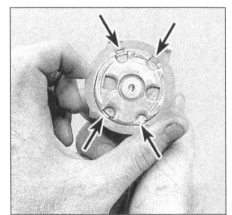

7.9b To unlock the retainer, turn it slightly in either direction until the tabs line up with the four small D-shaped detents (arrows) and lift off the ring

brake cable bracket is damaged and must be replaced, remove it now but do not install the new one until Step 17.

5   Using a wrench to rotate the actuator screw in a counterclockwise direction, push the piston out of the bore in the caliper. When the piston is protruding far enough from the caliper bore, press on the head of the actuator screw **(see illustration)** to break the inner seal loose from the base of the caliper bore.

6   A C-clamp can also be used to remove the piston from the caliper **(see illustration)**. **Caution:** *Even though it is permissible to use a C-clamp directly against the actuator screw to remove the piston when the caliper is out of the vehicle, it must never be used directly on the actuator screw when the caliper is installed on the vehicle or damage to the actuator screw could result.*

7   Remove the piston, balance spring and actuator screw from the piston bore. Remove the balance spring and actuator screw from the piston.

8   Remove the seal and thrust washer from the actuator screw **(see illustrations)**. Note

that the bearing surface (the copper side) of the washer faces the piston and the grayish surface faces the caliper bore. The thrust washer must be installed in exactly the same way when the caliper is reassembled.

9   To remove the inboard pad retaining ring from the caliper piston, turn the ring either clockwise or counterclockwise until the locking tangs are lined up with the four small D-shaped detents in the face of the piston and remove the ring **(see illustrations)**. **Note:** *Not all calipers are equipped with the same type of inboard pad retainer. Some vehicles have a wire clip type dampening spring that is secured to the end of the piston. To remove it, pop it off with a small screwdriver.*

10   Remove the piston dust boot from the caliper with a screwdriver **(see illustration)**. Use a wood or plastic tool to remove the piston seal from the bore.

11   If you have not already done so, remove the caliper mounting bolt dust boots and bushings **(see illustrations)**.

12   Remove the bleeder valve protector cap and the bleeder valve.

Inspection

13   Carefully inspect the caliper bore for scoring, nicks, corrosion and excessive wear. Light corrosion may be polished out with crocus cloth, but do not use anything more abrasive. If the bore cannot be fully restored with crocus cloth, replace the caliper with a new one.

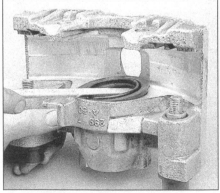

7.10 Pry the old piston dust boot out of the counterbore with a screwdriver

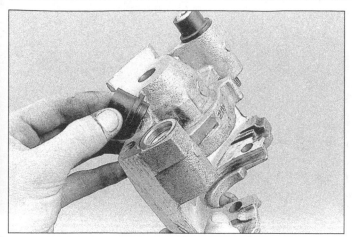

7.11a Pull off the old caliper mounting bolt dust boots

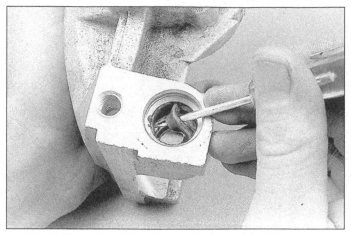

7.11b Pry out the old caliper mounting bolt bushings - be careful not to scratch the bolt holes or the sleeves will not be able to slide freely

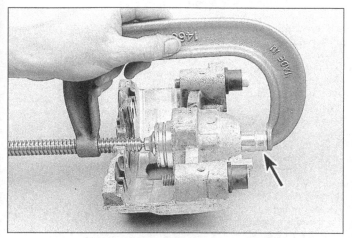

7.24a The piston can be pressed into the bore in the caliper with a C-clamp - note the socket placed between the right end of the clamp (arrow) and the caliper body to protect the actuator screw from damage

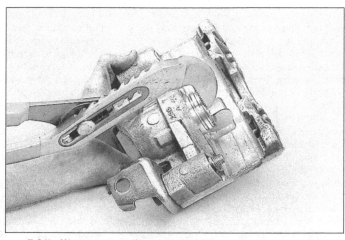

7.24b Water pump pliers can also be used to depress the piston into the caliper - just be sure that you don't touch the actuator screw with the jaws of the pliers

14   Clean the disassembled caliper, piston and other metal parts with denatured alcohol, brake cleaner or clean brake fluid. **Warning:** *Never use gasoline, kerosene or cleaning solvents on brake parts.*

15   Use compressed air to dry the parts. Be sure to blow out all passages in the caliper housing, including the bleeder valve. Lay the parts out to dry on a clean workbench.

Reassembly

16   Install the bleeder valve and tighten it to the specified torque. Install the bleeder valve protector.

17   If you removed the old parking brake cable bracket, install the new one at this time and tighten the bracket bolt.

18   Lubricate the new piston seal and the caliper housing bore with brake fluid. Install the piston seal into the groove in the caliper bore. Make sure that the seal is not twisted.

19   Install the thrust washer on the actuator screw with the bearing surface (the copper side) of the washer toward the piston assembly and the grayish surface toward the caliper housing.

20   Lubricate the shaft seal with clean brake fluid and install it on the actuator screw.

21   Lubricate the actuator screw with brake fluid and install it in the piston.

22   Install the balance spring in the piston bore.

23   Lubricate the piston and caliper bore with clean brake fluid and start the piston assembly into the caliper bore.

24   Using a piston compressor, C-clamp or a pair of water pump pliers **(see illustrations)**, push the piston in until it bottoms in the caliper bore.

25   Before removing the tool, lubricate the anti-friction washer and lever seal with silicone lubricant and install them over the end of the actuating screw, making sure that the sealing bead on the lever seal is against the housing. Install the lever onto the actuating screw, then rotate the lever slightly away from the stop on the housing and hold it while installing the lever retaining nut. Tighten the nut to the specified torque, then rotate the lever back to the stop.

26   Install the piston dust boot. The inside lip of the boot must seat into the piston groove and the open side of the boot must face the piston bore **(see illustration)**. If the

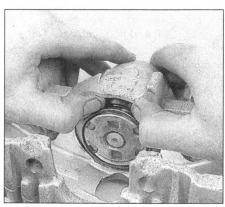

7.26 The piston dust seal can usually be pressed into the counterbore by hand - just be sure that the inside lip is seated in the groove in the piston wall

7.28 Make sure that the inboard brake pad retaining tangs on the retainer clip are aligned so that they will fit over the edges of the pad when it is installed into the caliper

8.2 To determine whether the brake disc has become warped, measure the runout by setting up a dial gauge like this - if runout is greater than the specified maximum allowable amount, replace the disc

boot can't be pushed into the counterbore by hand, use a socket slightly smaller than the inside diameter of the piston bore and drive the boot into place with a mallet, or carefully tamp the boot down with the tip of a screwdriver. Avoid pushing on the soft, inner face of the boot or you will damage it.

27   If the caliper is equipped with a wire clip pad dampening spring, install it in the groove in the piston end. It may be necessary to move the parking brake lever off its stop so that the piston can be extended enough to make the spring groove accessible. If so, be sure to push the piston back into the bottom of the caliper bore before installing the caliper.

28   If the caliper you are working on is equipped with the lock ring type pad retaining spring, install it on the face of the piston by turning it slightly until the four locking tangs are locked in place **(see illustration)**. **Note:** *The four outer tangs must be properly aligned to accept the inboard brake pad.*

29   Install the brake pads (Section 6).

30   Install the caliper on the vehicle (Section 5).

8   Disc brake rotor - inspection, removal and installation

Refer to illustration 8.2

1   Inspect the rotor surfaces. Light scoring or grooving is normal, but deep grooves or severe erosion is not. If pulsating has been noticed during application of the brakes, suspect disc runout.

2   Attach a dial indicator to the caliper mounting bracket, turn the rotor and note the amount of runout **(see illustration)**. Check both inboard and outboard surfaces. If the runout is more than the specified allowable maximum, the rotor must be removed from the vehicle and taken to an automotive machine shop for resurfacing.

3   Using a micrometer, measure the thickness of the rotor. If it is less than the speci-

fied minimum, replace the rotor with a new one. Also measure the disc thickness at several points to determine variations in the surface. Any variation over 0.0005-inch may cause pedal pulsations during brake application. If this condition exists and the disc thickness is not below the minimum, the rotor can be removed and taken to an automotive machine shop for resurfacing.

4   Refer to Chapter 1 for the removal and installation procedure for the front rotor/hub assembly.

5   The rear rotor is retained by the wheel itself. Once the wheel is removed, the caliper is the only thing holding the rotor on. Refer to Section 5 in this Chapter for the rear caliper removal procedure. **Note:** *It is not necessary to disconnect the brake hose. After removing the caliper mounting bolts, hang the caliper out of the way on a piece of wire. Never hang the caliper by the brake hose.*

9   Master cylinder - removal, overhaul and installation

Refer to illustrations 9.3, 9.7, 9.9, 9.10 and 9.22

Note: *Replace all components included in the rebuild kit for this master cylinder. Lubricate the rubber parts with clean, fresh brake fluid to ease assembly. If any hydraulic component is removed or brake line disconnected, bleed the brake system. The torque values specified are for dry, unlubricated fasteners.*

1   A master cylinder overhaul kit should be purchased before beginning this procedure. The kit will include all the replacement parts necessary for the overhaul procedure. The rubber replacement parts, particularly the seals, are the key to fluid control within the master cylinder. It is critical that they be installed securely and properly. Be careful during the rebuild procedure that no grease or mineral-based solvents come in contact with the rubber parts.

Removal

2   Completely cover the front fender and cowling area of the vehicle. Spilled brake fluid can ruin painted surfaces.

3   Place several shop rags or newspapers underneath the master cylinder to soak up the brake fluid that will drain from it when the brake lines are disconnected. Disconnect the brake line connections **(see illustration)**.

4   Remove the two master cylinder mounting nuts attaching the master cylinder mounting bracket to the booster assembly. Remove the master cylinder assembly.

Overhaul

5   Remove the reservoir cover and reservoir diaphragm, then discard any remaining fluid in the reservoir.

6   Inspect the reservoir cover and the diaphragm for tears, cracks and deformation. Replace any damaged parts.

7   Place the master cylinder in a vise and pry the reservoir loose with a screwdriver **(see illustration)**. Pry up the old reservoir grommets with a small screwdriver.

8   Inspect the reservoir and the grommets for tears, cracks and deformation. Replace any damaged parts.

9   Remove the primary piston lock ring by depressing the piston with a large screwdriver and prying the ring out with a small screwdriver **(see illustration)**.

10   You should be able to remove the primary piston assembly by tapping the master cylinder body into the palm of your hand. The primary piston assembly is under spring pressure and should pop out enough so that you can grasp it and extract it by hand **(see illustration)**.

11   Remove the secondary piston assembly by carefully tapping the master cylinder assembly on a block of wood.

12   If the primary and/or secondary piston assembly is difficult to remove, compressed air directed into the appropriate outlet, depending on which piston is stuck, will pop

9.3 The master cylinder is located in the left side of the front compartment and is attached to the brake booster

A   Front brake lining fitting
B   Rear brake line fitting

C   Master cylinder mounting nuts

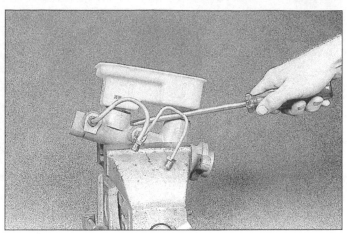

9.7 The easiest way to remove the reservoir from the master cylinder is to place the master cylinder in a vise and pry the reservoir loose with a screwdriver

the piston(s) loose. **Caution:** *If compressed air is used to remove either piston assembly, wear safety glasses to prevent brake fluid spray from causing eye injury.*

13   Before disassembling the primary and secondary pistons, note the direction in which each piston faces and the difference in size of the primary and secondary piston springs.

14   Once you have noted the above, disassemble the primary and secondary piston assemblies. Before you remove each old seal, note the direction in which the cupped side faces.

15   You may notice the quick take-up valve in the passage between the rear reservoir and the primary piston chamber. Do not attempt to remove this device from the cylinder body. It is neither removable nor serviceable.

16   Clean all parts with denatured alcohol or brake cleaner and dry them with compressed air. **Warning:** *Do not, under any circumstances, use petroleum-based solvents to clean brake parts. Lay the parts out in order to prevent confusion during reassembly.*

17   Inspect the master cylinder bore for scoring and corrosion. If either is present,

replace the master cylinder body. Abrasives cannot be used on the bore.

18   Lubricate the reservoir grommets with silicone lubricant and press them into the master cylinder body. Make sure they are properly seated.

19   Lay the reservoir, upside down, on a hard surface and press the master cylinder body onto the reservoir, using a rocking motion.

20   Install the new seals on the secondary piston. Make sure that the cupped sides face out, away from the secondary piston.

21   Attach the spring retainer to the secondary piston assembly.

22   Lubricate the cylinder bore with clean brake fluid and install the spring, spring retainer and secondary piston assembly in the same order in which they were removed **(see illustration).**

23   Lubricate the primary piston seals with clean brake fluid and install them on the primary piston. Be sure that the cupped face of the seal is facing into the bore.

24   Lubricate the cylinder bore and primary piston assembly with clean brake fluid and install the spring and piston assembly.

9.9 To free the primary piston assembly, depress it with a screwdriver and pry the lock ring loose with a small screwdriver

25   Depress the primary piston assembly and install the lock ring in the groove inside the master cylinder bore.

26   **Note:** *Every time the master cylinder is removed, the complete hydraulic system must be bled. The time required to bleed the*

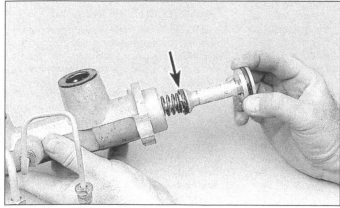

9.10 When removing the primary piston assembly, note that the wider side of the cupped seal is facing forward (arrow)

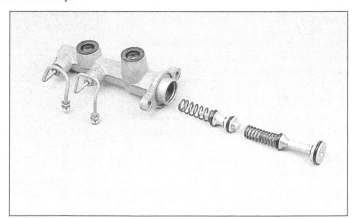

9.22 Make sure that the spring with fewer coils is installed on the secondary piston and the spring with more, tightly wound coils is installed on the primary piston

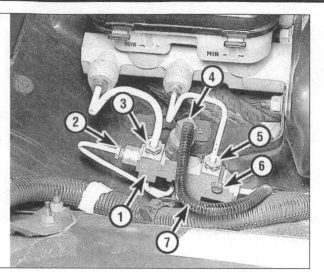

10.1a   The combination valve

1　*Left front brake line outlet*
2　*Right front brake line outlet*
3　*Front brake line outlet*
4　*Brake warning lamp switch connector*
5　*Rear brake line inlet*
6　*Mounting bolt*
7　*Rear brake line outlet (not visible)*

10.1b   To remove the brake warning light switch connector, depress the sides of connector (A) to release tabs (B)

system can be reduced if the master cylinder is filled with fluid and bench bled (refer to Steps 27 through 29) *before the master cylinder is installed on the vehicle.*

27　Insert threaded plugs of the correct size into the cylinder outlet holes and fill both reservoirs with brake fluid. The master cylinder should be supported in a level position so that brake fluid will not spill during the bench bleeding procedure.

28　Loosen one plug at a time and push the piston assembly into the bore to force air from the master cylinder. To prevent air from being drawn back into the cylinder, the appropriate plug must be replaced before allowing the piston to return to its original position.

29　Stroke the piston three or four times for each outlet to ensure that all air has been expelled.

30　Refill the master cylinder reservoirs and install the diaphragm and cover assembly. **Note:** *The reservoirs should only be filled to the top of the reservoir divider to prevent overflowing when the cover is installed.*

31　Install the diaphragm on the reservoir cover and install the cover on the reservoir to prevent spillage during installation.

Installation

32　Install the master cylinder on the brake booster and tighten the mounting nuts to the specified torque.

33　Install the brake lines and tighten the fittings securely. If a special adapter is available, use a torque wrench.

34　Bleed the brakes at the brake caliper bleeder valves (Section 13).

---

10   Combination valve - testing and replacement

Refer to illustrations 10.1a and 10.1b

Note: *The combination valve is not repairable and must be replaced as a complete assembly.*

Testing the electrical circuit

1　To remove the electrical wire connector

from the pressure differential switch **(see illustration)**, squeeze the elliptically shaped plastic locking ring sides and then pull up. This will move the locking tabs away from the switch **(see illustration)**. A pliers can be used to aid in the removal of the connector.

2　Use a jumper wire to connect the switch wire to a good ground.

3　Turn the ignition key to On. The warning lamp on the instrument panel should light. If the lamp does not light, either the bulb is burned out or the electrical circuit is defective. Replace the bulb or repair the electrical circuit as necessary.

4　When the warning lamp lights, turn the ignition switch Off. Disconnect the jumper and reconnect the wire to the switch terminal.

Testing the warning light switch

5　Attach a hose to one rear brake bleeder valve and immerse the other end of the hose in a container filled with clean brake fluid. Be sure that the master cylinder reservoirs are both full.

6　Turn the ignition switch to On. Open the bleeder valve while an assistant applies moderate pressure to the brake pedal. The warning lamp should light. Close the bleeder valve before the assistant releases the pedal. Reapply the brake pedal with moderate-to-heavy pressure. The lamp should go out.

7　Attach the hose to one front brake bleeder valve and repeat the above test. The warning lamp action should be the same as in Step 6. Turn the ignition switch off.

8　If the warning lamp does not light during Steps 6 and 7, but does light when a jumper is connected to ground, the warning light switch portion of the combination valve is defective. Do not attempt to disassemble the combination valve. If any part of the combination valve is defective, it must be replaced with a new combination valve.

Replacement

9　Disconnect the hydraulic lines at the combination valve. Plug the lines to prevent

loss and contamination of fluid.

10　Disconnect the warning switch wiring harness from the valve switch terminal.

11　Remove the combination valve mounting bolt and detach the combination valve.

12　Install the new combination valve. Tighten the mounting bolt securely.

13　Install the hydraulic line fittings and tighten them securely.

14　Plug in the warning switch wiring harness to the valve switch terminal.

15　Bleed the entire brake system (Section 13). **Warning:** *Do not move the vehicle until a firm brake pedal is obtained.*

---

11   Power brake booster - removal, installation and servicing

Refer to illustrations 11.4, 11.5, 11.9a and 11.9b

1　The power brake booster requires no special maintenance apart from periodic inspection of the vacuum hose, lines and air filter between the intake manifold and the booster unit.

2　Disassembly of the power brake booster requires special tools. If a problem develops, obtain a new or rebuilt unit.

Removal and installation

3　Remove the master cylinder mounting nuts from the booster assembly (Section 9). If possible, carefully push the master cylinder aside to allow enough room for removal of the power booster. **Caution:** *Do not bend or kink any of the hydraulic lines attached to the master cylinder. If you cannot gain enough room to remove the booster by moving the master cylinder out of the way, disconnect the hydraulic lines.*

4　Disconnect the vacuum hose leading to the front of the power brake booster **(see illustration)**. Cover the end of the hose.

5　Loosen the four nuts securing the booster to the firewall **(see illustration)**. Do not remove these nuts at this time.

6　Inside the vehicle, disconnect the power

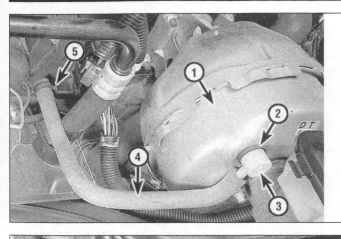

11.4 The power brake booster and vacuum hose assembly

1 Brake booster
2 Grommet
3 Elbow fitting
4 Vacuum hose
5 Clamp

11.5 The four mounting bolts (arrows - fourth bolt not visible) for the brake booster assembly are located on the left side of the rear bulkhead of the front engine compartment

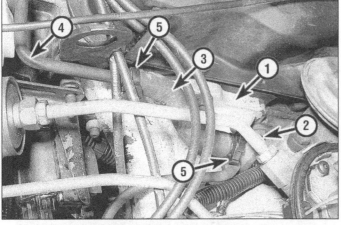

11.9a View of the power brake vacuum filter, tube and hose assembly on the 2.5L four cylinder engine

1 Intake manifold
2 Vacuum fitting
3 Vacuum hose

4 Vacuum tube
5 Clamps

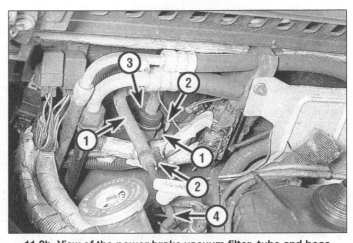

11.9b View of the power brake vacuum filter, tube and hose assembly on the 2.8L V-6 engine

1 Vacuum hose
2 Clamps

3 Filter assembly
4 Vacuum tube

brake pushrod from the brake pedal. Do not force the pushrod to the side when disconnecting it.

7   Remove the four booster mounting nuts and carefully lift the unit out.

8   When installing the booster, loosely install the four mounting nuts and connect the pushrod to the brake pedal. Tighten the nuts to the specified torque and reconnect the vacuum hose and master cylinder. If the hydraulic brake lines were disconnected, the entire brake system should be bled to eliminate any air which has entered the system (refer to Section 13).

Replacement of power brake vacuum filter, tube and hose

9   The power brake vacuum filter, tube and hose assembly is located in the engine compartment. It begins at the intake manifold **(see illustrations)** and is connected to the power brake booster assembly through a hose routed inside the center console.

10   Inspect the hose for cracks. If it shows signs of damage, remove and discard it. Install a new hose.

11   The filter is located on the forward firewall of the engine compartment. If it looks dirty, replace it.

12 Hydraulic brake hoses and lines - inspection and replacement

Refer to illustrations 12.3, 12.16 and 12.17

1   At least twice a year (when lubricating the chassis), inspect the flexible hydraulic brake hoses which connect the steel brake lines with the front and rear brake calipers for cracks, chafing of the outer cover, leaks, blisters and road damage. These hoses are important and vulnerable parts of the brake system. Your inspection should be thorough. A light and mirror are helpful for a complete inspection. If a hose exhibits any of the above conditions, replace it.

Front brake hose
Removal

2   Clean dirt and foreign material from both the hose and the fittings.

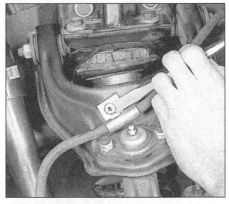

12.3 When removing the front brake hose, you'll have to cut the rivet securing the hose bracket to the upper A-arm

3   Remove the rivet attaching the brake hose clip to the upper A-arm **(see illustration)**.

4   Remove the U-clip from the female fitting at the support bracket immediately forward of the upper A-arm and remove the hose from the bracket.

5   Remove the bolt from the caliper end of the hose. Remove the hose from the caliper and discard the two copper washers on either side of the fitting block.

Installation

6   Use new copper washers on either side of the fitting block and lubricate the bolt threads with clean brake fluid.
7   With the fitting flange engaged with the caliper locating ledge, attach the hose to the caliper.
8   Without twisting the hose (there must be no kinks in the hose), install the female fitting in the hose bracket. It will fit the bracket in only one position.
9   Install the U-clip to the female fitting at the frame bracket.
10   Rivet the brake hose clip to the upper A-arm.
11   Using a backup wrench, attach the brake line to the hose fitting.
12   When the brake hose installation is complete, there should be no kinks in the hose. Make sure the hose does not contact any part of the suspension. Check this by turning the wheels to the extreme left and right positions. If the hose makes contact, remove the hose and correct the installation as necessary.
13   Bleed the brake system (Section 13).

Rear brake hose

Removal

14   Disconnect the brake hose from the brake line at the hose mounting bracket. Use a backup wrench and do not bend the bracket or steel lines.
15   Remove the U-clip at the hose mounting bracket.
16   Remove the bolt and disconnect the bracket holding the brake line to the strut **(see illustration)**.
17   Remove the bolt attaching the fitting block to the caliper **(see illustration)**.

Installation

18   Attach the hose assembly to the brake line.
19   Attach the spring clip to the hose mounting bracket.
20   Attach the brake hose fitting block to the caliper along with new copper washers.
21   Attach the brake line to the strut.
22   Fill and maintain the brake fluid level in the reservoirs and bleed the system.

Steel brake lines

Caution: *Never use copper tubing for brake lines. Copper is subject to fatigue cracking and corrosion which could result in brake failure. Use double-wall steel tubing only.*
23   Auto parts stores and brake supply houses carry various lengths of prefabricated brake line. These sections can be bent into the desired shape with a tubing bender. Obtain the recommended tubing and steel fitting nuts of the correct size (the outside diameter of the tubing is used to specify size).
24   With a tubing bender, bend the tubing to match the shape of the old brake line. **Cau-**

12.16 Disconnect the hose bracket from the strut by removing this bolt

tion: *Leave at least 3/4-inch clearance between the line and any moving parts.*

---

13   Hydraulic system - bleeding

Note: *Bleeding the hydraulic system is necessary to remove any air that manages to find its way into the system when it has been opened during removal and installation of a hose, line, caliper or master cylinder.*
1   It will probably be necessary to bleed the system at all four brakes if air has entered the system due to low fluid level, or if the brake lines have been disconnected at the master cylinder.
2   If a brake line was disconnected only at one wheel, then only that caliper must be bled.
3   If a brake line is disconnected at a fitting located between the master cylinder and any of the brakes, that part of the system served by the disconnected line must be bled.
4   If the master cylinder has been removed from the vehicle, refer to Section 9, Step 26 before proceeding.
5   If the master cylinder is installed on the vehicle but is known to have, or is suspected of having, air in the bore, then it must be bled before any caliper can be bled. Follow Steps 7 through 16 to bleed the master cylinder while it is installed on the vehicle.
6   Remove any residual vacuum from the brake power booster by applying the brake several times with the engine off.
7   Remove the master cylinder reservoir cover and fill the reservoir with brake fluid. Reinstall the cover.
8   **Note:** *Check the fluid level often during the bleeding operation and add fluid as necessary to prevent the fluid level from falling low enough to allow air bubbles into the master cylinder.*
9   Disconnect the forward brake line fitting at the master cylinder.
10   Fill the master cylinder with brake fluid until it begins to flow from the forward line connector port. Have a container and shop rags handy to catch spilled fluid.
11   Reconnect the forward brake line to the master cylinder.

12.17 Break loose the bolt securing the fitting block to the caliper and remove the hose

12   Have an assistant depress the brake pedal very slowly, one time only, and hold it down.
13   Loosen the forward brake line at the master cylinder to purge the air from the bore, tighten the connection, then have the brake pedal released slowly.
14   Wait 15 seconds (this is very important).
15   Repeat the sequence outlined in Steps 9 through 14, including the 15 second wait, until all air is removed from the bore.
16   After the forward port has been completely purged of air, bleed the rear port in the same manner.
17   Before bleeding the individual calipers, refer to Steps 6 and 7.
18   Have an assistant on hand, as well as a supply of new brake fluid, an empty clear plastic container, a length of clear tubing to fit over the bleeder valve and a wrench to open and close the bleeder valve. The vehicle may have to be raised and placed on jackstands for clearance.
19   Beginning at the right rear wheel, loosen the bleeder valve slightly, then tighten it to a point where it is snug but can still be loosened quickly and easily.
20   Place one end of the tubing over the bleeder valve and submerge the other end in brake fluid in the container.
21   Have the assistant pump the brakes a few times to get pressure in the system, then hold the pedal firmly depressed.
22   While the pedal is held depressed, open the bleeder valve just enough to allow a flow of fluid to leave the valve. Watch for air bubbles to exit the submerged end of the tube. When the fluid flow slows after a couple of seconds, close the valve and have your assistant release the pedal.
23   Repeat Steps 21 and 22 until no more air is seen leaving the tube, then tighten the bleeder valve and proceed to the left rear wheel, the right front wheel and the left front wheel, in that order, and perform the same procedure. Be sure to check the fluid in the master cylinder reservoir frequently.
24   Never use old brake fluid. It contains moisture which will deteriorate the brake sys-

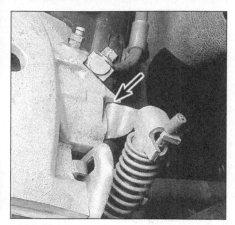

14.5 Make sure that each parking brake cable lever is against the stop on the caliper - if it isn't, the cable still has too much tension and must be loosened at the adjuster nut

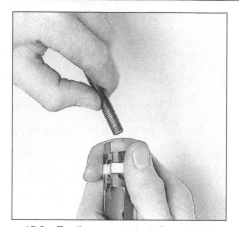

15.2a To disconnect the left and right cables, unscrew the threaded bolt from the adjuster nut

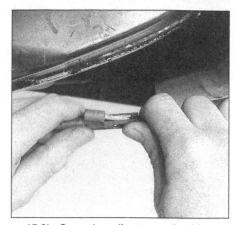

15.2b Once the adjuster nut is either backed off or disconnected, the right cable can be removed from the adjuster

tem components.

25   Refill the master cylinder with fluid at the end of the operation.

26   If any difficulty is experienced in bleeding the hydraulic system, or if an assistant is not available, a pressure bleeding kit is a worthwhile investment. If connected in accordance with the instructions, each bleeder valve can be opened in turn to allow the fluid to be pressure ejected until it is clear of air bubbles without the need to replenish the master cylinder reservoir during the process.

14 Parking brake - adjustment

Refer to illustration 14.5

Note: *If the hydraulic system operates with adequate reserve but the parking brake handle travel exceeds nine ratchet clicks, the*

parking brake cable must be adjusted. It must also be adjusted any time the rear brake cables have been disconnected.

1   Release the parking brake.

2   Raise the rear of the vehicle and support it securely on jackstands.

3   Before adjusting, make sure the equalizer nut groove is lubricated liberally with multi-purpose lithium base grease.

4   Hold the brake cable stud and tighten the equalizer nut until there is no cable slack.

5   Make sure that the caliper levers are against the stops on the caliper housing after tightening the equalizer nut **(see illustration)**.

6   If the levers are off the stops, loosen the cable until the levers return to the stops.

7   Operate the parking brake lever several times to check the adjustment. A properly adjusted brake cable will require five to eight notches of movement at the handle when enough force is applied to lock the

rear wheels.

8   Lower the vehicle. The levers must be on the caliper stops after completion of adjustment. Back off the parking brake equalizer if the levers are not against the stops when the vehicle is lowered.

15 Parking brake cable - removal and installation

Refer to illustrations 15.2a, 15.2b, 15.3a, 15.3b, 15.7 and 15.12

Removal

1   Raise the vehicle and support it securely on jackstands.

2   Loosen the adjusting nut at the equalizer and separate the cables **(see illustrations)**.

3   Remove the cable clips **(see illustrations)**.

15.3a Typical front parking brake cable routing details

1   *Front cable*   3   *Bolt*
2   *Retainer clamp*

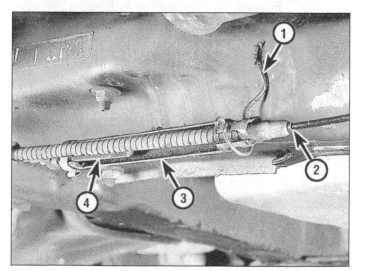

15.3b Typical parking brake cable and equalizer retainer clip details

1   *Retainer clip*   3   *Equalizer (adjuster)*
2   *Front cable*   4   *Left cable*

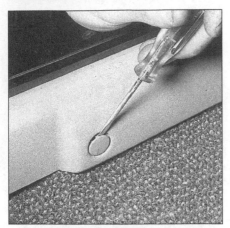

15.7 To get at the parking brake lever, remove the carpet trim finishing molding by prying the pop fasteners loose with a small screwdriver

4   Remove the retaining clip bolts in the left wheel well.

5   Lower the vehicle.

6   Unsnap the clip holding the parking brake boot to the lever.

7   Remove the seat belt bolt and carpet finishing molding **(see illustration)**.

8   Remove the shoulder harness retaining bolt.

9   Remove the quarter trim finishing molding.

10   Pull the carpet back and note how the cable is routed **(see illustration 15.3)**.

11   Remove the cable from the parking brake lever and push it through the body.

Installation

12   Route the new cable into place and attach the forward end to the parking brake

hand lever **(see illustration)**.

13   Lay the carpet back in place and install the quarter trim finish molding.

14   Install the shoulder harness retaining bolt and tighten it securely.

15   Install the seat belt bolt and carpet finish molding.

16   Install the parking brake boot on the parking brake hand lever.

17   Raise the vehicle and support it securely on jackstands.

18   Install the two retaining clip bolts in the left wheel well.

19   Install the cable clips.

20   Slide the right rear caliper cable into place in the cable adjuster.

21   Thread the equalizer nut into place and tighten it until the slack is removed from the cable (see Section 14).

16 Parking brake warning lamp switch - removal and installation

Refer to illustration 16.3

Note: *The parking brake warning lamp switch assembly, which is bolted to the hand brake mounting bracket and actuated by the hand brake lever, is* nonadjustable.

Removal

1   Remove the retainer holding the brake boot to the handle.

2   Remove the carpet trim finish molding (see Section 15).

3   Disconnect the electrical connection at the switch **(see illustration)**.

4   Remove the bolt at the switch.

5   Remove the switch.

Installation

6   Install the new switch.

7   Install the bolt at the switch.

8   Connect the electrical connection at the switch.

9   Install the carpet trim finish molding.

10   Install the retainer holding the brake boot to the handle.

17 Stop light switch - removal, installation and adjustment

Refer to illustration 17.1

Note: *The stop light switch also serves as the cutoff switch for the Torque Converter Clutch (TCC) (automatic transaxle equipped vehicles) as well as the kill switch for cruise control on models equipped with automatic transaxles. The switch is located on the clutch pedal bracket on vehicles with cruise control and manual transaxles.*

1   The stop light switch is located on the brake pedal support under the instrument panel **(see illustration)**.

2   With the brake pedal in the fully released position, the stop light switch plunger should be fully depressed against the pedal shank. When the pedal is pushed, the plunger releases and sends electrical current to the stop lights.

3   If the stop lights are inoperative, first determine whether the bulbs are burned out (see Chapter 12).

4   If the bulbs are okay, make certain that the tubular clip is seated firmly in the brake pedal mounting bracket.

5   With the brake pedal depressed, insert the stop light switch into the tubular clip until the switch body seats on the clip. Note the audible clicks made by the threaded portion of the switch as it is pushed through the tubular clip toward the brake pedal.

6   Pull the brake pedal all the way to the

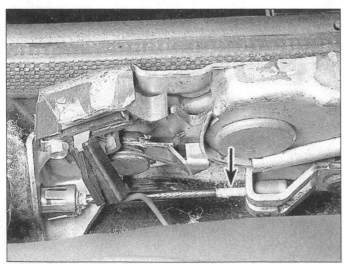

15.12 The parking brake cable (arrow) looks like this when it is properly installed in the slot at the base of the parking brake lever pivot

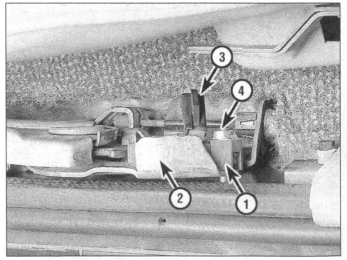

16.3 The parking brake warning switch

1   *Switch*
2   *Parking brake lever assembly*
3   *Parking brake switch connector*
4   *Switch mounting bolt*

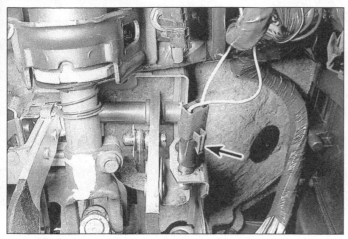

17.1 The stop lamp and/or Torque Converter Clutch (TCC) switch is located underneath the instrument panel to the right of the steering column, above the pedal assembly

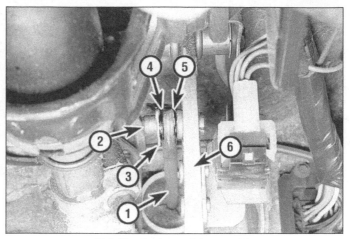

18.2 The brake pedal assembly

1	Power brake booster pushrod	4	Washer
2	Pedal pin	5	Spring washer
3	Wire clip	6	Pedal assembly

rear against the pedal stop until no further clicks can be heard. This will seat the switch in the tubular clip and provide the correct adjustment.

7   Release the brake pedal and repeat Step 6 to ensure that no further clicks can be heard.

8   Make sure that the stop lights are working.

9   If the lights are still not working, disconnect the electrical connectors at the stop light switch and remove the switch from its retainer.

10   Install a new switch and adjust it by per-

forming Steps 4 through 7, making sure the electrical connectors are hooked up.

18   Brake pedal - removal and installation

Refer to illustration 18.2

Removal

1   Disconnect the stop light/TCC/cruise control switch assembly from the pedal bracket (Section 17).

2   Remove the wire clip, flat washer, power brake booster pushrod and washer from the brake pedal pin **(see illustration)**.

3   Unbolt the brake pedal assembly from the dash panel and remove the assembly.

Installation

4   Place the new pedal assembly in position and install the mounting bolts.

5   Install the washer, power brake booster pushrod, flat washer and wire clip.

6   Attach the stop light/TCC/cruise control switch assembly to the pedal bracket (Section 17).

Notes

Chapter 10
Steering and suspension systems

Contents

Specifications

Torque specifications

	Ft-lbs (unless otherwise indicated)
Upper front shock mounting bolts/nut	
1984 through 1987	20
1988	96 in-lbs
Lower front shock mounting bolt	
1984 through 1987	51
1988	20
Stabilizer bar bushing clamp mounting bolts	
1984 through 1987	15
1988	20
Stabilizer bar link bolt	144 in-lbs
Upper front balljoint mounting bolts	28
Upper control arm balljoint-to-knuckle castellated nut	35
Upper control arm-to-crossmember pivot bolt	
1984 through 1987	66
1988	52
Lower control arm-to-crossmember pivot bolt nut	
1984 through 1987	52
1988	37 (plus 3/4-turn)
Lower control arm balljoint-to-knuckle castellated nut	
1984 through 1987	55
1988	26 (plus 1/2-turn)
Rear strut damper-to-knuckle mounting bolt nuts	140
Rear strut damper upper mounting nuts	18
Rear lower balljoint mounting bolt nuts	156 in-lbs
Rear lower balljoint stud clamp bolt nut	37
Rear lower control arm pivot bolts	66
Rear hub and bearing bolts	62

Torque specifications (continued)

Ft-lbs (unless otherwise indicated)

Rear driveaxle/hub nut	See Chapter 8
Rear hub and bearing shield mounting bolts..	55 to 70
Rear toe link rod nut	30 to 39
Wheel lug nuts	
Steel	80
Aluminum	100
Caliper mounting bolts	See Chapter 9
Steering wheel-to-shaft nut	30
Outer tie-rod-to-steering knuckle nut	
1984 through 1987	29
1988	15 (plus 1/2-turn)
Inner tie-rod-to-outer tie-rod jam nut	50
Steering damper-to-rack and pinion assembly nuts	32
Shock damper-to-boot support stud	35
Front brake splash shield-to-steering knuckle bolt	84 in-lbs
Steering damper mounting nuts	32
Flexible coupling pinch bolt...	46
Rack and pinion steering assembly mounting bolts...............................	21
Crossmember brace bolts	20
Wheel lug nuts	See Chapter 1

1   Suspension system - general information

1984 through 1987 models

Refer to illustrations 1.1 and 1.2

The Fiero features an independent front suspension with unequal length (shorter upper and longer lower) control arms, shock absorbers and coil springs **(see illustration)**. The shock absorbers are conventional sealed hydraulic units. The control arms are attached to the frame with bolts and rubber bushings and are connected to the steering knuckle/front wheel spindle assembly with balljoints.

The rear suspension is a MacPherson strut (combination spring and strut) design **(see illustration)**. The upper ends of the struts are connected to the vehicle with bolts and mounting plates and the lower ends are bolted to the knuckles. The lower control arms are attached to the engine cradle with bolts and rubber bushings.

1988 models

All 1988 models are equipped with a redesigned front suspension system. Although similar in configuration to the earlier design - unequal length (shorter upper and longer lower) control arms, shock absorbers and coil springs - the new system shares few parts with the older system, and some of the removal and installation procedures for various components vary slightly.

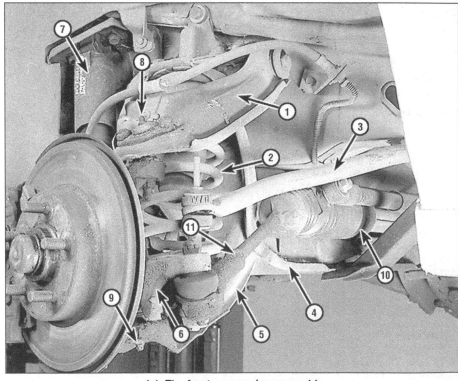

1.1 The front suspension assembly

1	Upper control arm		7	Shock absorber
2	Coil spring		8	Upper balljoint grease fitting
3	Stabilizer bar		9	Lower balljoint grease fitting
4	Frame/crossmember		10	Steering gear assembly
5	Lower control arm		11	Tie-rod end
6	Knuckle/spindle assembly			

1.2 The rear suspension assembly

1	Strut assembly	3	Spring	5	Driveaxle
2	Toe link rod	4	Rear control arm		

2 Suspension system - inspection

1   Suspension components subjected to normal wear will provide a long service life. However, handling and ride quality will deteriorate as the suspension wears, so it should be periodically inspected.

2   Check that the springs have not sagged. Park the vehicle on a level surface and note whether it leans to one side. This condition usually appears only after high mileage and is more likely to occur on the driver's side of the vehicle.

3   Put the transaxle in gear and take off the parking brake. Grip the steering wheel at the top with both hands and rock it back-and-forth. Listen for any squeaks or metallic noises. Feel for free play. If any of these conditions is noted, have an assistant move the wheel while the source of the trouble is located.

4   Check the shock absorbers, as these are the parts of the suspension system likely to wear out first. If there is any evidence of fluid leakage, replace the shocks. Bounce the vehicle up-and-down vigorously. It should feel stiff and well damped by the shock absorbers. As soon as the bouncing is stopped the vehicle should return to its normal position without excessive up-and-down movement. Do not replace the shock absorbers as single units, but rather in pairs.

5   Check all rubber bushings for signs of deterioration and cracking.

3 Front shock absorber - removal, check and installation

1984 through 1987 models

Refer to illustrations 3.4 and 3.5

Removal

1   Loosen the front wheel lug nuts.

2   Raise the front of the vehicle and place it securely on jackstands.

3   Remove the wheel and tire assembly.

4   Remove the two retaining bolts from the upper end of the shock absorber **(see illustration)**.

5   Remove the nut and bolt from the lower end of the shock absorber **(see illustration)**.

6   Remove the old shock absorber from the vehicle.

Checking

7   To test the shock absorber, hold it in an upright position and work the piston rod up-and-down its full length of travel. If you can feel a strong resistance because of hydraulic pressure, the shock absorber is functioning properly. If you feel no pronounced resistance or if there is a sudden lack of resistance, the shock absorber should be replaced.

8   If there is fluid leakage evident on the outside of the shock absorber body, the shock absorber must be replaced.

9   Even if only one shock shows signs that it has reached the end of its service life, replace both front shock absorbers to preserve uniform ride quality at both wheels and to maintain good handling.

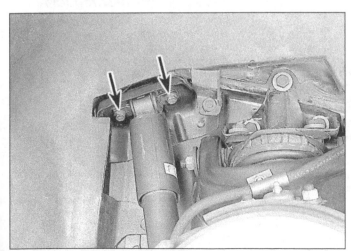

3.4 Remove the upper shock absorber mounting bolts (arrows)

3.5 The lower mounting bolt and nut for the front shock absorber

4.2 The nut (arrow) on the top of this long bolt must be removed in order to disconnect the stabilizer bar from the lower control arm

4.3 These bolts (A) secure the stabilizer bar clamps to the frame rails - note the relationship of the rubber bushing (B) to the clamp before disassembling it

Installation

10   With the lower end of the shock absorber in position, hand tighten the nut and bolt and extend the shock up into its support until the upper bolts can be installed. Tighten the upper bolts to the specified torque.
11   Tighten the lower nut and bolt to the specified torque.
12   Replace the wheel and tire assembly. Install the wheel lug nuts and tighten them finger tight.
13   Lower the vehicle and tighten the wheel lug nuts to the specified torque.

1988 models

Removal

14   Loosen the front wheel lug nuts.
15   Raise the front of the vehicle and place it securely on jackstands.

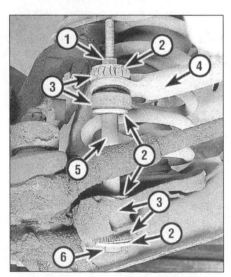

4.7 Stabilizer bar link bolt components

1	*Nut*	*4*	*Stabilizer bar*
2	*Washer*	*5*	*Spacer*
3	*Grommet*	*6*	*Bolt*

16   Remove the wheels.
17   Remove the upper retaining nut, washer and upper shock insulator.
18   Remove the two bolts from the lower end of the shock absorber.
19   Remove the shock absorber through the lower control arm.

Installation

20   Extend the shock absorber and insert it through the lower control arm with the lower shock insulator and washer in place.
21   Install the upper shock insulator, washer and retaining nut. Tighten the nut to the specified torque.
22   Install the lower retaining bolts and tighten them to the specified torque.
23   Install the wheel and tighten the lug nuts snugly.
24   Lower the vehicle to the ground.
25   Tighten the lug nuts to the specified torque.

---

4   Stabilizer bar - removal and installation

Refer to illustrations 4.2, 4.3 and 4.7

Removal

1   Raise the front of the vehicle and support it on jackstands.
2   Remove the nut from the top of the link bolt connecting the stabilizer bar to the lower control arm **(see illustration)**. Note the position of the various grommets and washers on the bolt, then remove the entire assembly.
3   Disconnect the stabilizer bar clamps from the frame rails **(see illustration)**.
4   Remove the stabilizer bar.
5   Carefully inspect the stabilizer bar rubber bushings and the rubber bushings on the link bolts connecting the bar to the lower control arm. Look for cracks, dryness and deterioration. If any damage is noted, replace the bushings as a set.

Installation

Note: *The eyelet on the left end of the stabilizer bar is pointed instead of round.*
6   Holding the stabilizer bar in place, install the stabilizer bar clamps. Make sure that the rubber bushings that insulate the clamps from the bar are correctly aligned with the clamps. Tighten the clamp bolts to the specified torque.
7   Install the link bolts connecting the stabilizer bar to the lower control arms. Make sure that the washers, grommets and spacers are installed onto each bolt in the correct order **(see illustration)**.
8   Tighten the nut on the top end of each link bolt until it bottoms on the end of the bolt threads, then tighten it to the specified torque.
9   Lower the vehicle.

---

5   Balljoints - removal and installation

1984 through 1987 models

Refer to illustrations 5.5, 5.6 and 5.8
Note: *This procedure applies only to the upper control arm balljoints. Refer to Section 7 for lower balljoint removal and installation procedures.*

Removal

1   Loosen the front wheel lug nuts.
2   Raise the front of the vehicle and support it on jackstands.
3   Remove the front tire and wheel assembly.
4   Support the lower control arm with a floor jack.
5   Remove the upper ball stud nut cotter pin and loosen but do not remove the castellated nut **(see illustration)**.
6   Install a pickle fork tool between the upper balljoint and the knuckle and split the balljoint **(see illustration)**.

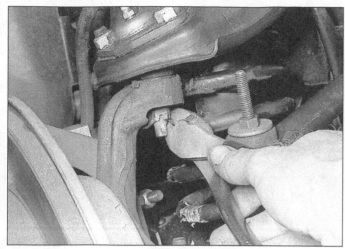

5.5 Remove the cotter pin from the upper ball stud castellated nut and loosen, but do not remove, the nut

5.6 Insert a pickle fork between the upper balljoint and the knuckle and split the balljoint

7   Remove the pickle fork and undo the castellated nut.

8   Remove the two nuts and bolts attaching the balljoint to the upper control arm **(see illustration)**. Note that the flat side of the balljoint flange faces out.

9   Remove the balljoint from the upper control arm.

10   Inspect the tapered hole in the steering knuckle. If it is contaminated by dirt or foreign particles, clean it out before reassembling. If out-of-roundness, deformation or damage is noted, the knuckle must be replaced.

Installation

11   Install the bolts and nuts that attach the balljoint to the upper control arm and tighten them to the specified torque.

12   Mate the upper control arm ball stud to the steering knuckle and install the ball stud castellated nut. Tighten it to the specified torque.

13   Turn the nut only the amount necessary to align the holes and install a new cotter pin.

14   Install the tire and wheel assembly.

15   Install the wheel lug nuts and tighten them finger tight.

16   Lower the vehicle to the ground.

17   Tighten the lug nuts to the specified torque.

1988 models

Note: *This procedure applies only to the upper control arm balljoints. Refer to Section 7 for lower balljoint removal and installation procedures.*

Removal

Note: *GM specifies a special balljoint separator tool, to disconnect the balljoint from the steering knuckle. If you don't want to buy this tool, and you can't borrow or rent it, you can make a tool that will work using a large bolt, nut, washer and socket.*

18   Loosen the wheel lug nuts.

19   Raise the vehicle and place it securely on jackstands.

20   Remove the wheel.

21   Place a floor jack under the lower control arm and raise the jack to compress the coil spring slightly. **Warning:** *Make sure the lower control arm is well supported, and there is no chance of the jack coming out, or you could be injured during this procedure.*

22   Remove the bolt and detach the brake line clip from the upper control arm.

23   Disconnect the tie rod end from the steering knuckle (Section 16) and swing the knuckle out of the way.

24   Remove the cotter pin, loosen the castellated nut on the upper balljoint stud and detach the upper balljoint from the steering knuckle with special tool. If the special tool is not available, you can make your own tool using a large bolt, nut, washer and socket. Whichever tool you use, leave the castellated nut in place to prevent the stud from separating violently from the knuckle.

25   Detach the upper balljoint from the upper control arm by removing the three attaching rivets.

Installation

Note: *You will need to obtain a service repair package from your dealer for this procedure. The package contains the nuts and bolts you will use to replace the rivets you removed. It also specifies the correct bolt torque for your vehicle.*

26   Attach the new upper balljoint to the control arm with nuts and bolts from the service repair package. Tighten them to the specifications provided in the package.

27   Inspect the tapered hole in the steering knuckle and remove any dirt. If you note any out-of-roundness, deformation or damage, replace the knuckle.

28   Insert the upper balljoint stud into the hole in the steering knuckle. Install the castellated nut and tighten it to the specified torque. Install a new cotter pin.

29   Connect the tie-rod end to the steering knuckle (see Section 16), then attach the brake line clip to the control arm.

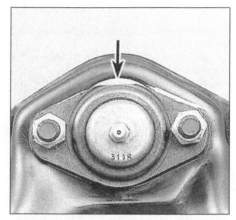

5.8 Note that the flat side (arrow) of the upper balljoint flange faces out, away from the vehicle

30   Install the wheel and tighten the lug nuts snugly.

31   Lower the vehicle to the ground and tighten the lug nuts to the specified torque.

---

6   Upper control arm and bushings - removal and installation

---

1984 through 1987 models

Refer to illustrations 6.4 and 6.8

Note: *The pivot bolt bushings in the upper control arm are nonserviceable, so if they are worn or damaged, the upper control arm assembly must be replaced.*

Removal

1   Loosen but do not remove the wheel lug nuts.

2   Raise the front of the vehicle and secure it on jackstands.

3   Remove the tire and wheel assembly.

4   Break off the rivet holding the brake hose clip to the upper control arm with a hammer and chisel **(see illustration)**.

6.4 Break off the rivet attaching the brake hose clip to the upper control arm with a hammer and chisel

6.8 Remove the upper control arm pivot bolt and nut - note the position of any spacers and/or washers when removing the bolt and install everything in exactly the same place

5   Remove the brake hose bracket from the frame rail.

6   Support the lower control arm with a floor jack.

7   Disconnect the upper balljoint from the steering knuckle (Section 5).

8   Remove the upper control arm pivot bolt **(see illustration)**. Note the proper positioning of the washers and shims before removing the pivot bolt assembly. Unless a change in geometry is desired, the washers and shims must be reinstalled in the same order in which they were removed.

9   Remove the upper control arm. Inspect the bushings for wear. If they are worn or damaged, replace the upper control arm.

10   Transfer the balljoint, if it's not damaged or worn, to the new upper control arm (Section 5).

Installation

11   Place the new upper control arm in position and install the pivot bolt assembly. **Note:** *The pivot bolt must be installed with the bolt head toward the front of the vehicle.*

12   Install the pivot bolt nut, put a back-up wrench on the bolt head and tighten the nut to the specified torque.

13   Attach the balljoint to the steering knuckle (Section 5).

14   Install the tire and wheel assembly. Tighten the wheel lug nuts finger tight.

15   Lower the vehicle and tighten the wheel lug nuts to the specified torque.

1988 models

Removal

16   Detach the upper balljoint from the steering knuckle (see Section 5).

17   Remove the two bolts that attach the pivot arm to the crossmember, then remove the paddle nut assembly. Remove the control arm from the vehicle.

Installation

18   Install the upper control arm on the

crossmember using a new paddle nut assembly. Tighten the bolts securely, but do not apply final torque until a wheel alignment is performed.

19   Inspect the tapered hole in the steering knuckle and remove any dirt. If you note any out-of-roundness, deformation or damage, replace the knuckle.

20   Attach the upper balljoint to the steering knuckle.

21   Install the wheel and tighten the lug nuts snugly, then lower the vehicle to the ground and tighten the lug nuts to the specified torque. Have the front wheels aligned by a properly equipped shop.

---

7   Front lower control arm and balljoint - removal and installation

1984 through 1987 models

Refer to illustrations 7.8 and 7.14

Warning: *This procedure involves removal of the coil spring, which is under load and could result in personal injury if it is released too quickly. Be sure to lower the jack slowly when dropping the lower control arm.*

Note: The lower balljoint is welded to the lower control arm and cannot be serviced separately. Replacement of the entire lower control arm will be necessary if the lower balljoint requires replacement.

Removal

1   Loosen but do not remove the wheel lug nuts.

2   Raise the front of the vehicle and support it on jackstands.

3   Remove the tire and wheel assembly.

4   Disconnect the stabilizer bar from the lower control arm. If only one lower control arm is being removed, disconnect only that end of the stabilizer bar. If both lower control arms are beeing removed, remove the entire

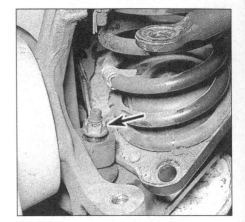

7.8 After removing the cotter pin from the balljoint stud, loosen but do not remove this castellated nut

stabilizer bar (Section 4).

5   Disconnect the tie-rod from the steering knuckle (Section 16).

6   Disconnect the shock absorber at the lower control arm.

7   Support the lower control arm with a floor jack.

8   Remove the cotter pin from the lower balljoint stud. Loosen but do not remove the castellated nut from the lower balljoint stud **(see illustration)**.

9   Insert a pickle fork tool between the knuckle and the balljoint and break the balljoint loose from the knuckle (see Section 5).

10   Remove the balljoint nut and swing the knuckle and hub out of the way.

11   Loosen the lower control arm pivot bolts.

12   Install a chain over the upper control arm and through the coil spring as a safety precaution.

13   Slowly lower the jack, remove the chain and withdraw the spring.

7.14 If the lower control arms on your vehicle have pivot bolts with forward facing heads, you will have to remove the steering rack to get the bolt out

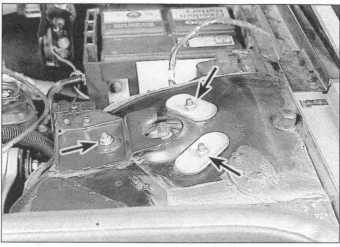

8.5 Remove the three nuts and washers (arrows) that secure the top end of the strut

14   Remove the pivot bolts at the frame rail and the crossmember and remove the lower control arm. **Note:** *Depending upon the direction in which the bolt head of the forward pivot bolt faces, removal of the forward pivot bolt at the crossmember may require the loosening or removal of the steering assembly mounting bolts* **(see illustration)**. *If the lower control arms on your vehicle have pivot bolts with forward facing heads, see Section 19 for steering rack removal.*

Installation

15   Place the lower control arm in position. Install the pivot bolts at the frame rail and crossmember and tighten them slightly.
16   Before installing the spring, inspect the condition of the rubber control arm bumper mounted on the crossmember. If it is worn or damaged, replace it.
17   Position the spring and install it into the upper pocket. Align the spring bottom with the lower control arm pocket.
18   Install the spring lower end onto the lower control arm. It may be necessary to have an assistant help you compress the spring far enough to slide it over the raised area of the lower control arm seat.
19   Use the floor jack to raise the lower control arm and compress the coil spring.
20   Install the balljoint in the steering knuckle. Install the nut on the balljoint stud and tighten the nut to the specified torque. Install a new cotter pin.
21   Connect the stabilizer bar (Section 4) and tighten the bolt to the specified torque.
22   Connect the tie-rod (Section 16) and tighten the nut to the specified torque.
23   Install the shock absorber on the lower control arm and tighten the bolt to the specified torque.
24   If the steering rack assembly bolts were removed or loosened, replace with new bolts and tighten them to the specified torque.
25   Tighten the lower control pivot bolts to the specified torque.

26   Install the tire and wheel assembly. Tighten the wheel lug nuts finger tight.
27   Lower the vehicle and tighten the wheel lug nuts to the specified torque.
28   It's a good idea to have the wheels aligned after reassembling the suspension.

1988 models

Note: *A balljoint separator, balljoint installer and adapters are required for this procedure. However, if the tools are not available, you can remove the lower control arm and have an automotive machine shop remove the old balljoint and install the new one.*
29   Loosen the wheel lug nuts.
30   Raise the vehicle and place it securely on jackstands.
31   Remove the wheel.
32   Place a floor jack under the lower control arm and raise the jack to compress the coil spring slightly. **Warning:** Make sure the lower control arm is well supported and there is no chance of the jack coming out or you could be injured during this procedure.
33   Disconnect the tie-rod end from the steering knuckle (see Section 16).
34   Detach the lower balljoint from the steering knuckle.
35   If you don't have access to this special tool, you can make your own tool from a large bolt, nut, washer and socket.
36   Swing the knuckle, rotor and caliper assembly out of the way.
37   Inspect the tapered hole in the steering knuckle and remove any dirt. If you note any out-of-roundness, deformation or damage, replace the knuckle.
38   If you have the special balljoint remover/installer and adapter, position the tool - with the indicated adapters for balljoint removal - on the lower control. Tightening the tool's screw will press the new balljoint into the control arm.
39   If you don't have the special tools, remove the lower control arm (see Section 7).

Take the control arm to an automotive machine shop and have them press out the old balljoint and press in the new one. Install the lower control arm following the procedure in Chapter 10, Section 7 (disregard the remainder of this procedure).
40   install the steering knuckle on the lower balljoint stud and install the stud nut. Tighten the nut to the specified torque and install a new cotter pin.
41   Attach the tie-rod end to the steering knuckle (see Section 16).
42   Remove the jack supporting the lower control arm.
43   Install the wheel. Tighten the wheel lug nuts snugly.
44   Lower the vehicle to the ground and tighten the lug nuts to the specified torque.
45   Have the front wheels aligned by a properly equipped shop.

---

8   Rear strut damper assembly - removal and installation

Refer to illustrations 8.5, 8.6 and 8.7
Note: *Although strut damper assemblies don't always reach the end of service life simultaneously, replace both left and right struts at the same time to prevent handling peculiarities and abnormal ride quality.*

Removal

1   Loosen but do not remove the rear wheel lug nuts.
2   Raise the rear of the vehicle and support it on jackstands.
3   Remove the rear wheel.
4   Support the lower control arm with a jack. Do not compress the spring.
5   Remove the three nuts and washers inside the engine compartment that secure the top of the strut to the vehicle **(see illustration)**.

8.6 Disconnect the brake hose clip (arrow) from the strut

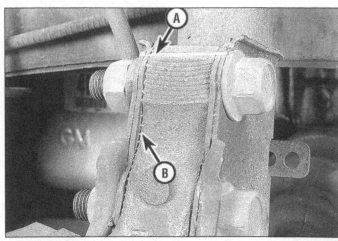

8.7 Scribing the strut and knuckle at points A and B is the best way to maintain proper camber adjustment

6   Disconnect the brake hose clip from the strut **(see illustration)**.

7   Scribe the strut and knuckle to assure that proper camber adjustment is maintained upon reassembly **(see illustration)**. **Caution:** *If you are replacing the struts, the new strut dampers will have no scribe marks, so the vehicle must be taken to a dealer or alignment shop immediately after reassembly. Failure to do so will result in uneven tire wear and abnormal handling.*

8   Remove the strut mounting nuts and bolts.

9   If necessary, lower the jack and remove the strut assembly and spacer plate.

Installation

10   Place the strut assembly and spacer plate in position.

11   Using the floor jack to raise the control arm if necessary, align the mounting holes in the knuckle with the holes in the strut bracket and install the strut mounting bolts. Install the nuts on the bolts finger tight.

12   Align the scribe marks on the strut bracket and the knuckle and tighten the strut mounting bolts to the specified torque.

13   Install the brake hose clip. Tighten the clip mounting bolt securely.

14   Install the tire and wheel assembly.

15   Lower the vehicle.

16   Tighten the wheel lug nuts to the specified torque.

17   Install the three upper strut washers and nuts. Tighten the nuts to the specified torque.

18   If the strut damper has been replaced, drive the vehicle to a dealer or alignment shop and have the camber and toe-in of the rear wheels adjusted.

9   Rear control arm balljoint - removal and installation

Refer to illustration 9.4

Note: *It isn't absolutely necessary to remove the rear control arm in order to drill out the balljoint mounting rivets. But unless you have*

access to a hydraulic hoist, it is easier to perform the following procedure with the control arm on a bench, since the balljoint mounting rivets are facing down. If you want to replace the balljoint while the control arm is installed, follow this procedure. If you want to do it on the bench, refer to the following Section to remove the control arm first.

Removal

1   Loosen but do not remove the rear wheel lug nuts.

2   Raise the vehicle and support it securely on jackstands.

3   Remove the clamp bolt from the balljoint stud.

4   Disconnect the balljoint from the knuckle. If necessary, pry the knuckle off the ball stud with a large screwdriver **(see illustration)**.

5   The balljoint is riveted to the lower control arm. To remove it, use a 1/8-inch drill bit to drill a pilot hole approximately 1/4-inch deep into the center of each rivet.

6   Use a 1/2-inch drill bit to drill just deep enough to remove the rivet heads.

7   Knock out the rivets with a hammer and punch.

9.4 If necessary, use a large screwdriver to pry the balljoint from the knuckle

Installation

8   Install the new balljoint, which comes with bolts and nuts to replace the rivets. Tighten the nuts to the specified torque.

9   Lower the knuckle onto the ball stud, install the clamp bolt and tighten it to the specified torque.

10   Install the tire and wheel assembly and tighten the wheel lug nuts finger tight.

11   Lower the vehicle.

12   Tighten the wheel lug nuts to the specified torque.

10   Rear control arm and bushings - removal and installation

Refer to illustration 10.5

Note: *Replacement of the pivot bolt bushings requires replacement of the entire lower control arm.*

1   Refer to Steps 1 through 4 in the previous Section.

2   Remove the control arm pivot bolts from the cradle.

3   Inspect the bushings. If they are worn out, replace the control arm.

4   Remove the balljoint from the old control arm and install it in the new one (Section 9).

5   Place the control arm in its installed position **(see illustration)** with the heads of the bolts facing away from each other and tighten the nuts to the specified torque.

6   Lower the knuckle onto the ball stud, install the clamp bolt and tighten it to the specified torque.

7   Install the wheel and tire.

8   Lower the vehicle and tighten the wheel lug nuts to the specified torque.

11   Rear hub and wheel bearings - removal and installation

Refer to illustrations 11.10a, 11.10b, 11.11 and 11.12

1   Loosen but do not remove the wheel lug nuts.

10.5 Make sure that the pivot bolts and nuts are facing each other when installing them

1 Left, rear control arm   3 Nuts
2 Pivot bolts

11.10a Remove the hub nut, then remove the three shield mounting bolts by rotating the flange so that the semicircular cutout (arrow) exposes each bolt to permit removal

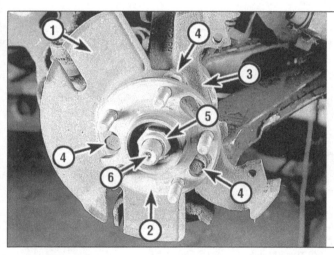

11.10b Exploded view of the relationship between the knuckle and the hub and bearing assembly

1 Shield
2 Hub and bearing assembly
3 Knuckle assembly
4 Shield/hub assembly mounting bolts
5 Hub nut
6 Axle

11.11 If you intend to reuse the same hub and bearing assembly, mark the relationship between the knuckle and the bearing flange (arrow) with a scribe or paint mark

11.12 Remove the hub with a puller

2   Raise the rear of the vehicle and secure it on jackstands.

3   Remove the tire and wheel assembly.

4   Remove the parking brake cable and spring from the lever on the backside of the caliper (see Chapter 9).

5   Remove the brake hose bracket from the strut (Chapter 9).

6   Remove the caliper mounting bolts and the caliper (Chapter 9).

7   Hang the caliper out of the way with a piece of wire.

8   Remove the brake disc.

9   Remove the hub nut and discard it (use a new one during installation). Wedge a screwdriver through one of the holes in the flange and position the tip behind the knuckle to hold the flange stationary while breaking the hub nut loose.

10   Remove the three shield mounting bolts and the shield **(see illustrations)**.

11   If the same hub and bearing assembly is going to be reused, mark the relationship of the hub and bearing assembly flange to the knuckle **(see illustration)**.

12   Remove the hub and bearing assembly with a puller **(see illustration)**.

13   Clean and inspect the bearing mating surfaces and the knuckle bore for dirt, nicks and burrs. **Note:** *The hub and bearing assembly must be replaced as a unit if the bearing surfaces are worn or damaged.*

14   If it is necessary to install a new hub and bearing assembly, remove the knuckle seal by prying it out with a screwdriver. Place the new knuckle seal in position and drive it into the knuckle bore with a large socket slightly smaller in diameter than the inside diameter of the knuckle bore.

15   Push the hub and bearing onto the axle-shaft. **Note:** *If the old hub and bearing assembly is being reused, make sure that the alignment marks between the hub flange and the knuckle are matched up.*

16   Install the shield and shield mounting bolts. Tighten the bolts to the specified torque.

17   Install the new hub nut. Apply the initial specified torque to the new hub nut. This will seat the hub and bearing assembly.

18   Install the disc and caliper assemblies and parking brake cable and spring (Chapter 9).

19   Install the tire and wheel assembly. If your vehicle is equipped with the standard steel wheels, do not install the plastic wheel

12.2 The toe link rod can be separated from the knuckle with a puller or with a special tool such as this

12.11 This nut secures the inboard end of the toe link rod to a bracket on the engine cradle

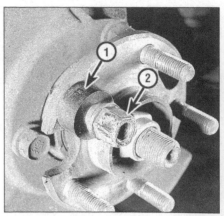

13.8 To install a new stud in the hub flange, insert the stud into the flange hole, install a spacer over the stud and tighten a wheel nut onto the stud until it is seated with its head against the backside of the flange

1   Spacer
2   Nut

cover at this time. Tighten the wheel lug nuts securely.
20   Lower the vehicle.
21   Tighten the hub nut to its final specified torque.
22   If your vehicle is equipped with the standard steel wheels, raise the vehicle again, secure it on jackstands, remove the wheel lugs nuts, install the plastic wheel cover and reinstall the wheel lug nuts finger tight. Lower the vehicle to the ground again and tighten the wheel lug nuts to the specified torque.

12   Rear knuckle and toe link rod - removal and installation

Refer to illustrations 12.2 and 12.11

Rear knuckle

Removal

1   Remove the rear wheel hub and bearing assembly (Section 11).
2   Remove the toe link rod at the knuckle **(see illustration)**.
3   Remove the clamp bolt from the knuckle and disconnect the knuckle from the ball stud (Section 9). **Caution:** *Whenever you separate the balljoint from the knuckle, be careful not to cut or tear the balljoint seal or damage to the balljoint could occur. If the seal is cut or torn, the balljoint must be replaced.*
4   Support the lower control arm with a jack, then remove the large through-bolts holding the strut to the knuckle (be sure to follow the scribing procedure outlined in Section 8).

Installation

5   Connect the knuckle to the balljoint (Section 9).
6   Align the scribe marks between the strut bracket and the knuckle, install the knuckle-to-strut mounting bolts and tighten them to the specified torque (Section 8).
7   Attach the toe link rod to the knuckle and tighten the nut to the specified torque.
8   Attach the hub and bearing assembly to

the knuckle (Section 11).
9   Drive the vehicle immediately to a dealer service department or alignment shop and have the camber and toe adjusted to specifications.

Toe link rod

Removal

10   Remove the toe link rod at the knuckle **(see illustration 12.2)**.
11   Remove the nut from the inboard end of the toe link rod at the cradle **(see illustration)** and remove the rod.
12   Measure the length of the exposed threads on the end of the rod. Use this figure to determine the position of the adjusting nut on the new rod.

Installation

13   Install the inboard end of the new toe link rod into the mounting bracket at the cradle and tighten the mounting nut securely.
14   Install the outboard end of the toe link rod at the knuckle and tighten the mounting nut to the specified torque.
15   Drive the vehicle immediately to a dealer service department or alignment shop and have the toe adjusted to specification.

13   Wheel studs - replacement

Refer to illustration 13.8

Rear

1   Loosen but do not remove the rear wheel lug nuts.
2   Raise the rear of the vehicle and support it on jackstands.
3   Remove the tire and wheel assembly.
4   Remove the brake caliper and disc (Chapter 9).
5   Remove the hub nut (Section 11).
6   Remove the rear hub and bearing assembly and the splash shield (Section 11).
7   Position a C-clamp over both ends of the stud with a socket between the head of the stud and the C-clamp. Turn the screw on

the C-clamp bolt until the stud is pressed out of the hub flange into the socket.
8   Insert the new stud through its hole in the flange, slip a spacer over the stud and install the wheel nut onto the stud with the flat side of the nut facing toward the flange **(see illustration)**. Tighten the wheel nut until the stud is seated. Remove the wheel nut and spacer.
9   If you don't have a C-clamp, place the hub and bearing assembly in a vise and tap out the damaged stud with a brass hammer.
10   Place the new wheel stud in position and tap it down with a drift and hammer until it seats in the bore. **Caution:** *If you are using a hammer to remove and/or install the stud, do not strike the flange itself. A hammer blow could damage it.*
11   Install the hub and bearing assembly (Section 11).
12   Install a new hub nut onto the end of the axleshaft (Section 11).
13   Install the brake disc and caliper (Chapter 9).
14   Install the tire and wheel assembly. Tighten the wheel lug nuts finger tight.
15   Lower the vehicle.
16   Tighten the wheel lug nuts to the specified torque.

Front

17   Loosen but do not remove the front wheel lug nuts.
18   Raise the front of the vehicle and support it on jackstands.
19   Remove the tire and wheel assembly.
20   Remove the brake caliper (Chapter 9).
21   Remove the dust cap from the spindle.
22   Remove the cotter key from the spindle.
23   Remove the spindle nut and spacer washer.
24   Carefully remove the wheel bearing (do not allow the bearing to drop on the ground).
25   Remove the brake disc (Chapter 9).

15.2 If the horn pad in your vehicle looks like this, pry it off with a small screwdriver - if it doesn't, look on the side of the wheel facing the instrument panel for a couple of retaining screws and remove them before removing the pad

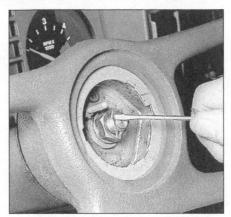

15.3 Pop off this steering wheel nut retainer clip with a small screwdriver

15.6 The steering wheel puller installed (make sure that the puller bolts don't damage the horn plunger)

26   Position a C-clamp over both ends of the stud with a socket between the head of the stud and the C-clamp. Turn the screw on the C-clamp bolt until the stud is pressed out of the hub into the socket. **Note:** *Be very careful not to damage the rotor faces while removing the old stud or installing the new one.*

27   Insert the new stud through the hole in the hub, slip four washers onto the stud and install the wheel nut onto the stud with the flat side of the nut facing toward the flange. Tighten the wheel nut until the stud is seated (it's seated when the stud head is against the hub). Remove the wheel nut and washers.

28   If you don't have a C-clamp, do not attempt to hammer the old stud out. Take the hub and disc assembly to a shop and have the old stud pressed out and the new one pressed in.

29   Install the hub and tighten the nut as described in Chapter 1.

30   Install the wheel and hand tighten the wheel lug nuts.

31   Lower the vehicle.

32   Tighten the wheel lug nuts to the specified torque.

14   Steering system - general information

The Fiero uses a rack and pinion steering system. When the steering wheel is turned, this motion is transferred via the steering column to the intermediate shaft and finally to the pinion gear. The pinion gear teeth mesh with the gear teeth of the rack, so the rack moves right or left in its housing when the pinion is turned. The movement of the rack is transmitted through the inner and outer tie-rods to the steering knuckles, which turn the wheels.

The steering column is a collapsible, energy absorbing unit designed to compress

in the event of a front end collision to minimize injury to the driver. The column also houses the ignition switch lock, key warning buzzer, turn signal controls, headlight dimmer control and windshield wiper controls. The ignition and steering wheel can both be locked to prevent theft when the vehicle is parked.

15   Steering wheel - removal and installation

Refer to illustrations 15.2, 15.3 and 15.6

Removal

1   Disconnect the negative battery cable.

2   Remove the horn pad **(see illustration)**. **Note:** *The design of the horn pad on the steering wheel in your vehicle may differ slightly from the one shown. Some pads or caps are secured by two screws; others can be simply pried off with a small screwdriver. If the pad or cap on your vehicle's steering wheel differs in appearance from the one shown, inspect the side of the wheel facing the instrument panel before attempting to pry off the cover.*

3   Pry off the steering wheel nut retainer **(see illustration)**.

4   Remove the steering wheel nut.

5   Scribe or paint a mark between the steering wheel and the steering column shaft to insure proper reinstallation.

6   Install a steering wheel puller and remove the steering wheel **(see illustration)**.

Installation

7   Align the mark on the steering wheel with the mark on the steering column shaft.

8   Install the nut and tighten it to the specified torque.

9   Install the retainer clip.

10   Install the horn pad. **Note:** *If the horn pad on the steering wheel in your vehicle is secured with two screws, install and tighten them securely.*

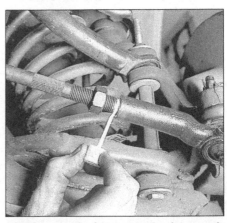

16.4 Loosen the jam nut and paint a mark on the threads of the inner tie-rod where it is screwed into the outer tie-rod - this will ensure that the toe-in adjustment is not altered when you remove and reinstall the outer tie-rod

16   Tie-rods - removal and installation

Refer to illustrations 16.4 and 16.6
Note: *When one or both tie-rods must be removed for any reason, the procedure can be performed without removing the rack and pinion assembly. You may wish, however, to conduct the removal and installation operation while the rack and pinion assembly is on the bench. If so, refer to Section 19 first.*

Removal

1   Loosen but do not remove the wheel lug nuts.

2   Raise the front of the vehicle and secure it on jackstands.

3   Remove the front wheel.

4   Back off the jam nut that locks the outer and inner tie-rods together, then paint an alignment mark onto the threads to insure that the new outer tie-rod is installed to exactly the same thread **(see illustration)**.

16.6 Loosen but do not remove the castellated nut from the outer tie-rod stud, install a puller and detach the outer tie-rod stud from the steering knuckle

5   Remove the cotter pin, then loosen but do not remove the castellated nut from the outer tie-rod stud.
6   Install a puller **(see illustration)** and disconnect the outer tie-rod from the knuckle.

Installation

7   Thread the outer tie-rod onto the inner tie-rod until the inner end of the outer tie-rod is matched up with the paint mark on the threaded portion of the inner tie-rod.
8   Install the outer tie-rod stud through the steering knuckle and tighten the castellated nut to the specified torque. Install a new cotter pin.
9   Tighten the jam nut securely.
10   It is a good idea to have the toe-in adjusted at a dealer service department or alignment shop after the outer tie-rod is installed.

17   Steering knuckle - removal and installation

Removal

1   Loosen but do not remove the wheel lug nuts.
2   Raise the front of the vehicle and support it on jackstands.
3   Remove the tire and wheel assembly.
4   Support the lower control arm with a floor jack (Section 7).
5   Remove the disc brake caliper (Chapter 9).
6   Remove the hub and disc (Chapter 9).
7   Remove the splash shield (Chapter 9).
8   Remove the tie-rod end from the steering knuckle (Section 16).
9   Disconnect the upper balljoint from the steering knuckle (Section 5).
10   Disconnect the lower balljoint from the steering knuckle (Section 7).
11   Remove the steering knuckle.

Installation

12   Place the steering knuckle in position and insert the upper and lower ball studs into the knuckle bosses.
13   Install the upper and lower ball stud nuts and tighten to the specified torque. Install new cotter pins.
14   Attach the splash shield to the steering knuckle (Chapter 9).
15   Attach the tie-rod end to the steering knuckle (Section 16).
16   Repack the wheel bearings (Chapter 1).
17   Install the hub and disc, bearings and nut (Chapter 1).
18   Install the brake caliper (Chapter 9).
19   Install the tire and wheel assembly. Tighten the wheel lug nuts finger tight.
20   Lower the vehicle.
21   Tighten the wheel lug nuts to the specified torque.

22   Take the vehicle to a dealer service department or alignment shop and have the toe-in checked.

18   Steering damper - removal and installation

Refer to illustration 18.1

1   Hold the steering damper mounting studs with a back-up wrench and remove the nuts and washers **(see illustration)**.
2   Remove the steering damper.
3   Installation is the reverse of removal.

19   Rack and pinion assembly - removal and installation

1984 through 1987 models

Refer to illustrations 19.2, 19.3 and 19.6

Removal

1   Raise the vehicle and support it securely on jackstands.
2   Remove both front crossmember braces **(see illustration)**.
3   Remove the flexible coupling pinch bolt from the pinion shaft **(see illustration)**.
4   Disconnect the outer tie-rods from the steering knuckles (see Steps 5 and 6 in Section 16).
5   Scribe an alignment mark between the pinch clamp bolt and the pinion shaft boss to ensure proper alignment between the steering wheel and the rack assembly during reinstallation.
6   Remove the four bolts holding the rack and pinion steering assembly to the crossmember. If the flexible coupling sticks to the pinion shaft, carefully pry the rack and pinion assembly loose with a screwdriver **(see illustration)**.

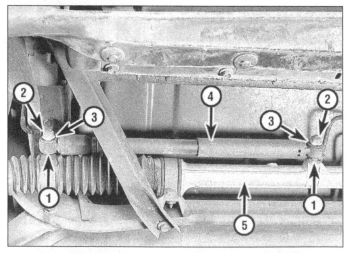

18.1 Exploded view of the steering damper assembly

1	*Shock damper stud*	*4*	*Steering shock damper*
2	*Nut*	*5*	*Steering gear assembly*
3	*Washer*		

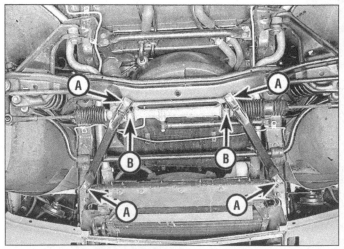

19.2 The rack and pinion steering assembly is removed from the vehicle by removing the crossmember brace bolts (A), the mounting bracket bolts (B), the flexible coupling pinch bolt (hidden from view by the crossmember) and disconnecting the outer tie-rods from the steering knuckles

19.3 The flexible coupling pinch bolt is just above and in front of the crossmember

Installation

7   Place the rack and pinion steering assembly in position. Make sure that the pinion shaft is properly seated in the flexible coupling.
8   Install four new steering assembly mounting bolts and tighten them to the specified torque.
9   Install the flexible coupling pinch bolt and tighten it to the specified torque.
10   Connect the outer ends of the tie-rods to the steering knuckles (Section 16).
11   Install both front crossmember braces and tighten the bolts to the specified torque.
12   Lower the vehicle.

1988 models

13   Because of the redesigned front suspension assembly, the rack and pinion assembly on all 1988 vehicles is easier to remove than the older unit. The procedure for removing and installing the new rack and pinion assembly is substantially the same as the above procedure, but it is no longer necessary to remove or install any crossmember braces.

20   Rack and pinion boot seals - removal and installation

Refer to illustrations 20.5a, 20.5b and 20.6
Note: *The boot seals protect the internals of*

the rack and pinion steering assembly from dirt and water. They should be checked periodically for holes, cracks and other damage. If any of these conditions are noted, replace the boot seals immediately or you may have to replace the entire steering assembly.

Removal

1   Remove the rack and pinion steering assembly (Section 19).
2   Remove the steering damper from the rack and pinion steering assembly (Section 18).
3   Remove the outer tie-rods from the steering assembly (Section 16).
4   Remove the jam nut from the inner tie-rod.
5   To remove the boot seal, cut off the inner boot clamp with a pair of diagonal cutters and discard it **(see illustration)**. If you are replacing the right boot seal, you must also remove the shock damper stud and the boot support clamp **(see illustration)**.
6   Place the tip of a small screwdriver underneath the outer lip of the boot seal **(see illustration)** and stretch the opening enough to slide the old seal off.

19.6 If the rack and pinion assembly pinion shaft sticks to the flexible coupling, carefully pry it loose with a screwdriver

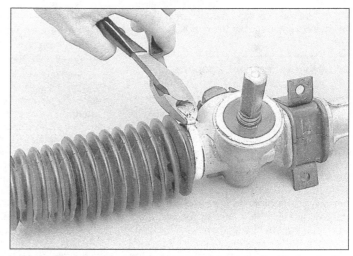

20.5a The inner boot seal clamp must be cut off with a pair of diagonal cutters and cannot be reused

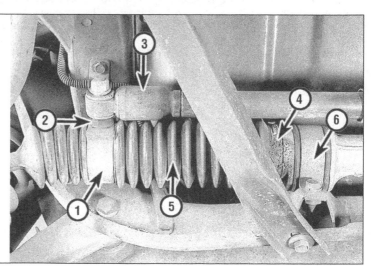

20.5b Remove the shock damper stud and boot support from the right boot seal before attempting to remove it from the right end of the rack and pinion assembly - make sure that the new boot is properly seated over the damper adapter before attempting to reinstall the boot support and shock damper stud

  1   Boot support
  2   Shock damper stud
  3   Shock damper
  4   Boot clamp
  5   Boot
  6   Steering gear clamp

Installation

7   Place a new boot clamp over the end of the new boot before installing it.

8   Slide the new boot onto the inner tie-rod until the outer end of the boot snaps into its groove. If you are replacing the right boot seal, be sure to line up the hole for the shock damper stud over the damper adapter. Make sure that the boot seal is properly seated over the adapter, then install the boot support and tighten the shock damper stud to the specified torque.

9   Secure the boot clamp with the recommended crimping tool (available at your dealer).

10   Install the steering damper onto the rack and pinion steering assembly (Section 18). Tighten the shock damper stud nuts until they are snug.

11   Install the rack and pinion steering assembly and steering damper onto the vehicle (Section 19). Tighten the steering assembly mounting bolts to the specified torque. Tighten the shock damper stud nuts to the specified torque.

12   Connect the outer tie-rods to the inner tie-rods and to the steering knuckle (Section 16).

13   Drive the vehicle to a dealer or an alignment shop and have the toe-in adjusted.

21   Wheel alignment - general information

Wheel alignment refers to the adjustments made to the front suspension and steering components to bring the front wheels into the proper angular relationship with the suspension and the road. Such variables as the angle of the steering knuckles from the vertical, the toe-in of the front wheels, the tilt of the front wheels from vertical and the tilt of the suspension members from vertical affect alignment. Front wheels that are out of proper alignment not only affect steering control, but also increase tire wear.

Rear alignment refers to the angular relationship between the rear wheels, the rear suspension attaching components and the road. Camber and toe-in are the only adjustments required.

Obtaining the proper wheel alignment is a very exacting process that requires complex and expensive machines to perform the

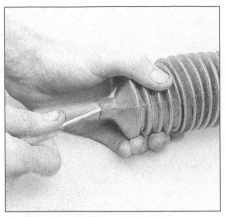

20.6 Inside the outer lip of the boot seal there is a ridge that fits into a groove in the tie-rod - pry the ridge out of this groove before attempting to slide off the seal

job properly. Therefore, it is advisable to have a properly equipped shop align the wheel immediately after you do any work on the front or rear suspension pieces for any reason.

22   Wheels and tires - general information

All Fieros are equipped with radial tires. When replacing the tires, make sure that the new ones have the same Tire Performance Criteria (TPC) specification number molded into their sidewalls as the original.

A specific TPC number is assigned to each tire size. This number means that the tire meets GM's performance standards for traction, endurance, dimensions, noise, handling, rolling resistance, etc. Tires of the same TPC number will also be the same size, load range and construction as the originals. Other types and/or sizes of tires may affect handling, ride quality, ground clearance, speedometer/odometer calibration and clearance between the tires and the vehicle.

It is recommended that tires be replaced in pairs on the same axle, but if only one tire is being replaced, be sure it is of the same size, structure and tread design as the other. Although they may appear different in tread design, tires manufactured by different companies can be intermixed on the same vehicle as long as they have the same TPC number.

Wheels must be replaced if they are bent, dented, leak air, have elongated bolt holes, are heavily rusted, out of vertical symmetry or if the lug nuts won't stay tight. Wheel repairs that use welding or peening are not recommended, as this can weaken the metal.

Tire and wheel balance is important in the overall handling, braking and performance of the vehicle . Unbalanced wheels can adversely affect handling and ride characteristics as well as tire life. Whenever a tire is installed on a wheel, the tire and wheel should be balanced by a shop with the proper equipment.

Because the compact spare is designed as a temporary replacement for an out-of-service standard wheel and tire, the compact spare should be used on the vehicle only until the standard wheel and tire are repaired or replaced. Continuous use of the compact spare at speeds of over 50 mph (80 kph) is not recommended. In addition, the expected tread life of the compact spare is only 3000 miles (4800 kilometers).

23   Wheels and tires - removal and installation

1   With the vehicle on a level surface, the parking brake on and the transaxle in gear (manual transaxles should be in Reverse, automatic transaxles should be in Park) remove the hub trim ring and loosen, but do not remove, the wheel lug nuts.

2   Using a jack positioned in the proper location on the vehicle, raise the vehicle just enough so that the tire clears the ground.

3   Remove the lug nuts.

4   Remove the wheel and tire.

5   If a flat tire is being replaced, make sure that there is adequate ground clearance for the new inflated tire, then mount the wheel and tire on the wheel studs.

6   Apply a light coat of spray lubricant or light oil to the wheel stud threads and install the lug nuts snugly with the cone-shaped end facing the wheel.

7   Lower the vehicle until the tire contacts the ground and the wheel studs are centered in the wheel holes.

8   Tighten the lug nuts evenly and in a criss-cross pattern to the specified torque.

9   Lower the vehicle and remove the jack.

10   Replace the hub trim ring.

Chapter 11 Body

Contents

Specifications

Torque specifications

	Ft-lbs
Door latch striker	34 to 46
Door lock assembly	80 to 100 in-lbs

1 General information

The Fiero is of unitized construction. The frame consists of a floorpan with front and rear frame side rails which support the body components, front and rear suspension systems and other mechanical components.

The exterior body panels of the Fiero are made from various kinds of plastic such as injection molded urethane (RIM), glass fiber reinforced RIM (RRIM), sheet molded compound (SMC) or thermoplastic olefin (TPO). All these materials are rustproof and can withstand minor impacts without damage.

Only general body maintenance practices and body panel repair procedures within the scope of the average home mechanic are included in this Chapter.

2 Body - maintenance

1   The condition of your vehicle's body is very important, because it is an important component of the resale value of the vehicle. It is much more difficult to repair a neglected or damaged body than it is to repair mechanical components. And the condition of the hidden areas of the body (the fender wells, the frame, and the engine compartment, to name a few) - even though they don't require as much attention as the rest of the body - is just as important as the condition of more obvious exterior panels.

2   Once a year, or every 12,000 miles, it is a good idea to have the underside of the body and the cradle steam cleaned. All traces of dirt and oil will be removed and the underside can then be inspected carefully for damaged brake lines, frayed electrical wiring, damaged cables, and other problems. The front suspension components should be greased after completion of this job.

3   At the same time, clean the engine and the engine compartment using either a steam cleaner or a water soluble degreaser.

4   The body should be washed once a week (or when dirty). Wet the vehicle thoroughly to soften the dirt, then wash it down with a soft sponge and plenty of clean soapy water. If the surplus dirt is not washed off very carefully, it will eventually wear down the paint.

5   Spots of tar or asphalt coating thrown up from the road should be removed with a cloth soaked in solvent.

6   Once every six months, give the body a thorough waxing.

3 Upholstery and carpets - maintenance

1   Every three months remove the carpets or mats and clean the interior of the vehicle (more frequently if necessary). Vacuum the upholstery and carpets to remove loose dirt and dust.

2   If the upholstery is soiled, apply upholstery cleaner with a damp sponge and wipe it off with a clean, dry cloth.

4 Body repair - minor damage

Note: *This section describes the minor surface repair of flexible ("Enduraflex") panels. Although a specific brand of material may be mentioned, it should be noted that equivalent products from other manufacturers may be used instead.*

Below is a list of the equipment and materials necessary to perform the following repair procedures:

Wax, grease and silicone removing solvent
Cloth back body tape
Sanding discs
Drill motor with three-inch disc holder
Hand sanding block
Rubber squeegees
Sandpaper
Non-porous mixing palette
Wood paddle or putty knife
Curved tooth body file
3M Flexible Parts Repair Material, or equivalent

These photos illustrate a method of repairing simple dents. They are intended to supplement *Body repair - minor damage* in this Chapter and should not be used as the sole instructions for body repair on these vehicles.

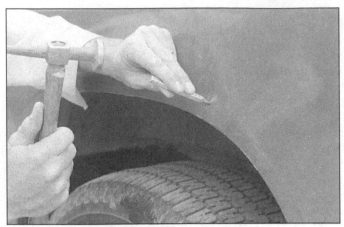

1 If you can't access the backside of the body panel to hammer out the dent, pull it out with a slide-hammer-type dent puller. In the deepest portion of the dent or along the crease line, drill or punch hole(s) at least one inch apart . . .

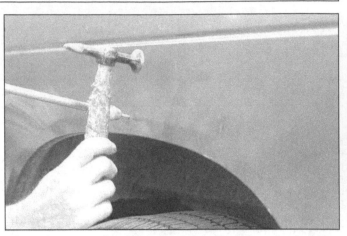

2 . . . then screw the slide-hammer into the hole and operate it. Tap with a hammer near the edge of the dent to help 'pop' the metal back to its original shape. When you're finished, the dent area should be close to its original contour and about 1/8-inch below the surface of the surrounding metal

3 Using coarse-grit sandpaper, remove the paint down to the bare metal. Hand sanding works fine, but the disc sander shown here makes the job faster. Use finer (about 320-grit) sandpaper to feather-edge the paint at least one inch around the dent area

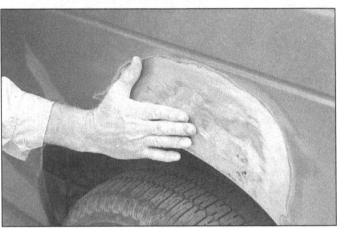

4 When the paint is removed, touch will probably be more helpful than sight for telling if the metal is straight. Hammer down the high spots or raise the low spots as necessary. Clean the repair area with wax/silicone remover

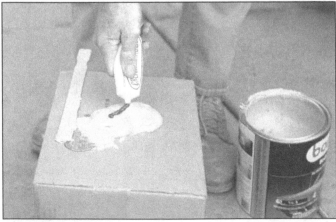

5 Following label instructions, mix up a batch of plastic filler and hardener. The ratio of filler to hardener is critical, and, if you mix it incorrectly, it will either not cure properly or cure too quickly (you won't have time to file and sand it into shape)

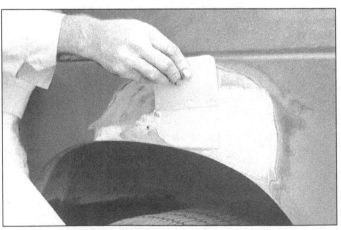

6 Working quickly so the filler doesn't harden, use a plastic applicator to press the body filler firmly into the metal, assuring it bonds completely. Work the filler until it matches the original contour and is slightly above the surrounding metal

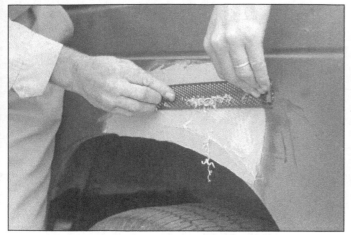

7 Let the filler harden until you can just dent it with your fingernail. Use a body file or Surform tool (shown here) to rough-shape the filler

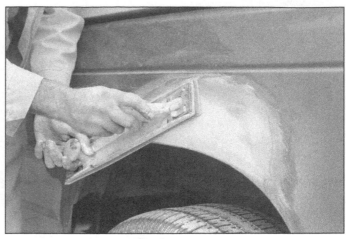

8 Use coarse-grit sandpaper and a sanding board or block to work the filler down until it's smooth and even. Work down to finer grits of sandpaper - always using a board or block - ending up with 360 or 400 grit

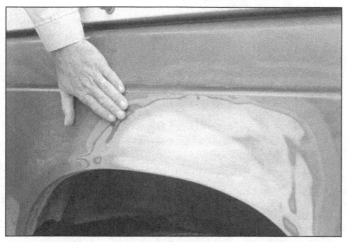

9 You shouldn't be able to feel any ridge at the transition from the filler to the bare metal or from the bare metal to the old paint. As soon as the repair is flat and uniform, remove the dust and mask off the adjacent panels or trim pieces

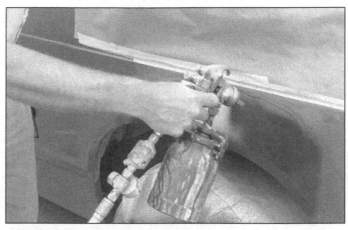

10 Apply several layers of primer to the area. Don't spray the primer on too heavy, so it sags or runs, and make sure each coat is dry before you spray on the next one. A professional-type spray gun is being used here, but aerosol spray primer is available inexpensively from auto parts stores

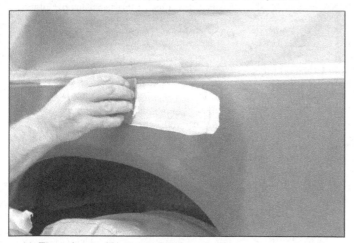

11 The primer will help reveal imperfections or scratches. Fill these with glazing compound. Follow the label instructions and sand it with 360 or 400-grit sandpaper until it's smooth. Repeat the glazing, sanding and respraying until the primer reveals a perfectly smooth surface

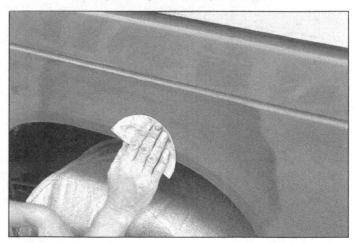

12 Finish sand the primer with very fine sandpaper (400 or 600-grit) to remove the primer overspray. Clean the area with water and allow it to dry. Use a tack rag to remove any dust, then apply the finish coat. Don't attempt to rub out or wax the repair area until the paint has dried completely (at least two weeks)

7.1a The front compartment lid and attaching hardware

A   *Hinge*
B   *Hinge-to-front compartment lid bolt*

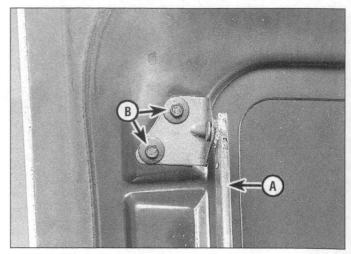

7.1b The front compartment lid and attaching hardware

A   *Front compartment lid support*
B   *Upper support bolt*

Flexible ("Enduraflex") panels

Note: *The following repair procedure applies to the bumpers, doors, fenders and lower rear quarter panels, all of which are made of this material.*

1   Remove the damaged panel if necessary or desirable. In most cases, repairs can be carried out with the panel installed.

2   Clean the area(s) to be repaired with a wax, grease and silicone removing solvent applied with a water-dampened cloth.

3   If the damage is structural, that is, if it extends through the panel, clean the backside of the panel area to be repaired as well. Wipe dry.

4   Sand the surface about 1-1/2 inches beyond the break.

5   Cut two pieces of fiberglass cloth large enough to overlap the break by about 1-1/2 inches. Cut only to the required length.

6   Mix the adhesive from the 3M #05900 kit according to the instructions included with the kit, and apply a layer of the mixture approximately 1/8-inch thick on the backside of the panel. Overlap the break by at least 1-1/2 inches.

7   Apply one piece of fiberglass cloth to the adhesive and cover the cloth with additional adhesive. Apply a second piece of fiberglass cloth to the adhesive and immediately cover the cloth with additional adhesive in sufficient quantity to fill the weave.

8   Allow the repair to cure for 20 to 30 minutes at 60° to 80°F.

9   If necessary, trim the excess repair material at the edge.

10   Remove all of the paint film over and around the area(s) to be repaired. The repair material should not overlap the painted surface.

11   With a drill motor and a sanding disc (or a rotary file), cut a "V" along the break line approximately 1/2-inch wide. Remove all dust and loose particles from the repair area.

12   Mix and apply the repair material. Apply a light coat first over the damaged area; then continue applying material until it reaches a level slightly higher than the surrounding finish.

13   Cure the mixture for 20 to 30 minutes at 60° to 80°F.

14   Roughly establish the contour of the area being repaired with a body file. If low areas or pits remain, mix and apply additional adhesive.

15   Block sand the damaged area with sandpaper to establish the actual contour of the surrounding surface.

16   If desired, the repaired area can be temporarily protected with several light coats of primer. Because of the special paints and techniques required for flexible body panels, it is recommended that the vehicle be taken to a paint shop for completion of the body repair.

Non-flexible panels

17   The non-flexible panels (primarily the roof skin) are made of a reinforced plastic material. Repair of these panels by the home mechanic should be limited to scratch repair.

18   If the scratch is superficial, lightly rub the scratched area with a fine rubbing compound to remove loose paint and built up wax. Rinse the area with clean water.

19   Apply touch-up paint to the scratch, using a small brush. Continue to apply thin layers of paint until the surface of the paint is level with the surrounding paint. Allow the new paint at least two weeks to harden, then blend it into the surrounding paint by rubbing with a very fine rubbing compound. Finally, apply a coat of wax to the scratch area.

20   If the scratch penetrates all the way through the paint and into the panel material, clean the area as above, then use a rubber or nylon applicator to coat the scratched area with a glaze-type filler. If required, the filler can be mixed with thinner to provide a very thin paste, which is ideal for filling very narrow scratches. Before the glaze filler in the scratch hardens, wrap a piece of small cotton around the tip of a finger. Dip the cloth in thinner and quickly wipe it along the surface of the scratch. This will ensure that the surface of the filler is slightly concave. The scratch can now be painted as described earlier.

5   Body repair - major damage

1   Unitized construction demands that the underbody components be properly aligned to assure correct suspension location. In the event of a collision, it is essential that the underbody be thoroughly checked and, if necessary, realigned by a body repair shop with the proper equipment.

2   Because each underbody component contributes directly to the overall strength of the body, proper materials and techniques must be used during body repair service. Again, because of the specialized materials, machinery, tools and techniques required for this kind of work, it should be performed by a professional paint and body shop.

3   Because all of the exterior panels are separate and replaceable units, they should be replaced rather than repaired if major damage occurs. Some of these components can be found in a wrecking yard that specializes in used vehicle components, often at considerable savings over the cost of new parts.

6   Hinges and locks - maintenance

Once every 3000 miles, or every three months, the hinges, locks and latch assemblies on the doors and the front and engine compartment lids should be given a few drops of light oil or lock lubricant. The door latch strikers should also be lubricated with a thin coat of grease to reduce wear and ensure free movement.

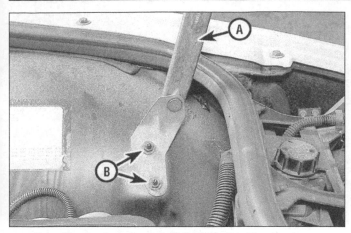

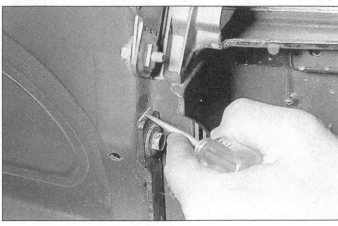

7.1c The front compartment lid and attaching hardware

A   *Front compartment lid support*
B   *Lower support nut*

7.2 Scribe a mark along the part of the hinge that bolts to the front compartment lid to assure proper alignment when the lid is reinstalled

7   Front compartment lid - removal, installation and adjustment

Refer to illustrations 7.1a, 7.1b, 7.1c and 7.2

Removal

1   Remove the two upper front compartment lid support attaching bolts **(see illustrations)**.
2   Scribe or paint alignment marks along the edges of the front compartment lid hinge upper flange **(see illustration)** and remove the hinge-to-lid mounting bolts.
3   Remove the front compartment lid.

Installation

4   Place the front compartment lid in position.
5   Place the upper mounting flange of the front compartment lid support in position, install the upper attaching bolts and tighten them until they are finger tight.

Adjustment

6   Slotted holes are provided at all front compartment lid hinge attaching points for proper vertical and fore and aft adjustment.

The best way to adjust the front compartment lid is to make one adjustment at a time.
7   If the front compartment lid is too high or too low at the front corners, loosen the hinge mounting nuts, reposition the lid and tighten the nuts.
8   If the lid is too high or too low at the rear corners, estimate the amount and direction of adjustment needed and adjust the front compartment lid bumpers.
9   If the lid is too far fore or aft, loosen the hinge-to-lid bolts, reposition the lid and tighten the bolts.
10   After the above three adjustments have been performed satisfactorily, make sure that all front compartment lid hinge and support nuts and bolts are tightened securely.

8   Door trim panel - removal and installation

Refer to illustrations 8.1, 8.3, 8.5, 8.6 and 8.9

Removal

1   Remove the wire clip from the window regulator handle **(see illustration)**.
2   Pry off the forward screw cover.
3   Remove the forward bezel mounting

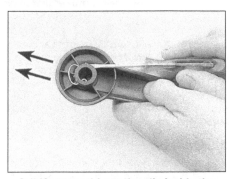

8.1 If you could see the clip inside the handle, this is what it would look like from the other side - note that the clip must be pushed away from the handle knob to release

screw **(see illustration)**.
4   Pry off the rear screw cover.
5   Remove the rear bezel mounting screw **(see illustration)**.
6   Pull the remote control handle bezel away from the door trim panel. Using a small flat-bladed tool like a small screwdriver, insert the blade between the backside of the locking knob and the lock rod and pry the rod loose from the knob **(see illustration)**.

8.3 Pry off the round screw cover behind the door remote control handle and remove the remote control bezel forward attaching screw

8.5 Pry off the square screw cover behind the door remote control handle and remove the remote control bezel rear attaching screw

8.6 Pull the remote control bezel away from the door and disconnect the locking rod from the backside of the inside locking knob

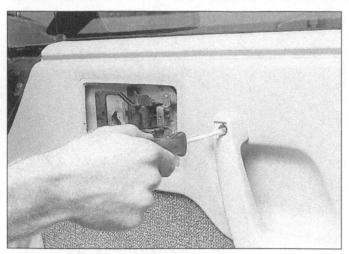

8.9 Pry off the square screw cover from the armrest and remove the upper armrest attaching screw

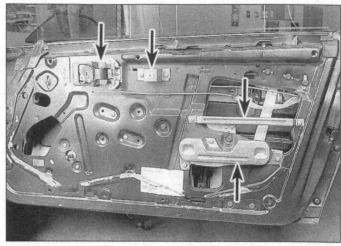

9.2 These parts (arrows) are riveted and cannot easily be removed, so you will have to cut the water deflector loose from them since it was installed at the factory between them and the door

7   Remove the remote control handle bezel.
8   Pry off the screw cover from the armrest.
9   Remove the armrest mounting screws **(see illustration)**.
10   Carefully pry along the edges of the door trim panel until it is free of all pop fasteners and remove it.

Installation

11   Place the door trim panel in position against the door and push on it around the edges until all the pop fasteners are seated into the door.
12   Place the armrest in position, install the mounting screws and tighten them moderately. **Caution:** *If you overtighten these screws, you may strip out the clips into which they are threaded.*
13   Install the screw cover into the armrest.
14   Connect the lock rod to the back of the inside locking knob and place the remote control handle bezel in position.
15   Install the forward and rear bezel mounting screws and tighten them moderately.
16   Install the forward and rear screw covers.
17   Install the window regulator handle (use the handle on the opposite door as a guide for proper positioning of the handle knob when the window is in the fully closed posi-

tion). Push the locking clip back into position until it locks into the groove in the splined window regulator shaft.

9   Door window glass - adjustment

Refer to illustrations 9.2 and 9.3

1   Remove the door trim panel (Section 8).
2   Remove the inner panel water deflector (the brown paper liner). **Note:** *Since complete removal of the deflector necessitates removal of the door inner panel stiffener and armrest hanger plates, both of which are riveted to the door, use an sharp knife or a razor blade to carefully trim around the non-removable parts* **(see illustration)**. The deflector is secured around its edges by a string loaded sealing material and by sealing tape. A flat-bladed putty knife can be used to release the sealer around the edges of the deflector. Keep the blade between the inner panel and the string that is imbedded in the sealer.
3   If the upper edge of the window does not line up with the roof side rail weatherstrip **(see illustration)**:

a) *Loosen the rear up-stop to support bolt and the front up-stop to inner door bolts*

b) *Adjust the window so that the upper edge of the glass is parallel with the roof side rail weatherstrip*
c) *Adjust the up-stops*
d) *Tighten the attaching bolts*
4   If the upper edge of the window is inboard or outboard:

a) *Loosen the front retainer to support bolt*
b) *Loosen the rear cam guide to support bolts*
c) *Loosen the rear up-stop support to inner door bolt*
d) *Loosen the front and rear stabilizer to inner door bolts*
e) *Adjust the vertical guide and rear up-stop support in or out as required*
f) *Tighten the attaching bolts*
5   If the window is too far forward or rearward:

a) *Loosen the front glass run channel support to inner door bolt and the lower front glass run channel to inner door bolt*
b) *Loosen the lower rear cam guide to inner door bolts and rear cam support to inner door bolts*
c) *Align the glass in the correct up position*
d) *Tighten the front glass run channel support to inner door bolt*

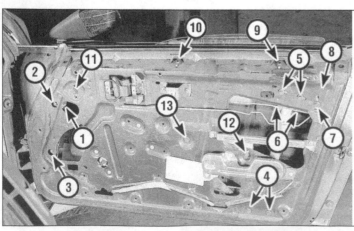

9.3 Door glass adjusting hardware

1   *Front retainer-to-support*
2   *Front glass run channel support-to-inner door*
3   *Lower front glass run channel-to-inner door*
4   *Lower rear cam guide-to-inner door*
5   *Rear cam support-to-inner door*
6   *Rear cam guide-to-support*
7   *Rear up-stop support-to-inner door*
8   *Rear up-stop-to-support*
9   *Rear stabilizer-to-inner door*
10   *Front stabilizer-to-inner door*
11   *Front up-stop-to-inner door*
12   *Rear inner panel cam-to-inner door*

e) *Tighten the rear cam support to inner door bolts*
f) *Lower the glass*
g) *Tighten the lower front glass run channel to inner door bolt*
h) *Tighten the lower rear cam guide to inner door bolts*

6   If the window is too high or low in the up position:

a) *Adjust the rear up-stop support to inner door and front up-stop to inner door as required*

10   Door latch striker - adjustment

Refer to illustrations 10.1 and 10.2

1   The door latch striker consists of a single metal bolt and washer assembly which is threaded into a tapped, floating cage plate in the body pillar **(see illustration)**. The door is secured in the closed position when the door lock fork bolt snaps and engages the striker bolt.

2   To check for fore and aft adjustment:

a) *Check for proper door alignment*
b) *Apply modeling clay or body caulking to the lock bolt opening* **(see illustration)**
c) *Close the door only as far as necessary for the striker to form an impression in clay or compound. Completely closing the door will make clay removal difficult*

3   If the striker is not centered in accordance with these dimensions, remove it with a special striker wrench.

4   Install a spacer or spacers as required to obtain the correct alignment. Spacers are available in 3/32-inch and 5/32-inch sizes.

5   Reinstall the striker and tighten it to the specified torque.

6   Check the striker for up or down and in or out adjustment.

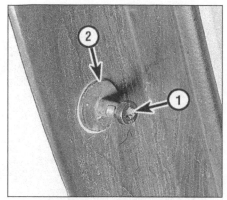

10.1 The door lock striker is a bolt with a spacer washer which allows it to be positioned fore and aft and a tapped cage plate that can be moved up or down and in and out to adjust the striker

1   *Striker bolt*
2   *Spacer/washer*

7   If the striker is too high or low or if it is too far in or out, remove again and enlarge the hole in the direction required with a rotary file. **Caution:** *It is vital that a flat end rotary file be used so that no damage is done to the tapped cage plate.*

8   Install the striker and tighten it to the specified torque.

11   Door latch - removal and installation

Refer to illustrations 11.3a and 11.3b

Removal

1   Remove the trim panel (Section 8).
2   Remove the water deflector (Section 9).
3   Disconnect the inside remote handle

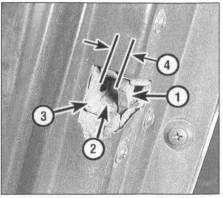

10.2 To check fore and aft alignment of the striker, apply a little modeling clay to the lock fork bolt opening, close the door until the striker forms an impression in the clay, then note whether the striker impression is centered fore and aft

1   *Lock fork bolt*
2   *Approximate location of the striker bolt impression*
3   *Modeling clay location (fill area)*
4   *Measurement*

lock rod and the lock cylinder-to-latch rod at the lock assembly **(see illustrations)**. **Note:** *Spring clips are used to secure remote control connecting rods and inside locking rods to levers and handles. A slot in the clip provides for disengagement.*

4   If your vehicle is equipped with a door ajar switch electrical lead in the door itself (later models have the switch in the pillar instead of the door), disconnect it.

5   Remove the latch assembly screws.

6   Lower the latch assembly to disconnect the outside handle lock rod from the latch.

7   Remove the latch assembly.

11.3a View showing the relationship between the door latch assembly, the door lock cylinder and the outside handle

1   *Outside handle studs*
2   *Handle attaching nuts*
3   *Outside handle-to-latch rod*
4   *Lock cylinder*
5   *Clip*
6   *Lock cylinder-to-latch rod*

11.3b View of the routing of the connecting rods between the remote handle assembly and the latch assembly

1   *Latch assembly*
2   *Inside locking rod*
3   *Outside handle-to-latch rod*
4   *Outside lock cylinder-to-latch rod*
5   *Inside door handle-to-latch rod*

Installation

8   Install the spring clips on the new latch assembly.
9   Place the latch assembly into position and install the lock rods.
10   Install the latch assembly mounting screws and tighten them to the specified torque.

12 Door lock cylinder - removal and installation

Refer to illustration 12.4

Removal

1   Remove the door trim panel (Section 8).
2   Remove the water deflector (Section 9).
3   Remove the lock rod retaining clip from the lock cylinder lever by prying between the lever and the clip in an outward and downward motion with a small screwdriver.
4   Remove the cylinder lock assembly retainer by pushing it forward (away from the lock cylinder) with the tip of a large screwdriver. It may be necessary to grip the leading edge of the retainer with a pair of needle nose pliers in order to prevent the tip of the screwdriver blade from slipping off the forward edge of the retainer **(see illustration)**.
5   Remove the lock cylinder.

Installation

6   Place the new lock cylinder in position.
7   Install the retainer by sliding it rearward (toward the lock cylinder). Make sure that the retainer is positioned between the small guide bosses on the edge of the cylinder lock casting and the steel stamping which supports the lock cylinder.
8   Install the lock rod and its retaining clip onto the end of the lock cylinder lever.
9   Check the operation of the new lock cylinder and the lock rod to make sure that they are working properly.
10   Install the water deflector (Section 9).
11   Install the door trim panel (Section 8).

13 Door outside handle - removal and installation

Refer to the illustrations in Section 11 for this procedure

Removal

1   Remove the door trim panel (Section 8).
2   Remove the water deflector (Section 9).
3   Remove the nuts at the door handle.
4   Remove the lock cylinder retainer and outside locking rod.
5   Remove the handle assembly.

Installation

6   Install the new handle assembly.
7   Install and tighten securely the door handle nuts.
8   Install the outside handle locking rod and lock cylinder retainer.

12.4 To free the lock cylinder, push forward on the retainer (arrow) with a large screwdriver (to prevent the screwdriver from slipping off the retainer flange, grip the flange with a pair of needle nose pliers and push on the pliers)

9   Install the water deflector (Section 9).
10   Install the door trim panel (Section 8).

14 Bumpers and energy absorbers - general information

The bumpers of all Fieros are designed so that the vehicle can withstand a collision into a fixed barrier at 2.5 mph. After absorbing the energy of a collision, the bumpers return to their original position.
The front and rear bumper face bars (fascias) are made of urethane. Urethane will withstand minor impact and return to its original shape. The front bumper fascia is integral with the front end panel.
The absorbing capability for both front and rear bumper systems is achieved through honeycombed energy absorbing devices in each bumper.
If, after a collision, there is obvious damage to either energy absorbing unit, replace it. If the collision is not severe enough to obviously damage the energy absorber but the bumper nevertheless fails to return to its original position, replace the energy absorber.

15 Front bumper - removal and installation

Removal

1   Remove the mounting screws from each sidemarker light and disconnect the bulb holder from each sidemarker light assembly.
2   Remove the two mounting screws from each sidemarker light assembly indent attaching the front bumper fascia to the front fender panel.
3   Remove the three mounting screws in each inner fender panel attaching the front bumper fascia to the inner fender panel.
4   Remove the seven screws attaching the front bumper fascia to the body front end assembly.
5   From underneath the vehicle, remove the three push retainers from the underside of the grille area of the fascia.
6   Remove the front bumper fascia.
7   Pull the center section of each push

retainer out and then drill it out with a 1/4-inch drill. Remove the energy absorber.
8   Install the new energy absorber and new push retainers.

Installation

9   Place the front bumper fascia in position and install the seven mounting screws that attach the fascia to the body front end assembly. Tighten the bolts finger tight.
10   Install the three push retainers in the underside of the front bumper fascia.
11   Install the three mounting screws to each inner fender panel. Tighten them finger tight.
12   Install the two mounting screws that attach the front bumper fascia to the front fender panel in each sidemarker light assembly indent. Tighten them finger tight.
13   Check the alignment between the front bumper fascia and the front compartment lid. Carefully manipulate the fascia until it lines up with the front edge of the front compartment lid and tighten all the mounting screws.
14   Plug in the bulb holder and install the sidemarker light assembly.

16 Rear bumper - removal and installation

Refer to illustrations 16.1, 16.4, 16.7 and 16.11

1   Open the engine compartment lid and remove the six screw covers from the trailing edge of the rear panel **(see illustration)**.
2   Remove the six taillight retaining screws from the rear panel.
3   Slide out each taillight lens far enough to reach the bulb holders.
4   Disconnect all four bulb holders from the backside of each taillight lens **(see illustration)**. **Note:** *To remove the inboard three bulb holders, push in the lock tab and rotate the holder counterclockwise; to remove the outboard bulb holder, rotate it counterclockwise and pull it out. Carefully set the bulb holders aside.*
5   Remove the seven pop fasteners from the top of the rear bumper fascia.
6   Remove the sidemarker retaining screws and disconnect the sidemarker bulb

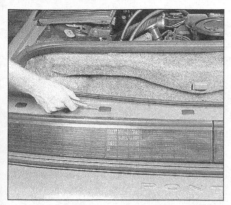

16.1 Remove the six plastic screw covers from the rear panel

holder from each sidemarker. Remove the sidemarker strips from both sides.

7   Remove the screw attaching the rear bumper fascia to the rear panel from each sidemarker strip indent **(see illustration)**.

8   Remove the four screws in each wheel well.

9   Remove the seven plastic push retainers from the underside of the rear bumper fascia.

10   Pull the rear bumper fascia away from the vehicle far enough to disconnect the two license plate lamp bulb holders.

11   If a new energy absorber is to be installed, pull out the pop fastener push retainers and remove the energy absorbing device **(see illustration)**. **Note:** *If you are unable to pull the retainers from the energy absorbing device on your vehicle, drill them out instead (they must be replaced anyway).*

12   Place a new energy absorbing device in position and install new push retainers.

13   Place the rear bumper fascia in position. Be sure to connect the two license plate light bulb holders to the backside of the license plate well.

14   Install the seven plastic push retainers in the underside of the fascia.

15   Install the four screws that attach the fascia to each wheel well and tighten them securely.

16   Install the screw that attaches the side of the fascia to the rear panel at the sidemarker strip indent.

17   Connect the bulb holder to the backside of each sidemarker, place the sidemarker in position over its indent and install the two sidemarker mounting screws.

18   Install the seven pop fastener push retainers into the top of the fascia.

19   Connect the four bulb holders to each taillight lens and push the taillight lenses into place.

20   Install and tighten securely the taillight lens mounting screws.

21   Install the taillight lens mounting screw plastic covers.

---

17 Fender - removal and installation

Refer to illustrations 17.3, 17.4a, 17.4b and 17.6

Removal

1   If you are removing the right front fender panel and it is fitted with a radio antenna, remove the antenna by unscrewing the hex nut base.

2   Remove the sidemarker light assembly and disconnect the bulb holder from the backside of the sidemarker.

3   Remove the two screws from the indent for the sidemarker light between the front bumper fascia and the front fender panel **(see illustration)**.

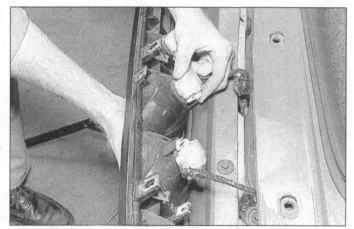

16.4 Slide each taillight lens out of the recess between the rear bumper fascia and the rear panel far enough to disconnect the bulb holders by pushing in the locking tabs and turning the bulb holders counterclockwise

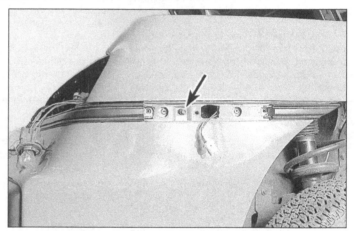

16.7 Remove the mounting screw (arrow) that attaches the rear bumper fascia to the rear panel

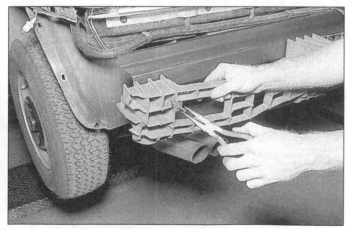

16.11 You should be able to pull out the push retainers attaching the energy absorber to the bumper

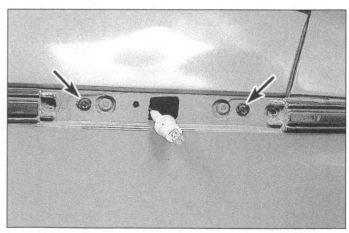

17.3 Remove the two mounting screws (arrows) that attach the front fender panel to the front bumper fascia

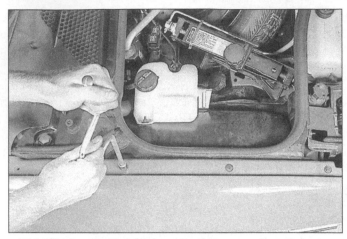

17.4a Remove the four larger mounting screws from the upper flange of the front fender panel . . .

17.4b . . . and the smaller fifth mounting screw in the extreme forward end of the front fender panel upper flange

4   Remove the five attaching screws from the top mounting flange of the front fender panel **(see illustrations)**.

5   Remove the five screws attaching the fender panel to the inner wheelhouse panel. **Note:** *The forward screw is smaller than the other four. Do not attempt to install one of the bigger screws into this hole during reassembly.*

6   Pry the forward end of the rocker panel away from the vehicle and, with a hammer and punch, knock off the head of the single rivet which attaches the lower rear edge of the fender panel to the vehicle **(see illustration)**.

7   Remove the front fender panel.

Installation

8   Place the new front fender panel in position.

9   Install the five attaching screws in the top mounting flange of the fender panel. **Note:** *Be sure to install the small screw at the forward hole. Tighten the screws finger tight.*

10   Install the five screws that attach the fender panel to the inner wheelhouse panel and tighten them finger tight.

11   Install the two sidemarker indent screws that attach the front bumper fascia to the front fender panel.

12   Install a nut and bolt of the appropriate size into the hole that was riveted at the bottom rear corner of the front fender panel.

13   Check the alignment of the front fender panel in relation to the front compartment lid and then tighten all attaching screws securely.

14   Plug in the bulb holder to the backside of the sidemarker light and install the sidemarker assembly. Tighten the two mounting screws.

15   Install the antenna.

---

18   Engine compartment lid - removal, installation and adjustment

Refer to illustrations 18.3 and 18.4

Removal

1   Disconnect the compartment lid ground strap.

2   If the engine compartment lid on your vehicle is equipped with a remote control release, disconnect the electrical lead at the left hinge.

3   Scribe or paint alignment marks around the edge of the hinges to assure proper alignment during reinstallation **(see illustration)**.

4   Remove the engine compartment lid mounting bolts from each hinge plate. **Warning:** *The torque rod bolts - the ones in the middle, between the two engine compartment lid mounting bolts on each hinge plate - are under tension. Do not disconnect them - personal injury or damage to the vehicle could result* **(see illustration)**.

5   Remove the engine compartment lid.

Installation

6   Place the engine compartment lid in position, making sure that the scribe marks line up along the edges of the hinge plates, and install the engine compartment lid mounting bolts.

7   Close the lid carefully and make sure that it is aligned properly with the rear panel. The lid can be manipulated by pushing or pulling it to the left or right. When both edges

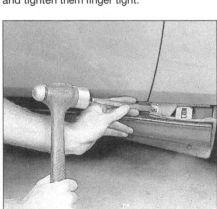

17.6 Pry the rocker panel away from the vehicle and remove the head of the rivet that attaches the lower rear corner of the front fender panel with a hammer and chisel

18.3 Before loosening the engine compartment lid mounting bolts, scribe or paint alignment marks along the edges of the hinge plates to assure proper alignment during reinstallation

18.4 Remove the engine compartment lid mounting bolts but do not remove the middle bolt (the one with the recessed hex in the head) because it is under tension from the torque rod

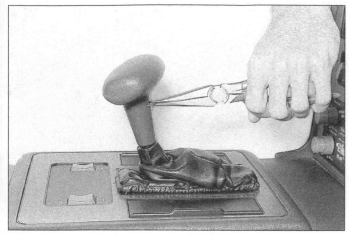

19.2 The shifter handle on automatic transaxle equipped vehicles cannot be removed until the wire retaining clip in the front of the handle is extracted

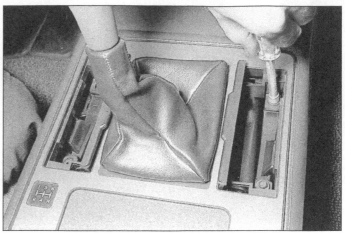

19.3 There are four screws - two in each ashtray receptacle - that must be removed before the shifter trim plate assembly can be removed

of the lid line up with the edges of the rear panel, carefully open the lid and tighten the mounting bolts.

8   If the engine compartment lid on your vehicle is equipped with a remote control release, connect the electrical connector.

9   Reconnect the ground strap.

19 Console - removal and installation

Refer to illustrations 19.2, 19.3, 19.12, 19.15, 19.17, 19.21, 19.23 and 19.28

1   Place the shifter in first gear and disconnect the cable from the negative terminal of the battery.

2   If your vehicle is equipped with a manual transaxle, unscrew and remove the shift knob. If it is equipped with an automatic transaxle, grasp the wire clip in the front of the shifter handle with needle nose pliers and pull it out **(see illustration)**. Remove the handle.

3   Remove both ashtrays from the shift plate assembly and remove the two screws underneath each ashtray receptacle **(see illustration)**. Remove the shift plate assembly. If your vehicle is equipped with an automatic transaxle, disconnect the gear indicator bulb holder from the underside of the shift plate assembly before removing it.

4   If your vehicle is equipped with power windows, remove the four screws securing the power window toggle switch trim plate to the console assembly and disconnect both connectors from the backside of the plate.

5   Remove both screws from the cigarette lighter trim plate. Remove both screws from the cigarette lighter flange.

6   Unplug the cigarette lighter, bulb holder and ground wire electrical connections.

7   Remove the four screws securing the rear console pad assembly. There are two screws at the front and two in the glovebox.

8   Remove the rear console pad assembly from between the seats.

9   Remove the two screws securing the Assembly Line Communications Link (ALCL)

diagnostics connector to the console support assembly.

10   A strip of electrical tape secures the cigarette lighter/ash tray lights/ground wire harness and the ALCL wire harness to the top of the console support assembly. Peel it away and set both harnesses aside.

11   Remove the trim plate for the radio and heater control units from the front pad assembly.

12   Remove the screws securing the radio **(see illustration)**, pull it out of the console support assembly, disconnect the electrical connectors and antenna lead from the rear and remove it.

13   Remove the screws securing the heater control assembly, pull it out of the dashboard console, disconnect the electrical connectors (one large connector and one smaller one) from the backside, disconnect the heater control cable (Chapter 3) and remove the heater control assembly.

14   Remove the screws securing the front pad assembly to the dashboard and remove the front pad assembly.

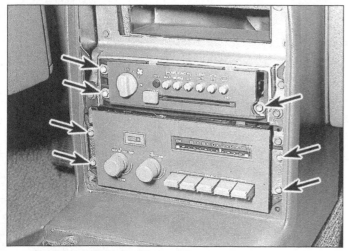

19.12 Three screws (arrows) secure the heater control assembly and four screws (arrows) support the radio to the console support frame

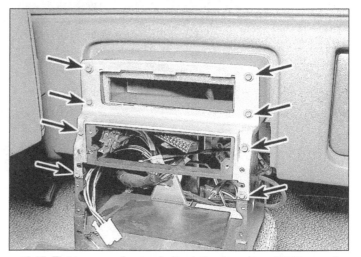

19.15 Eight screws (arrows) attach the console reinforcement frame to the dashboard

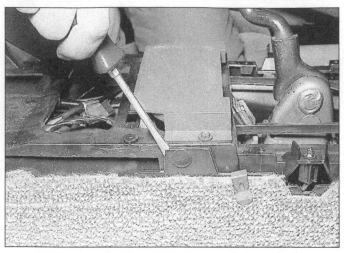

19.17 Four pop fasteners - two per side - can be pried loose with a small screwdriver

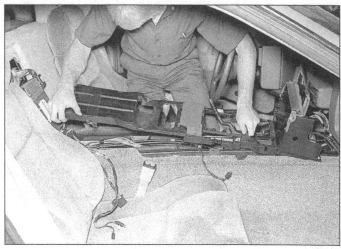

19.21 To remove the console support assembly, lift it up at the rear and pull it back from the dashboard

15   Remove the four screws securing the front reinforcement frame to the dashboard and remove the frame **(see illustration)**.

16   Remove the three carpet clips on each side of the console support assembly.

17   Remove the four plastic pop fasteners from each side of the console support assembly **(see illustration)**.

18   Remove the screw from the bracket between the shifter assembly and the console support assembly.

19   Rotate the two ashtray light bulb holders clockwise 1/4-turn and remove them from their plastic retainers.

20   Pop off the plastic clips securing the ash tray light bulb wiring harnesses to the console support assembly.

21   Remove all remaining attaching screws from the console support assembly, lift it up at the rear, pull it back from the instrument panel pad and remove it **(see illustration)**.

Installation

22   Lay the console support assembly in position.

23   Referring to the accompanying illustration, install the console support assembly

mounting screws in the following order:

a) *Support frame to console support assembly screws*
b) *Support frame to dashboard screws*
c) *Right rear console support assembly to body screw*
d) *Left rear console support assembly to body screw*
e) *Forward console support assembly to body screws*
f) *Forward side console support assembly to body screws*
g) *Shifter assembly bracket bolt*

Note: *Do not tighten any screws or bolts until all of the above have been installed finger tight, then tighten all fasteners securely.*

24   Install the plastic clips that secure the ash tray light bulb wiring harnesses to the console support assembly.

25   Install the ashtray light bulb holders by pushing them in and turning them 1/4-turn counterclockwise.

26   Install the four plastic pop fasteners (two on each side of the console support assembly).

27   Install the three carpet clips to each side

of the console support assembly.

28   Install the front pad assembly and tighten the mounting screws securely **(see illustration)**.

29   Connect the heater control cable (see Chapter 3) and the two electrical connectors to the backside of the heater control assembly, install the assembly and tighten the bolts securely.

30   Connect the electrical connectors and antenna lead to the backside of the radio, install the radio assembly and tighten the bolts securely.

31   Install the radio and heater control trim plate onto the front pad assembly and tighten the bolts securely.

32   Install the two screws that attach the ALCL connector to the top of the console support assembly.

33   Install the rear console pad assembly between the seats and tighten the four attaching screws (two front and two rear) securely.

34   Plug in the cigarette lighter, bulb holder and ground wire electrical connectors.

35   Tape the cigarette lighter/ashtray lights/ground wire harness and the ALCL

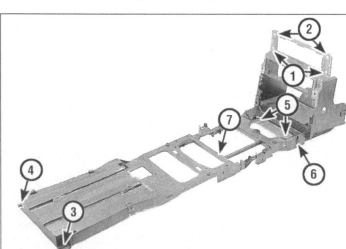

19.23 Install the console support assembly mounting screws in the following order

1   *Support frame-to-console support assembly screws*
2   *Support frame-to-dashboard screws*
3   *Right rear console support assembly-to-body screw*
4   *Left rear console support assembly-to-body screw*
5   *Forward console support assembly-to-body screws*
6   *Forward side console support assembly-to-body screws*
7   *Shifter assembly bracket bolt*

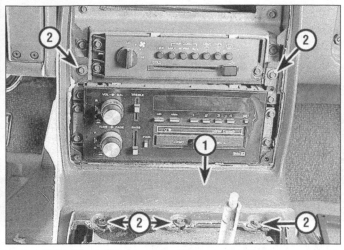

19.28 View of the console pad and mounting screws

1   Console pad
2   Mounting screws

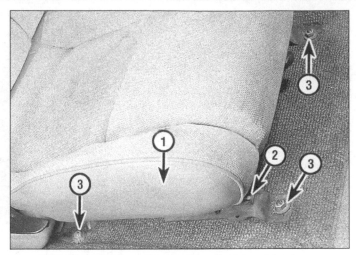

20.2 View of the seat assembly adjuster mechanisms

1   Seat
2   Seat adjuster assembly
3   Nuts (3 of 4 shown)

connector wire harness to the top of the console support assembly.

36   Install the cigarette lighter and tighten the two screws securely. Install the cigarette lighter trim plate and tighten the two screws securely.

37   If your vehicle is equipped with power windows, plug in the toggle switch electrical connectors to the underside of the trim plate, install the trim plate and tighten the four mounting screws securely.

38   Install the shifter plate assembly and tighten the four mounting screws (two in each ashtray receptacle) securely. **Note:** *If your vehicle is equipped with an automatic transaxle, be sure to plug in the gear position indicator bulb holder before placing the shifter plate assembly in position.*

39   Install the shifter knob. If your vehicle is equipped with a manual transaxle, screw the knob onto the shifter. If it is equipped with an automatic transaxle, slip the knob onto the shifter and install the wire retaining clip.

40   Connect the cable to the negative terminal of the battery.

20   Seats - removal and installation

Refer to illustration 20.2

Removal

1   Move the seat to the forward position.
2   Remove the adjuster-to-floor pan attaching nuts **(see illustration)**.
3   Move the seat to the rearward position.
4   Remove the adjuster-to-floor pan front attaching nuts.
5   Remove the seat assembly.

Installation

6   Place the seat assembly in position.
7   Move the seat to the rearward position.
8   Install the adjuster-to-floor pan front attaching nuts and tighten them securely.
9   Move the seat to the full-forward position.
10   Install the adjuster-to-floor pan rear attaching nuts and tighten them securely.
11   Check the seat assembly adjuster for proper operation.

21   Headlight door assembly - removal and installation

Refer to illustration 21.1
Note: *The headlight doors have slotted mounting points which insure proper clearance between the headlight door and the front compartment lid. The entire headlight door assembly can be adjusted to achieve the desired appearance and fit. Care should be exercised when adjusting the headlight door assembly so as not to damage any components.*

1   Remove the headlight door assembly-to-hinge assembly mounting bolts and spacers **(see illustration)**.
2   Remove the cover panel and door assembly as a unit.
3   If you are replacing the door assembly, but not the cover panel, you will need to remove the cover panel from the old door and install it onto the new one. To do so, remove the pair of retainers from the underside of the door assembly. Then, holding the

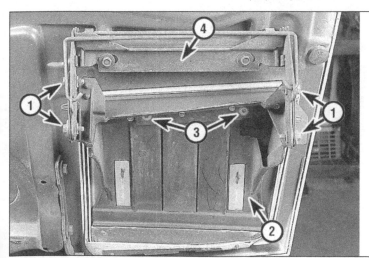

21.1 View of the headlamp door assembly

1   Door-to-hinge bolts and spacers
2   Door and cover panel
3   Cover panel-to-door retainers
4   Hinge

assembly open, lift the rear edge of the headlight cover panel and slide it forward to remove it.

4   Slide the cover panel into place on the new door assembly and install two new retainers.

5   Place the headlight cover panel/door assembly in position and install the door assembly-to-hinge assembly bolts and spacers.

22   Headlight door - emergency (manual) operation

Refer to illustration 22.1

    In the event that the headlight door should fail to open when the headlights are turned on, a manually operated crank assembly can be used to raise the door **(see illustration)**.

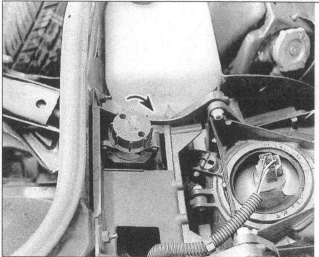

22.1 In the event of electrical or mechanical failure of the power headlamp door operating mechanism, turn this knob to manually raise or lower the door assembly

Chapter 12
Chassis electrical system

Contents

Specifications

Torque specifications

	In-lbs
Windshield wiper motor transmission-to-cowl panel screws	49 to 80
Windshield wiper motor attaching screws	51 to 73
Headlight mounting bracket bolts	
1984 and 1985	84
1986 and later	80

1 General information

The electrical system is a 12-volt, negative ground type. Power for the lights and all electrical accessories is supplied by a lead/acid-type battery which is charged by the alternator.

This Chapter covers repair and service procedures for the various electrical components not associated with the engine. Information on the battery, alternator, distributor and starter motor can be found in Chapter 5.

It should be noted that whenever portions of the electrical system are worked on, the negative battery cable should be disconnected at the battery to prevent electrical shorts and/or fires.

2 Electrical troubleshooting - general information

A typical electrical circuit consists of an electrical component, any switches, relays, motors, etc. related to that component and the wiring and connectors that connect the component to both the battery and the chassis. To aid in locating a problem in any electrical circuit, wiring diagrams are included at the end of this book.

Before tackling any troublesome electrical circuit, first study the appropriate diagrams to get a complete understanding of what makes up that individual circuit. Trouble spots, for instance, can often be narrowed down by noting if other components related to that circuit are operating properly or not. If several components or circuits fail at one time, chances are the problem lies in the fuse or ground connection, as several circuits are often routed through the same fuse and ground connections.

Electrical problems often stem from simple causes, such as loose or corroded connections, a blown fuse or a melted fusible link. Always visually inspect the condition of the fuse, wires and connections in a problem circuit before troubleshooting it.

If testing instruments are going to be utilized, use the diagrams to plan ahead of time where you will make the necessary connections in order to accurately pinpoint the trouble spot.

The basic tools needed for electrical

3.1 The fuse block is located under the left side of the dash

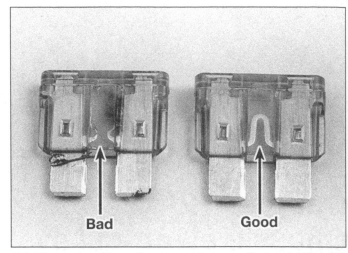

3.3 When a fuse blows, the element between the terminals melts - the fuse on the left is blown, the fuse on the right is good

troubleshooting include a circuit tester or voltmeter (a 12-volt bulb with a set of test leads can also be used), a continuity tester, which includes a bulb, battery and set of test leads, and a jumper wire, preferably with a circuit breaker incorporated, which can be used to bypass electrical components.

Voltage checks should be performed if a circuit is not functioning properly.

Connect one lead of a circuit tester to either the negative battery terminal or a known good ground. Connect the other lead to a connector in the circuit being tested, preferably nearest to the battery or fuse. If the bulb of the tester lights up, voltage is present, which means that the part of the circuit between the connector and the battery is problem free. Continue checking the rest of the circuit in the same fashion. When you reach a point at which no voltage is present, the problem lies between that point and the last test point with voltage. Most of the time the problem can be traced to a loose connection. **Note:** *Keep in mind that some circuits receive voltage only when the ignition key is in the Accessory or Run position.*

One method of finding shorts in a circuit is to remove the fuse and connect a test light or voltmeter in its place to the fuse terminals. There should be no voltage present in the circuit. Move the wiring harness from side-to-side while watching the test light. If the bulb goes on, there is a short to ground somewhere in that area, probably where the insulation has rubbed through. The same test can be performed on each component in the circuit, even a switch.

Perform a ground test to check whether a component is properly grounded. Disconnect the battery and connect one lead of a selfpowered test light, known as a "continuity tester", to a known good ground. Connect the other lead to the wire or ground connection being tested. If the bulb goes on, the ground is good. If the bulb does not go on, the ground is not good.

A continuity check his done to determine if there are any breaks in a circuit - if it is passing electricity properly. With the circuit off (no power in the circuit), a self-powered continuity tester can be used to check the circuit. Connect the test leads to both ends of the circuit (or to the "power" end and a good ground), and if the test light comes on the circuit is passing current properly. If the light doesn't come on, there is a break somewhere in the circuit. The same procedure can be used to test a switch, by connecting the continuity tester to the power in and power out sides of the switch. With the switch turned On, the test light should come on.

When diagnosing for possible open circuits, it is often difficult to locate them by sight because oxidation or terminal misalignment are hidden by the connectors. Merely wiggling a connector on a sensor or in the wiring harness may correct the open circuit condition. Remember this when an open circuit is indicated when troubleshooting a circuit. Intermittent problems may also be caused by oxidized or loose connections.

Electrical troubleshooting is simple if you keep in mind that all electrical circuits are basically electricity running from the battery, through the wires, switches, relays, fuses and fusible links to each electrical component (light bulb, motor, etc.) and back to ground, from which it is passed back to the battery. Any electrical problem is an interruption in the flow of electricity to and from the battery.

3   Fuses - general information

Refer to illustrations 3.1 and 3.3

1   The electrical circuits of the vehicle are protected by a combination of fuses, circuit breakers and fusible links. The fuse block is located under the instrument panel on the left side of the dashboard **(see illustration)**.

2   Each of the fuses is designed to protect a specific circuit, and the various circuits are

identified on the fuse panel itself. Miniaturized fuses are employed in the fuse block. These compact fuses, with blade terminal design, allow fingertip removal and replacement.

3   If an electrical component fails, always check the fuse first. The easiest way to check fuses is with a test light. Check for power at the exposed terminal tips of each fuse. If power is present on one side of the fuse but not the other, the fuse is blown. A blown fuse can also be confirmed by visually inspecting it **(see illustration)**.

4   Be sure to replace blown fuses with the correct replacement. Fuses of different ratings are physically interchangeable, but only fuses of the proper rating should be used. Replacing a fuse with one of a higher or lower value than specified is not recommended. Each electrical circuit needs a specific amount of protection. The amperage value of each fuse is molded into the fuse body. **Caution:** *At no time should a fuse be bypassed with pieces of metal or foil. Serious damage to the electrical system could result.*

5   If the replacement fuse immediately fails, do not replace it again until the cause of the problem is isolated and corrected. In most cases, this will be a short circuit in the wiring caused by a broken or deteriorated wire.

4   Fusible links - general information

Refer to illustration 4.1

1   Some circuits are protected by fusible links **(see illustration)**. These links are used in circuits which are not ordinarily fused, such as the ignition circuit.

2   Although the fusible links appear to be a heavier gauge than the wire they are protecting, the appearance is due to the thick insulation. All fusible links are four wire gauges smaller than the wire they are designed to

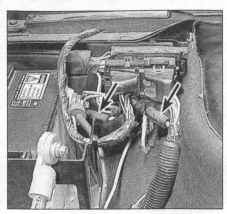

4.1 These two fusible links (arrows) in the main engine wiring harness of the Fiero are typical

protect. The location of the fusible links on your particular vehicle may be determined by referring to the wiring diagrams at the end of this book.

3　Fusible links cannot be repaired, but a new link of the same size wire can be put in its place. The procedure is as follows:

a) *Disconnect the negative cable at the battery.*

b) *Disconnect the fusible link from the wiring harness.*

c) *Cut the damaged fusible link out of the wiring just behind the connector.*

d) *Strip the insulation approximately 1/2-inch.*

e) *Position the connector on the new fusible link and crimp it into place.*

f) *Use rosin core solder at each end of the new link to obtain a good solder joint.*

g) *Use plenty of electrical tape around the soldered joint. No wires should be exposed.*

h) *Connect the battery ground cable. Test the circuit for proper operation.*

5　Circuit breakers - general information

Circuit breakers, which are located in the main fuse block, protect accessories such as power windows, power door locks and the rear window defogger.

The headlight wiring is also protected by a circuit breaker. An electrical overload in the system will cause the lights to go off and come on or, in some cases, to remain off. If this happens, check the headlight circuit immediately. The circuit breaker will function normally once the overload condition is corrected. *Refer to the wiring diagrams at the end of this book for the location of the circuit breakers in your vehicle.*

6　Turn signal and hazard flashers - check and replacement

Refer to illustration 6.10

Turn signal flasher unit

1　The turn signal flasher, a small canister shaped unit located on the left steering column bracket under the instrument panel, next to the main fuse panel, flashes either the left or right front and rear turn signal lamps.

2　When the flasher unit is functioning properly, an audible click can be heard during its operation. If the turn signals fail on one side or the other and the flasher unit does not make its characteristic clicking sound, a faulty turn signal bulb is indicated.

3　If both turn signals fail to blink, the problem many be due to a blown fuse, a faulty flasher unit, a broken switch or a loose or open connection. If a quick check of the fuse box indicates that the turn signal fuse has blown, check the wiring for a short before installing a new fuse.

4　Remove the instrument panel steering column cover.

5　Disconnect the electrical connection at the flasher.

6　Pull the flasher from its retainer.

7　Install the flasher at the retainer.

8　Plug the electrical connection into the flasher.

9　Install the instrument panel steering column cover.

Hazard flasher

10　The emergency four-way flasher, located in the convenience center on the right side of the heater or A/C module under the instrument panel **(see illustration)**, is checked in a fashion similar to the turn signal flasher above.

11　If the hazard flasher must be replaced, release the tab lock on the side of the flasher and pull the unit straight out from the convenience center.

12　Installation is the reverse of removal. **Note:** *When replacing either of the above flashers, make sure that the replacement unit is identical in specification to the unit it replaces. Compare the old one to the new one before installing it.*

7　Key lock cylinder - replacement

Refer to illustrations 7.4, 7.5, 7.9 and 7.11

Removal

1　Disconnect the negative cable at the battery.

2　Remove the steering wheel (Chapter 10).

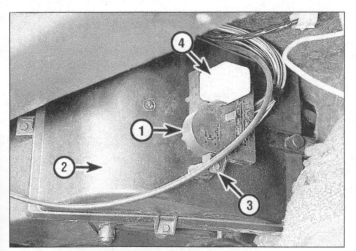

6.10 The convenience center containing the hazard flasher is attached to the heater/air-conditioning assembly underneath the right side of the dashboard

1	Convenience center	
2	Heater and air conditioning module	
3	Screw	
4	Horn relay	

7.4 Install a shaft lock depressor and depress the shaft lock just far enough to get at the retaining ring

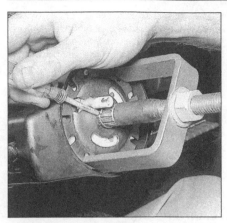

7.5 Remove the retaining ring from the shaft with a small screwdriver

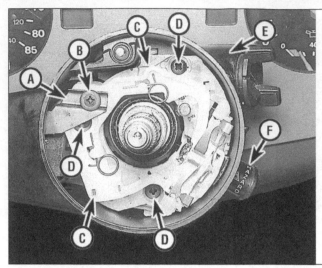

7.9 This is what you see when you remove the shaft lock

A   Switch actuator arm
B   Switch actuator arm retaining screw
C   Turn signal switch assembly
D   Turn signal switch assembly retaining screws
E   Key lock cylinder
F   Hazard flasher switch knob

3   Pry off the shaft lock cover.
4   Install a shaft lock depressor and tighten the nut until the tool slightly depresses the shaft lock **(see illustration)**.
5   Pry the retaining ring from its groove in the steering shaft **(see illustration)**.
6   Remove the steering shaft lock plate and the canceling cam assembly.
7   Remove the hazard flasher switch by removing the Phillips screw that attaches it to the turn signal switch.
8   Remove the switch actuator arm.
9   Remove the three screws attaching the turn signal switch **(see illustration)**.
10   Disconnect the turn signal switch wire harness connector at the base of the steering column cover.
11   Remove the lock cylinder retaining screw **(see illustration)**.
12   Turn the lock cylinder to the On position and remove it.

Installation

13   Turn the lock cylinder to the On position and install it.
14   Install the lock retaining screw and tighten it securely.
15   Pull the turn signal wire harness connec-

tor and wire harness back through the steering column cover and plug it into the connector at the base of the steering column cover.
16   Install and tighten the three screws retaining the switch actuator arm assembly.
17   Install the switch actuator arm assembly and tighten the retaining screw securely.
18   Install the hazard flasher switch knob and tighten the Phillips retaining screw securely.
19   Install the canceling cam assembly and the shaft lock.
20   Install the steering shaft lock plate retaining ring into the groove in the steering shaft.
21   Install the shaft lock cover.
22   Install the steering wheel (Chapter 10).
23   Reconnect the cable to the negative terminal of the battery.

8   Headlight - removal and installation

Refer to illustrations 8.1, 8.4, 8.6, 8.9, 8.12, 8.13, 8.14, 8.16a and 8.16b
Note: *If you are replacing both headlights,*

doing only one of them at a time will simplify reassembly because you will have the other headlight assembly as a guide.

Removal

1   Raise the front compartment lid and disconnect the headlight connectors **(see illustration)**.
2   Turn the headlight switch on to raise the headlamp assemblies.
3   Close the front compartment lid.
4   Remove the two Torx screws in the upper left and right corners of the black plastic bezel surrounding the front of the headlight **(see illustration)**.
5   Raise the front compartment lid.
6   Remove the two Torx screws from each side of the black plastic bezel **(see illustration)**.
7   Remove the bezel.
8   Turn the headlight switch Off.
9   Using a hooked tool or a pair of pliers, release the retaining spring by pulling it upwards and unseating it **(see illustration)**.
10   Turn the headlight switch to the On position.
11   Close the front compartment lid.
12   Disconnect the headlight assembly from

7.11 The key lock cylinder retaining bolt (arrow) must be removed to get the key lock cylinder out of the steering column

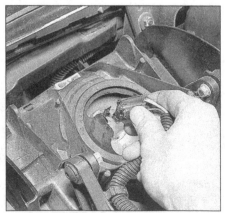

8.1 Raise the front compartment lid and disconnect the headlight electrical connector from the backside of the headlight

8.4 Remove the two headlight bezel retaining screws (arrows) from the front of the bezel . . .

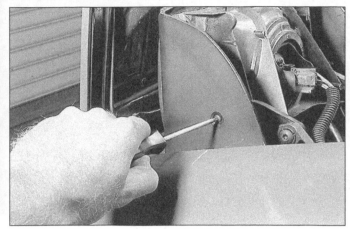

8.6 . . . then remove the retaining screws from each side of the bezel

8.9 Turn the headlight switch off to lower the headlight assembly and release the retaining spring by pulling it upward

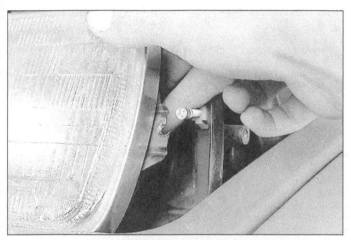

8.12 Turn the headlight switch on again, lower the front compartment lid and disconnect the headlight from its adjuster screws

8.13 Remove the four screws (arrows) from the headlight retaining ring

the adjuster screws and withdraw from the actuator, taking care not to lose the retaining spring **(see illustration)**.

13 Remove the four screws from the chrome headlight retaining ring **(see illustration)**, and remove the headlight.

Installation

14 With the front compartment lid open and the headlight switch turned to Off, position the retaining spring hook in its seat on the actuator **(see illustration)**.

15 Close the front compartment lid and turn the headlight switch to On.

16 Pull the headlight retaining spring forward (away from the front of the vehicle) with a pair of pliers or a hooked tool **(see illustration)**. Position the end of the headlight retaining spring in its seat on the inside lower corner of the headlight assembly **(see illustration)**.

8.14 With the front compartment lid open and the headlight switch off, place the retaining spring into its seat on the actuator

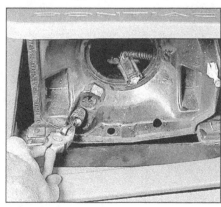

8.16a Pull the hooked end of the retaining spring forward with a pair of pliers . . .

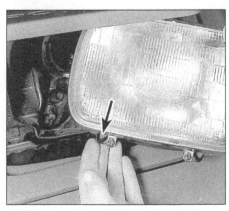

8.16b . . . and seat it in the dimpled tang (arrow) on the lower side of the headlight

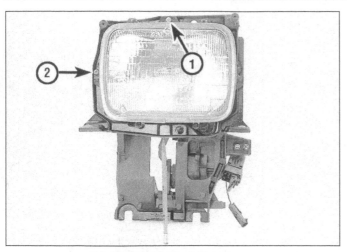

9.1 The headlight aiming screws

1   Vertical adjustment screw
2   Horizontal adjustment screw

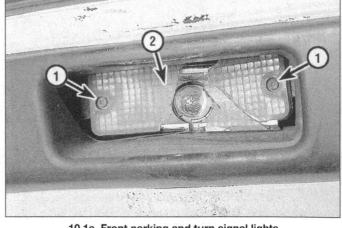

10.1a Front parking and turn signal lights

1   Screw
2   Turn signal/parking light lens

17   Position the headlight ring slots over the adjuster screws.
18   Raise the front compartment lid.
19   Reinstall the black plastic bezel and the two outboard Torx screws.
20   Plug in the headlight connector.
21   Close the front compartment lid.
22   Reinstall the two remaining Torx screws in the upper front face of the plastic bezel.
23   Test for proper operation.

9   Headlights - adjustment

Refer to illustration 9.1

Note: *It is important that the headlights be aimed correctly. If adjusted incorrectly they could blind an oncoming car and cause a serious accident or seriously reduce the your ability to see the road. The headlights should be checked for proper aim every 12 months and any time a new sealed beam headlight is installed or front end body work is performed.*
1   Headlights have two spring loaded adjusting screws, one on the top controlling up and down movement and one on the side

controlling left and right movement **(see illustration)**.
2   There are several methods of adjusting the headlights. The simplest method uses an empty wall 25 feet in front of the vehicle and a level floor.
3   Park the vehicle on a known level floor 25 feet from the wall.
4   Position masking tape vertically on the wall in reference to the vehicle centerline and the centerlines of both headlights.
5   Position a horizontal tape line in reference to the centerline of the headlights.
Note: *It may be easier to position the tape on the screen with the vehicle parked only a few inches away.*
6   Adjustment should be made with the vehicle sitting level, the gas tank half-full and no unusually heavy load in the vehicle.
7   Starting with the low beam adjustment, position the high intensity zone so it is two inches below the horizontal line and two inches to the side of the headlight vertical line, away from oncoming traffic. Adjustment is made by turning the top adjusting screw clockwise to raise the beam and counter-

clockwise to lower the beam. The adjusting screw on the side should be used in the same manner to move the beam left or right.
8   With the high beams on, the high intensity zone should be vertically centered with the exact center just below the horizontal line.
Note: *It may not be possible to position the headlight aim exactly for both high and low beams. If a compromise must be used, keep in mind that the low beams are the most used and have the greatest effect on driver safety.*

10   Bulb replacement

Refer to illustrations 10.1a, 10.1b, 10.6, 10.11, 10.18, 10.25 and 10.28

Front end

Parking and turn signal

1   Remove the two lens retaining screws and the lens **(see illustrations)**.
2   Remove the bulb socket from the lens and the bulb from the socket.
3   Installation is the reverse of removal.

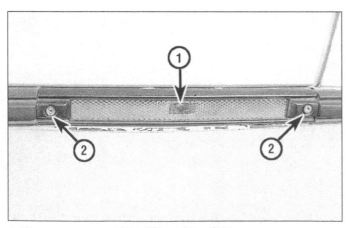

10.1a Side marker lights

1   Front side parking light lens   2   Screws

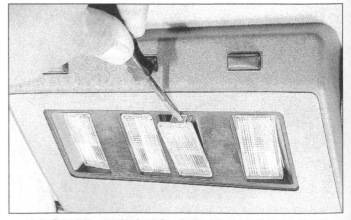

10.6 Carefully pry the lens from the overhead light assembly

Side marker lights (front and rear)

4   Remove the two Torx screws from the lens and remove the lens. Pull the bulb straight out of the socket.
5   Installation is the reverse of removal.

Interior

Reading (overhead) light bulbs

6   Remove the reading lamp lens **(see illustration)**. It can be pried out with a small screwdriver. Remove the bulb.
7   Installation is the reverse of removal.

Third brake light

Note: *1986 and 1987 vehicles are equipped with a third brake light, a quartz halogen unit located just inside the rear window.*
8   Pop the housing off, disconnect the electrical connector and pull out the quartz halogen bulb.
9   Installation is the reverse of removal.

Instrument panel light bulbs

10   Remove the rear cluster cover (Section 13).
11   Rotate the bulb counterclockwise and pull it straight out of the circuit board **(see illustration)**.
12   Replace the bulb and install the rear cluster cover (Section 13).

Console shift trim plate light bulb (automatics only)

13   Grasp the wire clip in the front of the shifter handle with needle nose pliers, pull it out and remove the handle.
14   Remove both ashtrays from the shift plate assembly and remove the two screws underneath each ashtray receptacle.
15   Lift up the trailing edge of the trim plate and rotate the console shift trim plate light bulb holder clockwise 1/4-turn to disconnect it from its plastic retainer. Remove the bulb by pulling it straight out of the holder.
16   Installation is the reverse of removal.

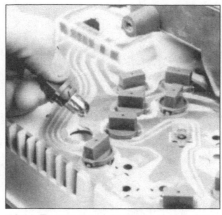

10.11 To remove an instrument bulb from the circuit board, simply turn it counterclockwise and pull it straight out (although the circuit board and cluster assembly have been removed for the sake of clarity in this photo, it is only necessary to remove the rear cluster cover to get at the circuit board)

Ashtray light bulbs

17   Remove the shift trim plate.
18   Rotate the two ashtray light bulb holders clockwise 1/4-turn and remove them from their plastic retainers **(see illustration)**.
19   Installation is the reverse of removal.

Cigarette lighter bulb replacement

20   Remove both Torx screws from the cigarette lighter trim plate and remove the plate.
21   Remove both Phillips screws from the cigarette lighter flange and pull the cigarette lighter far enough out of the console to get at the bulb holder retainer on the side of the lighter barrel.
22   Rotate the bulb holder 1/4-turn and remove it from its retainer. Pull the bulb straight out of the holder.
23   Installation is the reverse of removal.

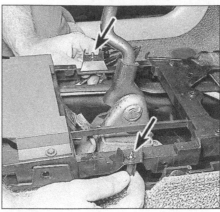

10.18 To remove either ashtray bulb (arrows), remove the light bulb holder by rotating it clockwise 1/4-turn and pulling it out of its retainer, then pull the bulb straight out of its holder

Rear end

Courtesy light

24   The rear compartment courtesy light bulb is contained in a plastic housing in the bulkhead between the trunk and the engine compartment. To replace it, pry off the plastic lens with a screwdriver, pull the bulb straight out, install a new bulb and pop the lens back into place.

License plate bulb

25   Remove the two bolts at the lamp assembly and remove the lamp assembly **(see illustration)**.
26   Remove the bulb socket from the lamp assembly and replace the bulb.
27   Installation is the reverse of removal.

Rear taillight bulbs

28   Open the deck lid. There are three plugs for each taillight lens. Remove the three black plastic plugs on the side of the vehicle with the burned out bulb **(see illustration)**.

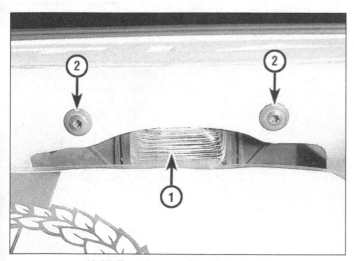

10.25 Rear license plate light details

1   Light assembly   *2   Screws*

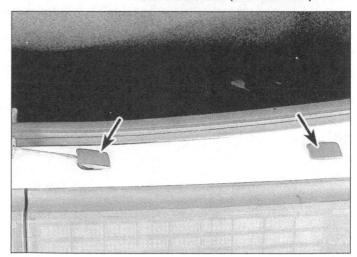

10.28 To disconnect the taillight lens from the body, raise the engine compartment lid, remove the three black plastic plugs in the top of the rear bumper fascia (arrows) and remove the recessed screw below each plug

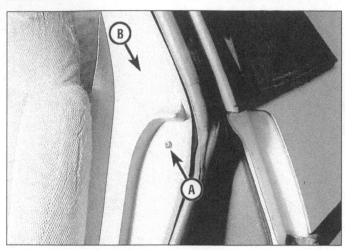

11.3 Rear quarter trim panel

A   Trim panel retaining screw   B   Trim panel

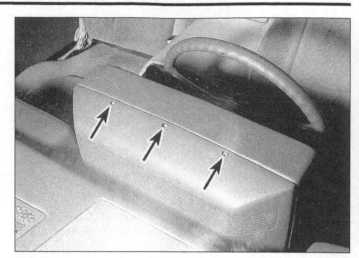

13.2 Remove these three Torx screws (arrows) to remove the rear cluster cover from the cluster pad assembly

29   Remove the screw in the bottom of each plug receptacle.

30   Slide the taillight lens out far enough to reach the bulb holder.

31   Disconnect the bulb holder from the backside of each taillight lens. To remove the inboard three bulb holders, push in the lock tab and rotate the holder counterclockwise; to remove the outboard bulb holder, rotate it counterclockwise and pull it out.

32   Rotate the burned out taillight bulb counterclockwise and remove it from the holder.

33   The installation procedure is the reverse of removal.

11   Radio and speakers - removal and installation

Refer to illustration 11.3

Caution: *All radios in this vehicle are the bridge audio type, using two wires to each speaker. It is very important when changing speakers or performing any radio work to avoid pinching any wires, as this will cause damage to the output circuit in the radio.*

Radio

1   Refer to *Console - removal and installa-tion* in Chapter 11 for the removal and instal-lation procedure for the radio.

Speakers

Sail panel

Note: *Replacement of either the left or right sail panel speaker involves removal of the rear quarter trim panel (the one piece plastic assembly which fits into the seatback-to-engine compartment panel).*

2   Disconnect the upper shoulder belt anchor assembly.

3   Remove the screw that fastens the for-ward edge of the quarter trim panel to the body **(see illustration).**

4   Pop off the panel. Unseat the retainer

clip at the top of the panel by grasping the panel with your hands and pulling it away from the body.

5   Slide the seat belt webbing through the slot on the panel and remove the panel.

6   Remove the three speaker mounting screws and pull the speaker away from the body.

7   Disconnect the electrical connector from the back of the speaker and remove the speaker.

8   Plug in the electrical connector to the backside of the new speaker.

9   Install the speaker and tighten the three mounting screws.

10   Route the shoulder belt webbing through the slot in the quarter trim panel and install the panel.

11   Push the retainer back into its slot and install and tighten the mounting screw in the leading edge of the trim panel.

12   Install the upper shoulder belt anchor assembly and tighten securely.

Sub-woofer speaker

Note: *If your vehicle is equipped with a sub-woofer assembly, it is installed underneath the dashboard on the left and right sides.*

13   Disconnect the sub-woofer wiring con-nector.

14   Remove the courtesy light bulb near the instrument panel support bracket, if applica-ble.

15   Remove the instrument panel support bracket bolt (and the plastic retaining pin, if applicable).

16   Remove the convenience center mount-ing bracket bolt.

17   Pull the instrument panel slightly toward the rear of the vehicle if necessary. Slide the sub-woofer assembly to the left, then pull it straight down to remove it.

18   Installation is the reverse of removal.

Front speaker

19   Remove the grille by popping it loose from its rubber mounting grommets with a small screwdriver.

20   Remove the four speaker mounting screws and lift the speaker up from the dash-board enough to disconnect the electrical connector from its underside. Remove the speaker.

21   Installation is the reverse of removal.

12   Radio antenna - removal and installation

Note: *The antenna on the right front fender is fixed. That is, it cannot be adjusted up or down. If it becomes slightly bent, you can straighten it by hand. If the antenna is severely bent, it must be replaced.*

1   The antenna mast can be removed by loosening the retaining nut and unscrewing the antenna.

2   The antenna body and cable assembly can be removed after the mast has been removed by unbolting it from the fender.

13   Instrument panel - removal and installation

Refer to illustrations 13.2, 13.5, 13.6, 13.7a and 13.7b

1   Disconnect the negative cable at the battery.

2   Remove the rear cluster cover **(see illustration).**

3   Remove the trim plate.

4   Unscrew the four Torx screws and pull out the headlight switch assembly from the left side of the cluster pad assembly.

5   Unplug the two electrical connectors from the backside of the headlight switch assembly and remove it **(see illustration).**

6   Remove the lower cover assembly **(see illustration).**

7   Remove the two bolts on the top that attach the cluster assembly to the steering column and the two on the bottom that attach the cluster assembly to the dashboard **(see illustrations).** Pull the cluster pad and

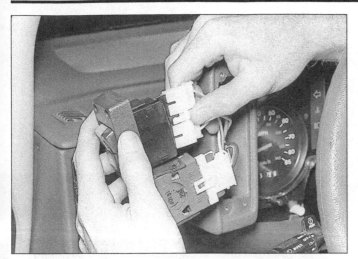

13.5 Disconnect these two electrical connectors from the backside of the headlight switch assembly and remove the switch

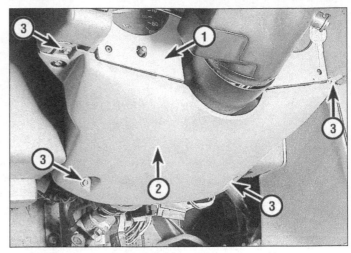

13.6 The instrument panel cluster trim plate and lower cover

1 Instrument panel cluster fascia trim
2 Lower cover
3 Screws

the cluster assembly away from the dashboard so that you can get at the electrical connectors on the backside and underneath.
8 Disconnect the large electrical connectors from the underside and backside of the cluster assembly.
9 Disconnect the wire harness clips.
10 Remove the cluster pad and cluster assembly from the vehicle.
11 For bulb replacement procedures, see Section 10.
12 For instrument replacement procedures, see Section 14. **Note:** *While the flexible circuit board is removed, check it for cracks. If damage or deterioration is evident, replace the circuit board.*
13 Place the cluster assembly in the cluster pad enclosure.
14 Place the cluster assembly and cluster pad in position over the steering column.
15 Plug in the large electrical connectors to

the backside and underside of the cluster assembly.
16 Install the wire harness retaining clips to the sides of the cluster assembly.
17 Install and tighten securely the two mounting bolts on top that secure the cluster assembly to the steering column and the two on the bottom that attach the cluster assembly to the dashboard.
18 Pull the headlight harness through the opening in the left side of the instrument panel and plug in the two electrical connectors to the backside of the headlight switch assembly.
19 Push the harness back through the opening and install the headlight switch assembly. Tighten the four Torx mounting screws.
20 Install the rear cluster cover (the upper cover) and tighten the three Torx retaining screws securely.

21 Place the lower cover in position and install and tighten the two screws that attach the lower cover to the cluster assembly and the two screws that attach it to the dashboard.

14 Instruments - removal and installation

Refer to illustration 14.13

Removal

1 Remove the instrument panel cluster assembly and cluster pad from the dashboard (Section 13).
2 Remove the cluster assembly from the cluster pad (Section 13).
3 Remove the four screws which attach the clear plastic cover to the instrument panel fascia and remove the cover.

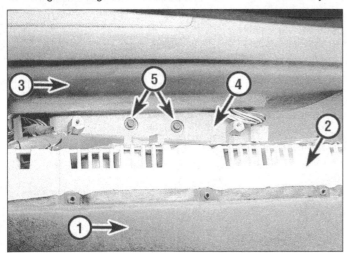

13.7a Instrument cluster upper mounting details

1 Cluster cover
2 Cluster assembly
3 Dashboard
4 Steering column support fixture
5 Upper cluster mounting bolts

13.7b Instrument cluster lower mounting details

1 Instrument cluster
2 Dashboard
3 Lower instrument cluster mounting screws

14.13 To service either gauge in the optional instrument package, remove the four Torx screws from the fascia trim plate then disconnect the electrical connectors from the bottom of the gauges

15.2 The governor and speedometer sensor assembly is located at the right rear corner of the automatic transaxle housing

4   Remove the single screw on the left side of the instrument fascia and lift off the fascia.

5   The speedometer, tachometer/oil pressure or fuel gauge/water temperature gauges can now be removed from the cluster assembly after removal of the gauge retaining screws.

Installation

6   Install the new gauge in the cluster assembly and tighten the gauge retaining screws.

7   Install the instrument fascia and the single fascia retaining screw.

8   Install the clear plastic instrument fascia cover and the four screws which retain it.

9   Install the cluster assembly in the cluster pad (Section 13).

10   Install the cluster pad and cluster assembly onto the dashboard (Section 13).

Optional gauge package

Note: *Some vehicles are equipped with an optional gauge package atop the center console.*

11   Remove the four Torx screws from the gauge fascia trim plate.

12   Remove the twin gauge cluster retaining screws, one per side, and pull the cluster out far enough to remove the electrical connectors.

13   Disconnect the electrical connectors from the underside of the gauge cluster **(see illustration)**.Remove the cluster.

14   Instrument replacement is similar to the procedure outlined in Section 14 for the main instrument panel.

15   Installation is the reverse of removal.

15   Quartz electric speedometer drive unit - removal and installation

Refer to illustration 15.2

1   The speedometer is an electrically driven unit, rather than a conventional cable-driven unit. The sender for the speedometer is mounted on the transaxle in the location normally reserved for the speedometer cable drive.

2   Disconnect the cable from the negative terminal of the battery and disconnect the electrical connection at the sensor **(see illustration)**.

3   Remove the sensor assembly to transaxle retainer.

4   Remove the sensor assembly and gear.

Installation

5   Install the sensor assembly and drive gear. Don't forget the O-ring seal.

6   Install the sensor assembly to transaxle retainer.

7   Attach the sensor electrical connector.

8   Attach the negative battery cable.

16   Headlight actuator switch - removal and installation

Refer to illustration 16.5

Removal

1   Before removing the headlight mounting bracket from the front panel compartment, mark its position by marking around the two upper attaching bolts and onto the headlight mounting bracket.

2   Disconnect the cable from the negative terminal of the battery.

3   Disconnect the electrical connection at the light.

4   Remove the bolts from each side of the light assembly.

5   Remove the bolt from the link assembly **(see illustration)**.

6   Remove the light, light assembly and bezel as a unit.

7   Remove the four attaching bolts from the headlight mounting bracket.

8   Disconnect the electrical connectors.

9   Remove the headlight mounting bracket assembly.

10   Remove the switch cover plate.

11   Remove the rubber slot filler.

12   Remove the switch and harness assembly.

Installation

13   Install the switch and harness assembly.

14   Install the rubber slot filler.

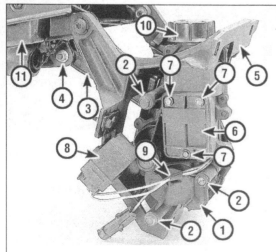

16.5 View of the headlight actuator motor assembly and components

1   *Headlight actuator motor*
2   *Actuator motor mounting bolts*
3   *Link*
4   *Link-to-headlight housing pivot bolt*
5   *Headlight mounting bracket*
6   *Switch cover plate*
7   *Switch cover screws*
8   *Relay*
9   *Switch harness*
10   *Manual control knob (headlights)*
11   *Headlamp housing*

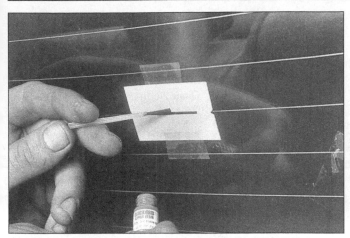

17.11 To repair a broken grid on the defogger, apply masking tape to the inside of the window at the damaged area, then brush on the special conductive coating

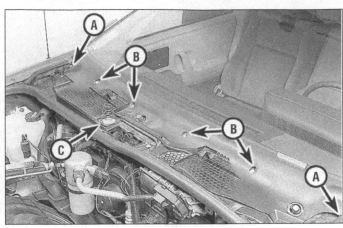

18.2 The top vent grille is secured to the body by several kinds of fasteners

A	Push-in connectors	
B	Screws	C Rivet

15   Install the switch cover plate.
16   Plug in the electrical connectors.
17   Install the headlight mounting bracket assembly. Be sure to align the marks on the headlight mounting bracket. Tighten the headlight bracket mounting bolts to the specified torque.
18   Install the light, light assembly and bezel as a unit.
19   Install and tighten the bolts at each side of the light assembly.
20   Install the bolt at the link assembly.
21   Plug in the electrical connector at the light.

17   Rear window defogger - check and repair

Refer to illustration 17.11

1   This option consists of a rear window with a number of horizontal elements baked into the glass surface during the glass forming operation.
2   Small breaks in the element can be successfully repaired without removing the rear window.
3   To test the grids for proper operation, start the engine and turn on the system.
4   Ground one lead of a test light and carefully touch the other lead to each element line.
5   The brilliance of the test light should increase as the lead is moved across the element. If the test light glows brightly at both ends of the lines, check for a loose ground wire. All of the lines should be checked in at least two places.
6   To repair a break in a line, it is recommended that a repair kit specifically for this purpose be purchased from a GM dealer. Included in the repair kit will be a decal, a container of silver plastic and hardener, a mixing stick and instructions.
7   To repair a break, first turn off the system and allow it to de-energize for a few minutes.
8   Lightly buff the element area with fine steel wool, then clean it thoroughly with alcohol.

9   Use the decal supplied in the repair kit or apply strips of electrician's tape above and below the area to be repaired. The space between the pieces of tape should be the same width as the existing lines. This can be checked from outside the vehicle. Press the tape tightly against the glass to prevent seepage.
10   Mix the hardener and silver plastic thoroughly.
11   Using the wood spatula, apply the silver plastic mixture between the pieces of tape, overlapping the undamaged area slightly on either end **(see illustration)**.
12   Carefully remove the decal or tape and apply a constant stream of hot air directly to the repaired area. A heat gun set at 500 to 700 degrees Fahrenheit is recommended. Hold the gun one inch from the glass for two minutes.
13   If the new element appears off color, tincture of iodine can be used to clean the repair and bring it back to the proper color. This mixture should not remain on the repair for more than 30 seconds.
14   Although the defogger is now fully operational, the repaired area should not be disturbed for at least 24 hours.

18.3 To disconnect the transmission drive link-to-windshield wiper motor connection, loosen the two nuts (arrows) and pull the wiper motor arm and transmission drive link apart

18   Windshield wiper transmission (linkage) - removal and installation

Refer to illustrations 18.2, 18.3, 18.4 and 18.5

1   Remove the wiper arms from the transmission spindle shafts by prying upwards on the underside of the wiper arm where it mates to the spindle shaft.
2   Remove the four screws, two push connectors and plastic rivet that secure the top vent grille to the cowl **(see illustration)**. Remove the vent grille by prying up along the forward edge, unseating the plastic push-in mounting studs. The single plastic rivet near the hood spring may be dislocated using the same method. Disconnect the washer nozzle hoses and lift the vent grille from the vehicle.
3   Loosen but do not remove the two drive link-to-windshield wiper motor crank arm nuts **(see illustration)**. Separate the drive link from the crank arm.
4   Remove the transmission-to-cowl panel screws **(see illustration)**.
5   Push the transmission spindle shafts through the cowl and guide the assembly out through the opening in the cowl on the left

18.4 There are three transmission-to-cowl mounting screws on each side which must be removed before the transmission can be removed

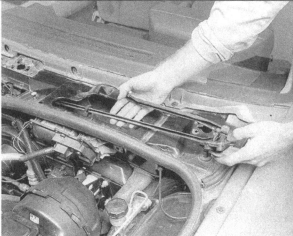

18.5 Push the transmission spindle shafts through the cowl and guide the assembly out through the opening in the cowl on the left side

3   Disconnect the electrical leads and remove the motor attaching screws.
4   Rotate the motor up and out to remove it.

Installation

5   Install the motor by placing the crank arm through the opening in the body **(see illustration)**.
6   Replace the motor attaching screws and tighten them to the specified torque.
7   Install the transmission drive link to the crank arm, with the motor in the park position **(see illustration)**.
8   Replace the shroud top vent grille and wiper arms.
9   Check the operation of the wiper system.

side **(see illustration)**.
6   Inspect the swivel joints and spindle shafts for binding or excessive play. Repair or replace if necessary.
7   Feed the transmission and drive link assembly through the left side opening in the cowl. The spindle shaft housings must be positioned above and behind the drive link during installation.
8   Push the spindle shaft assemblies through their openings and install the screws.
9   Attach the drive link to the wiper motor crank arm.
10   Place the top vent grille over the cowl, reconnect the washer hoses and install the screws and push-in connectors.
11   Install the wiper arms with the wiper

blades positioned as near as possible to the top edge of the blackout line on the glass, without actually touching it.

19 Windshield wiper motor - removal and installation

Refer to illustrations 19.5 and 19.7

Removal

1   Loosen but do not remove the transmission drive link to the motor crank arm attaching nuts.
2   Detach the drive link from the motor crank arm.

20 Clutch operated Neutral start switch - removal and installation

Removal

1   Disconnect the electrical connection at the switch.
2   Remove the bolt attaching the switch to the clutch bracket.
3   Rotate the switch slightly to disconnect the shaft from the clutch pedal hole.

Installation

4   Install the shaft at the clutch pedal hole.
5   Install the bolt at the switch.
6   Plug in the electrical connection.

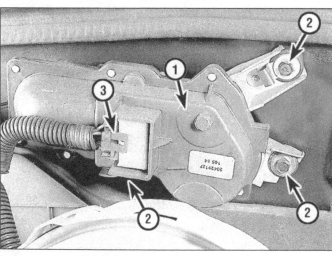

19.5 Windshield wiper motor details

1   *Wiper motor*
2   *Mounting bolts*
3   *Electrical connector*

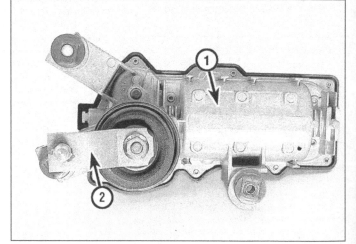

19.7 The windshield wiper motor crank arm in the park position

1   *Wiper motor*
2   *Crank arm*

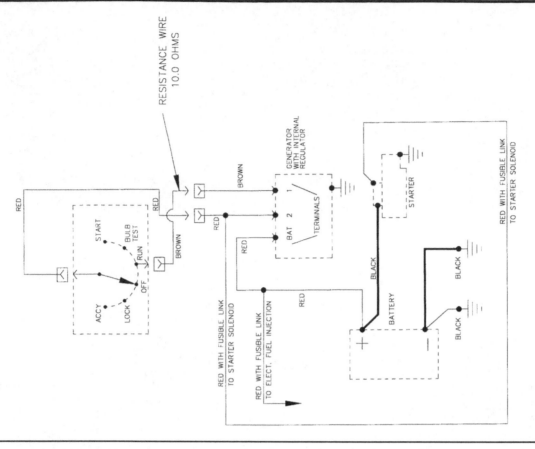

Charging system

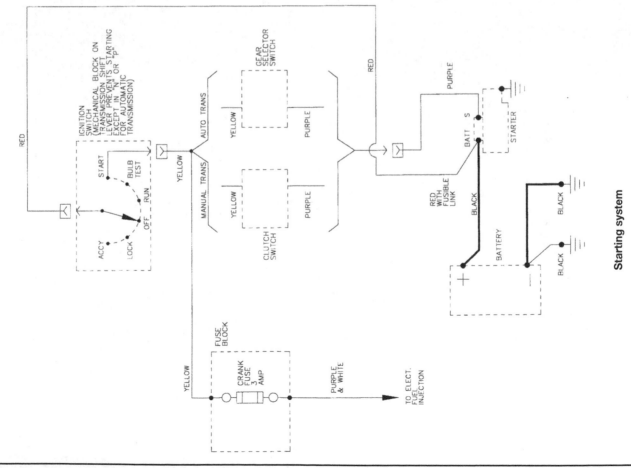

Starting system

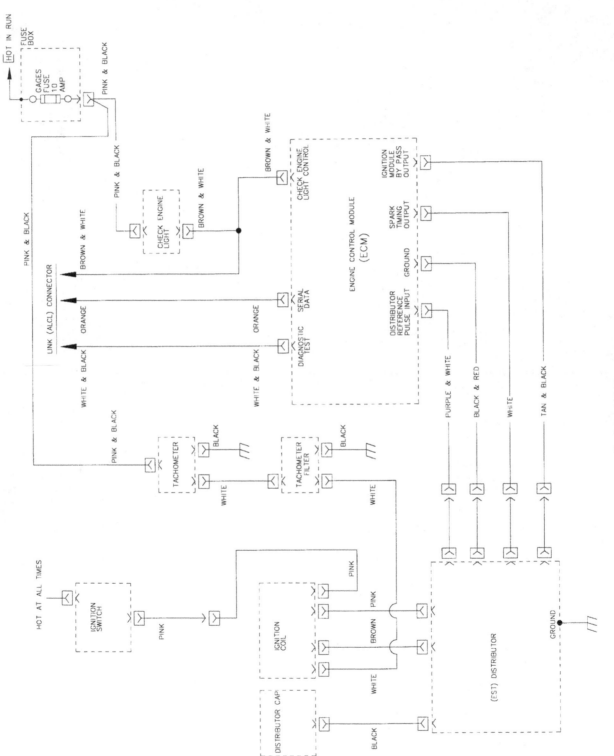

Engine control system

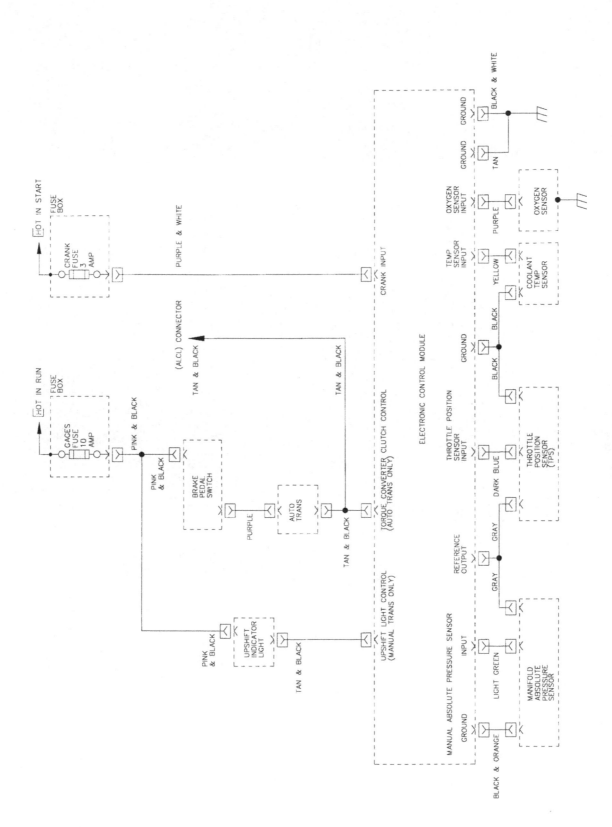

Transaxle control system

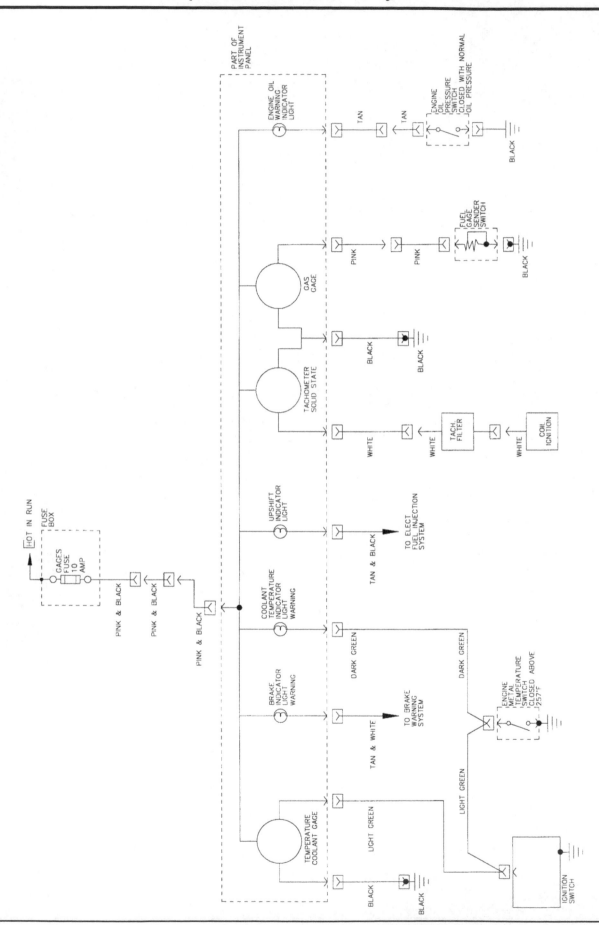

Engine warning lights

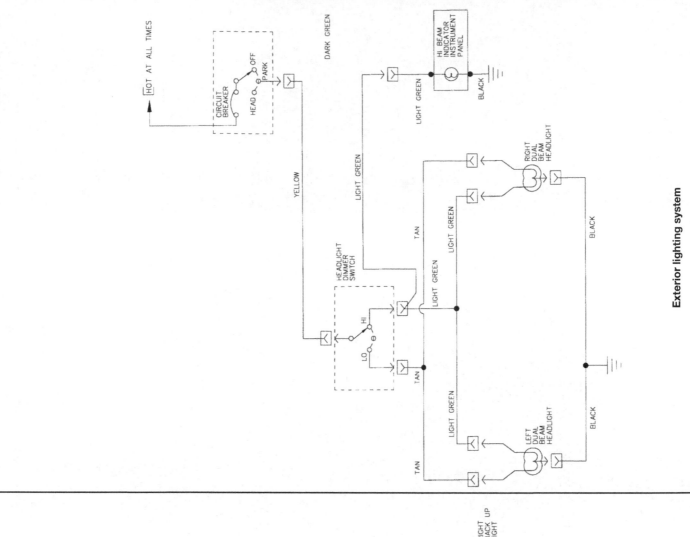

Exterior lighting system

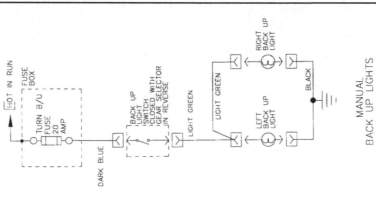

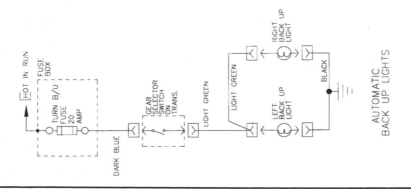

Exterior lighting system

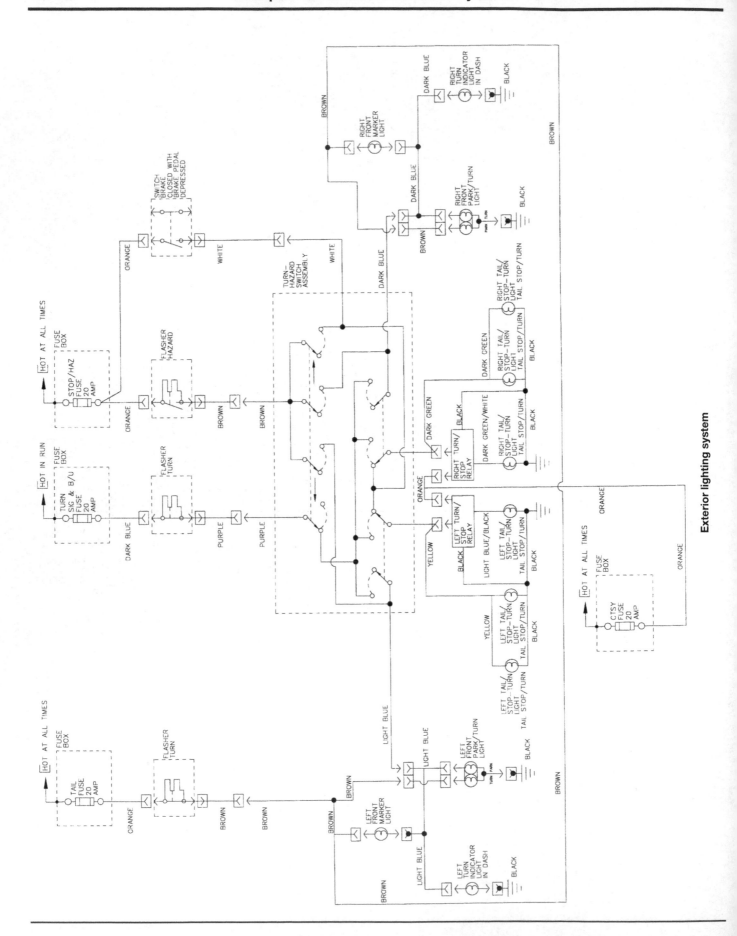

Exterior lighting system

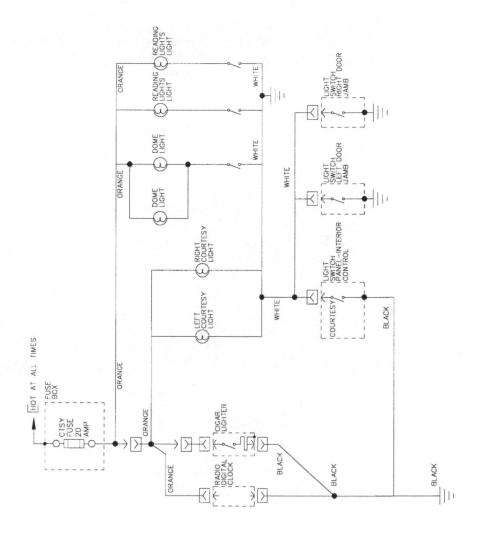

Interior lighting system

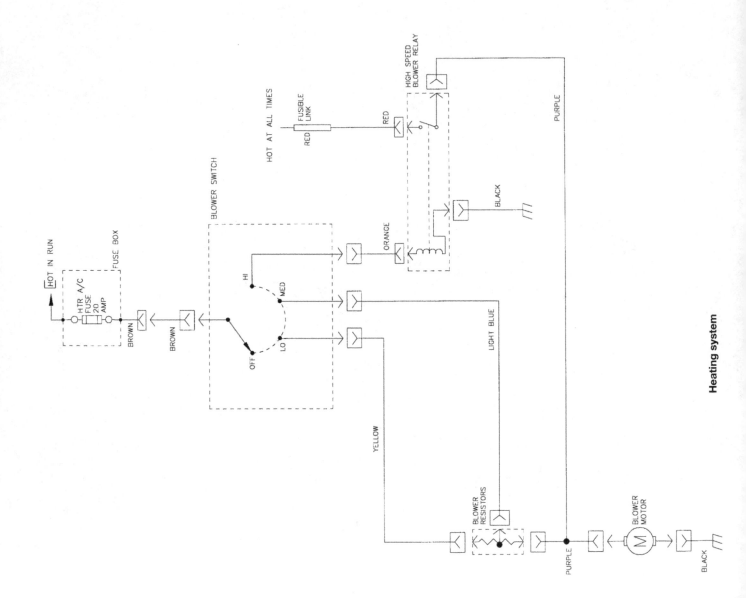

Heating system

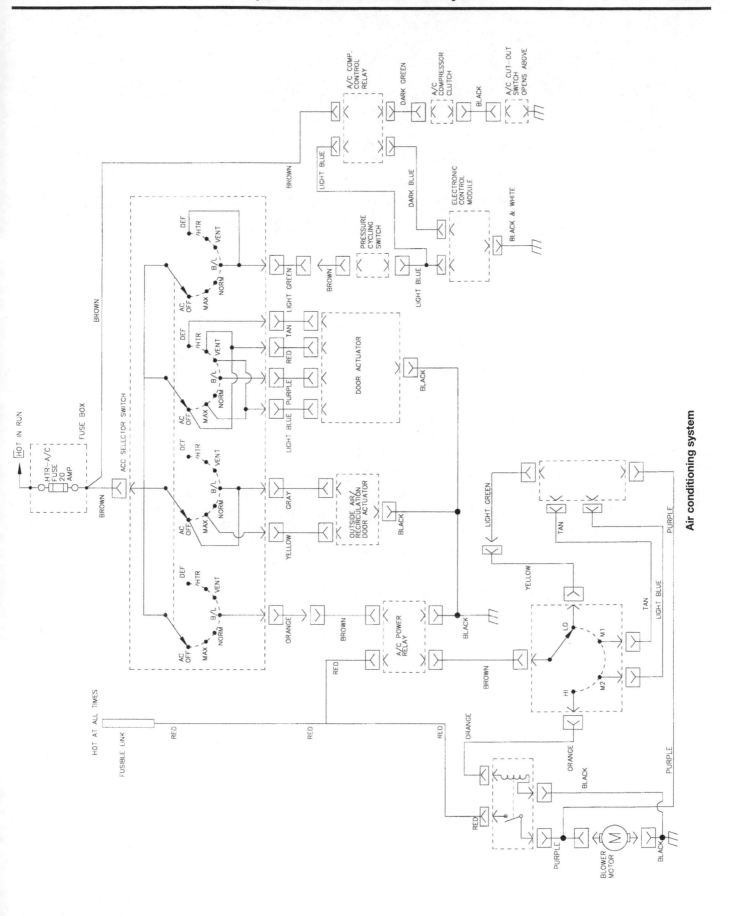

Air conditioning system

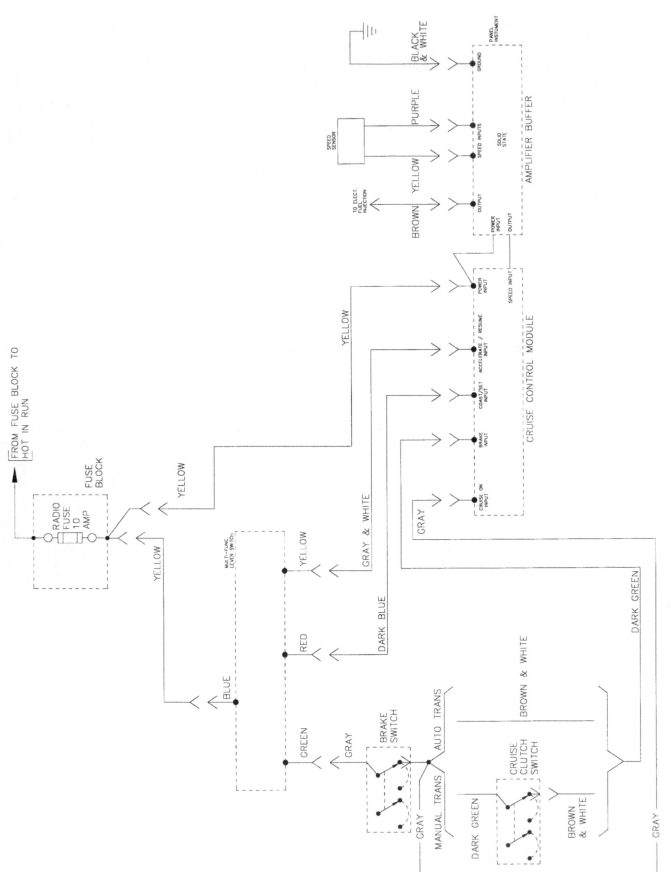

Cruise control system (1 of 2)

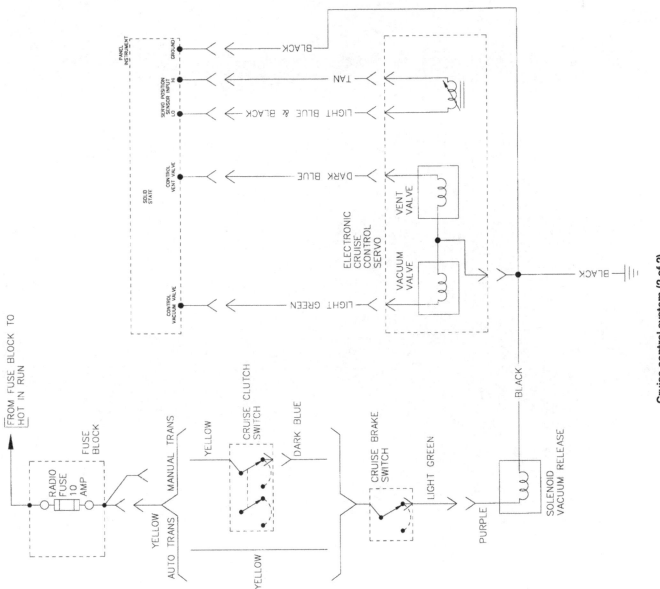

Cruise control system (2 of 2)

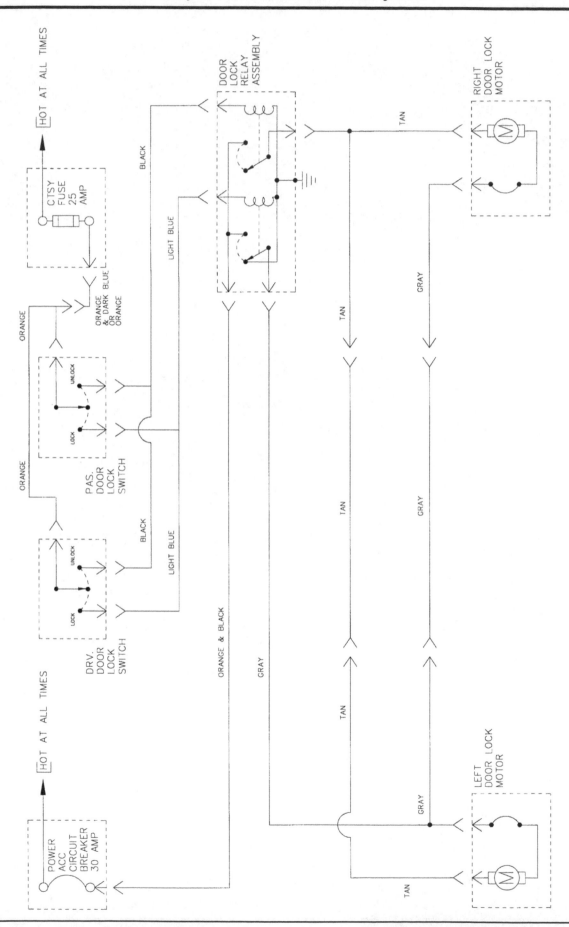

Power door lock system

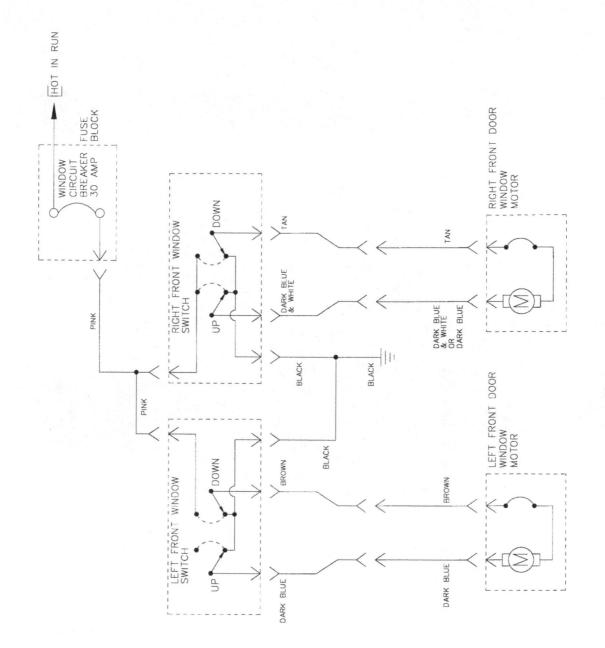

Power window system

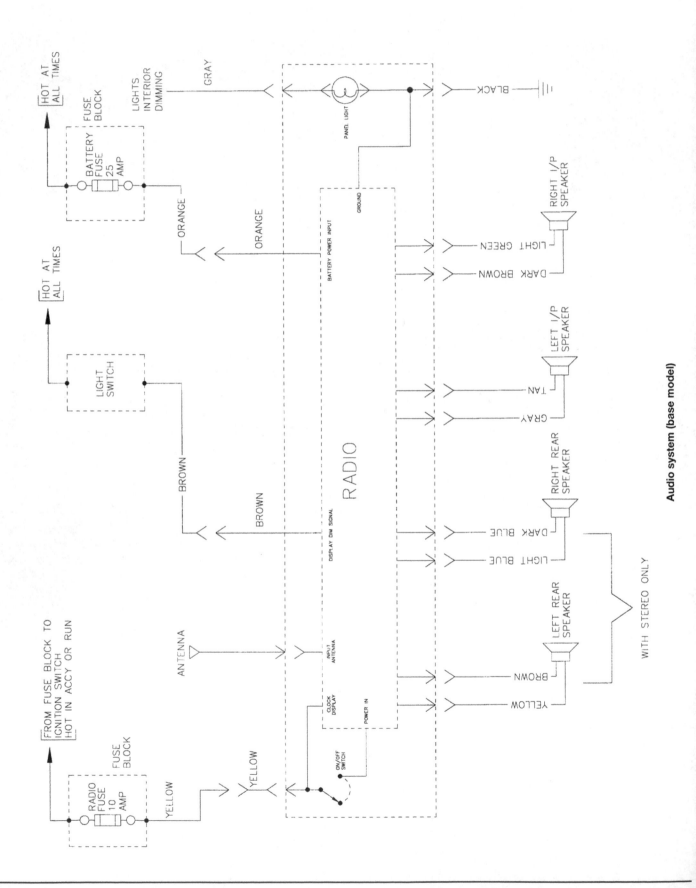

Audio system (base model)

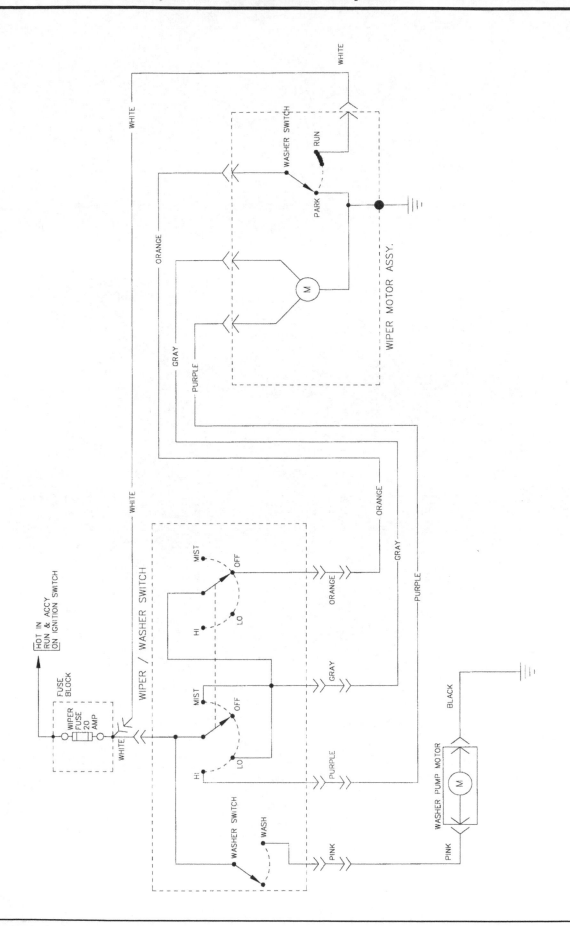

Windshield wiper and washer system

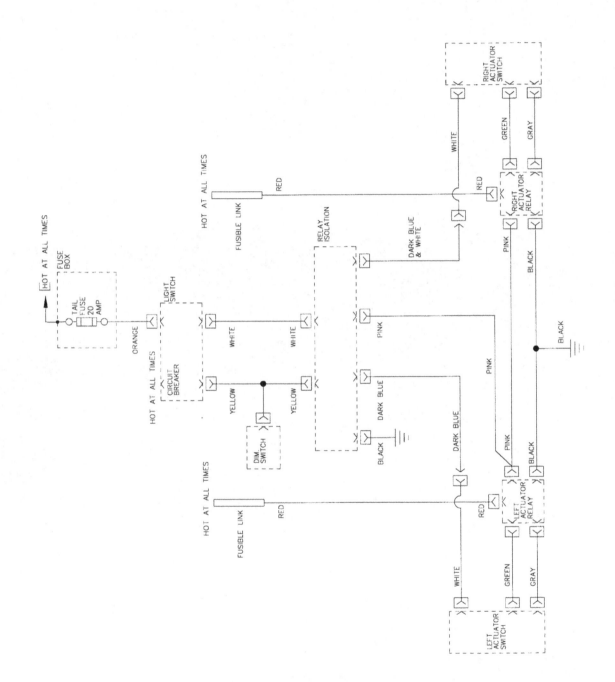

Headlight door actuator system

Index

Haynes Automotive Manuals

NOTE: If you do not see a listing for your vehicle, please visit **haynes.com** for the latest product information and check out our **Online Manuals!**

ACURA
12020	Integra '86 thru '89 & Legend '86 thru '90
12021	Integra '90 thru '93 & Legend '91 thru '95
	Integra '94 thru '00 - see HONDA Civic (42025)
	MDX '01 thru '07 - see HONDA Pilot (42037)
12050	Acura TL all models '99 thru '08

AMC
14020	Concord/Hornet/Gremlin/Spirit '70 thru '83
14025	(Renault) Alliance & Encore '83 thru '87

AUDI
15020	4000 all models '80 thru '87
15025	5000 all models '77 thru '83
15026	5000 all models '84 thru '88
	Audi A4 '96 thru '01 - see VW Passat (96023)
15030	Audi A4 '02 thru '08

AUSTIN-HEALEY
	Sprite - see MG Midget (66015)

BMW
18020	3/5 Series '82 thru '92
18021	3-Series including Z3 models '92 thru '98
18022	3-Series including Z4 models '99 thru '05
18023	3-Series '06 thru '14
18025	320i all 4-cylinder models '75 thru '83
18050	1500 thru 2002 except Turbo '59 thru '77

BUICK
19010	Buick Century '97 thru '05
	Century (front-wheel drive) - see GM (38005)
19020	Buick, Oldsmobile & Pontiac Full-size (Front wheel drive) '85 thru '05
19025	Buick, Oldsmobile & Pontiac Full-size (Rear wheel drive) '70 thru '90
19027	Buick LaCrosse '05 thru '13
	Regal - see GENERAL MOTORS (38010)
	Skyhawk - see GM (38015)
	Skylark - see GM (38020, 38025)
	Somerset - see GENERAL MOTORS (38025)

CADILLAC
21015	CTS & CTS-V '03 thru '14
21030	Cadillac Rear Wheel Drive '70 thru '93
	Cimarron, Eldorado & Seville - see GM (38015, 38030, 38031)

CHEVROLET
10305	Chevrolet Engine Overhaul Manual
24010	Astro & GMC Safari Mini-vans '85 thru '05
24015	Camaro V8 all models '70 thru '81
24016	Camaro all models '82 thru '92
	Cavalier - see GM (38016)
	Celebrity - see GM (38005)
24017	Camaro & Firebird '93 thru '02
24018	Camaro '10 thru '15
24020	Chevelle, Malibu, El Camino '69 thru '87
24024	Chevette & Pontiac T1000 '76 thru '87
	Citation - see GENERAL MOTORS (38020)
24027	Colorado & GMC Canyon '04 thru '12
24032	Corsica & Beretta all models '87 thru '96
24040	Corvette V8 models '68 thru '82
24041	Corvette all models '84 thru '96
24042	Corvette all models '97 thru '13
24044	Cruze '11 thru '19
24045	Full-size Sedans Caprice, Impala, Biscayne, Bel Air & Wagons '69 thru '90
24046	Impala SS & Caprice and Buick Roadmaster '91 thru '96
	Impala '00 thru '05 - see LUMINA (24048)
24047	Impala & Monte Carlo all models '06 thru '11
	Lumina '90 thru '94 - see GM (38010)
24048	Lumina & Monte Carlo '95 thru '05
	Lumina APV - see GM (38035)
24050	Luv Pick-up all 2WD & 4WD '72 thru '82
24051	Malibu '13 thru '19
24055	Monte Carlo all models '70 thru '88
	Monte Carlo '95 thru '01 - see LUMINA (24048)
24059	Nova all V8 models '69 thru '79
24060	Nova/Geo Prizm '85 thru '92
24064	Pick-ups '67 thru '87 - Chevrolet & GMC
24065	Pick-ups '88 thru '98 - Chevrolet & GMC
24066	Pick-ups '99 thru '06 - Chevrolet & GMC
24067	Chevy Silverado & GMC Sierra '07 thru '14
24068	Chevy Silverado & GMC Sierra '14 thru '19
24070	S-10 & S-15 Pick-ups '82 thru '93
24071	S-10 & Sonoma Pick-ups '94 thru '04
24072	Chevrolet TrailBlazer, GMC Envoy & Oldsmobile Bravada '02 thru '09
24075	Sprint '85 thru '88, Geo Metro '89 thru '01
24080	Vans - Chevrolet & GMC '68 thru '96
24081	Chevrolet Express & GMC Savana Full-size Vans '96 thru '19

CHRYSLER
10310	Chrysler Engine Overhaul Manual
25015	Chrysler Cirrus, Dodge Stratus, Plymouth Breeze, '95 thru '00
25020	Full-size Front-Wheel Drive '88 thru '93
	K-Cars - see DODGE Aries (30008)
25025	Chrysler LHS, Concorde & New Yorker, Dodge Intrepid, Eagle Vision, '93 thru '97
25026	Chrysler LHS, Concorde, 300M, Dodge Intrepid '98 thru '04
25027	Chrysler 300 '05 thru '18, Dodge Charger '06 thru '18, Magnum '05 thru '08 & Challenger '08 thru '18
25030	Chrysler & Plymouth Mid-size '82 thru '95
	Rear-wheel Drive - see DODGE (30050)
25035	PT Cruiser all models '01 thru '10
25040	Chrysler Sebring '95 thru '06, Dodge Stratus '01 thru '06 & Dodge Avenger '95 thru '00
25041	Chrysler Sebring '07 thru '10, 200 '11 thru '17 & Dodge Avenger '08 thru '14

DATSUN
28005	200SX all models '80 thru '83
28012	240Z, 260Z & 280Z Coupe '70 thru '78
28014	280ZX Coupe & 2+2 '79 thru '83
	300ZX - see NISSAN (72010)
28018	510 & PL521 Pick-up '68 thru '73
28020	510 all models '78 thru '81
28022	620 Series Pick-up all models '73 thru '79
	720 Series Pick-up - see NISSAN (72030)

DODGE
30008	Aries & Plymouth Reliant '81 thru '89
30010	Caravan & Plymouth Voyager '84 thru '95
30011	Caravan & Plymouth Voyager '96 thru '02
30012	Challenger & Plymouth Sapporo '78 thru '83
30013	Caravan, Chrysler Voyager & Town & Country '03 thru '07
30014	Grand Caravan & Chrysler Town & Country '08 thru '18
30020	Dakota Pick-ups all models '87 thru '96
30021	Durango '98 & '99 & Dakota '97 thru '99
30022	Durango '00 thru '03 & Dakota '00 thru '04
30023	Durango '04 thru '09 & Dakota '05 thru '11
30025	Dart, Challenger/Plymouth Barracuda & Valiant 6-cylinder models '67 thru '76
30030	Daytona & Chrysler Laser '84 thru '89
	Intrepid - see Chrysler (25025, 25026)
30034	Neon all models '95 thru '99
30035	Omni & Plymouth Horizon '78 thru '90
30036	Dodge & Plymouth Neon '00 thru '05
30040	Pick-ups full-size '74 thru '93
30041	Pick-ups full-size '94 thru '01
30042	Pick-ups full-size '02 thru '08
30043	Pick-ups full-size '09 thru '18
30045	Ram 50/D50 Pick-ups & Raider and Plymouth Arrow Pick-ups '79 thru '93
30050	Dodge/Plymouth/Chrysler RWD '71 thru '89
30055	Shadow & Plymouth Sundance '87 thru '94
30060	Spirit & Plymouth Acclaim '89 thru '95
30065	Vans - Dodge & Plymouth '71 thru '03

EAGLE
	Talon - see MITSUBISHI (68030, 68031)
	Vision - see CHRYSLER (25025)

FIAT
34010	124 Sport Coupe & Spider '68 thru '78
34025	X1/9 all models '74 thru '80

FORD
10320	Ford Engine Overhaul Manual
10355	Ford Automatic Transmission Overhaul
11500	Mustang '64-1/2 thru '70 Restoration Guide
36004	Aerostar Mini-vans '86 thru '97
36006	Contour & Mercury Mystique '95 thru '00
36008	Courier Pick-up all models '72 thru '82
36012	Crown Victoria & Mercury Grand Marquis '88 thru '11
36016	Escort & Mercury Lynx '81 thru '90
36020	Escort & Mercury Tracer '91 thru '02
36022	Escape '01 thru '17, Mazda Tribute '01 thru '11 & Mercury Mariner '05 thru '11
36024	Explorer & Mazda Navajo '91 thru '01
36025	Explorer & Mercury Mountaineer '02 thru '10
36026	Explorer '11 thru '17
36028	Fairmont & Mercury Zephyr '78 thru '83
36030	Festiva & Aspire '88 thru '97
36032	Fiesta all models '77 thru '80
36034	Focus all models '00 thru '11
36035	Focus '12 thru '14
36045	Ford Fusion '06 thru '14 & Mercury Milan '06 thru '10
36048	Mustang V8 all models '64-1/2 thru '73
36049	Mustang II 4-cylinder, V6 & V8 '74 thru '78
36050	Mustang & Mercury Capri '79 thru '93
36051	Mustang all models '94 thru '04
36052	Mustang '05 thru '14
36054	Pick-ups and Bronco '73 thru '79
36058	Pick-ups and Bronco '80 thru '96
36059	F-150 '97 thru '03, Expedition '97 thru '17, F-250 '97 thru '99, F-150 Heritage '04 & Lincoln Navigator '98 thru '17
36060	Super Duty Pick-up & Excursion '99 thru '10
36061	F-150 full-size '04 thru '14
36062	Pinto & Mercury Bobcat '75 thru '80
36063	F-150 full-size '15 thru '17
36064	Super Duty Pick-ups '11 thru '16
36066	Probe all models '89 thru '92
36070	Ranger & Bronco II gas models '83 thru '92
36071	Ranger '93 thru '11 & Mazda Pick-ups '94 thru '09
36074	Taurus & Mercury Sable '86 thru '95
36075	Taurus & Mercury Sable '96 thru '07
36076	Taurus '08 thru '14, Five Hundred '05 thru '07, Mercury Montego '05 thru '07 & Sable '08 thru '09
36078	Tempo & Mercury Topaz '84 thru '94
36082	Thunderbird & Mercury Cougar '83 thru '88
36086	Thunderbird & Mercury Cougar '89 thru '97
36090	Vans all V8 Econoline models '69 thru '91
36094	Vans full size '92 thru '14
36097	Windstar '95 thru '03, Freestar & Mercury Monterey Mini-van '04 thru '07

GENERAL MOTORS
10360	GM Automatic Transmission Overhaul
38005	Buick Century, Chevrolet Celebrity, Olds Cutlass Ciera & Pontiac 6000 '82 thru '96
38010	Buick Regal, Chevrolet Lumina, Oldsmobile Cutlass Supreme & Pontiac Grand Prix front wheel drive '88 thru '07
38015	Buick Skyhawk, Cadillac Cimarron, Chevrolet Cavalier, Oldsmobile Firenza Pontiac J-2000 & Sunbird '82 thru '94
38016	Chevrolet Cavalier/Pontiac Sunfire '95 thru '05
38017	Chevrolet Cobalt & Pontiac G5 '05 thru '11
38020	Buick Skylark, Chevrolet Citation, Olds Omega, Pontiac Phoenix '80 thru '85
38025	Buick Skylark & Somerset, Olds Achieva, Calais & Pontiac Grand Am '85 thru '98
38026	Chevrolet Malibu, Olds Alero & Cutlass, Pontiac Grand Am '97 thru '03
38027	Chevrolet Malibu '04 thru '12
38030	Cadillac Eldorado, Seville, Oldsmobile Toronado & Buick Riviera '71 thru '85
38031	Cadillac Eldorado, Seville, DeVille, Fleetwood, Oldsmobile Toronado & Buick Riviera '86 thru '93
38032	Cadillac DeVille '94 thru '05, Seville '92 thru '04 & Cadillac DTS '06 thru '10
38035	Chevrolet Lumina APV, Olds Silhouette & Pontiac Trans Sport all models '90 thru '96
38036	Chevrolet Venture, Olds Silhouette, Pontiac Trans Sport & Montana '97 thru '05
38040	Chevrolet Equinox '05 thru '17, GMC Terrain '10 thru '17 & Pontiac Torrent '06 thru '09

GEO
	Metro - see CHEVROLET Sprint (24075)
	Prizm - '85 thru '92 see CHEVY (24060), '93 thru '02 see TOYOTA Corolla (92036)
40030	Storm all models '90 thru '93
	Tracker - see SUZUKI Samurai (90010)

GMC
	Vans & Pick-ups - see CHEVROLET

HONDA
42010	Accord CVCC all models '76 thru '83
42011	Accord all models '84 thru '89
42012	Accord all models '90 thru '93
42013	Accord all models '94 thru '97
42014	Accord all models '98 thru '02
42015	Accord '03 thru '12 & Crosstour '10 thru '14
42016	Accord '13 thru '17
42020	Civic 1200 all models '73 thru '79
42021	Civic 1300 & 1500 CVCC '80 thru '83
42022	Civic 1500 CVCC all models '75 thru '79
42023	Civic all models '84 thru '91
42024	Civic & del Sol '92 thru '95
42025	Civic '96 thru '00, & CR-V '97 thru '01 & Acura Integra '94 thru '00
42026	Civic '01 thru '11 & CR-V '02 thru '11
42027	Civic '12 thru '15 & CR-V '12 thru '16
42030	Fit '07 thru '13
42035	Odyssey all models '99 thru '10
	Passport - see ISUZU Rodeo (47017)
42037	Honda Pilot '03 thru '08, Ridgeline '06 thru '14 & Acura MDX '01 thru '07
42040	Prelude CVCC all models '79 thru '89

HYUNDAI
43010	Elantra all models '96 thru '19
43015	Excel & Accent all models '86 thru '13
43050	Santa Fe all models '01 thru '12
43055	Sonata all models '99 thru '14

INFINITI
	G35 '03 thru '08 - see NISSAN 350Z (72011)

ISUZU
	Hombre - see CHEVROLET S-10 (24071)
47017	Rodeo '91 thru '02, Amigo '89 thru '94 & '98 thru '02 & Honda Passport '95 thru '02
47020	Trooper '84 thru '91 & Pick-up '81 thru '93

JAGUAR
49010	XJ6 all 6-cylinder models '68 thru '86
49011	XJ6 all models '88 thru '94
49015	XJ12 & XJS all 12-cylinder models '72 thru '85

JEEP
50010	Cherokee, Comanche & Wagoneer Limited all models '84 thru '01
50011	Cherokee '14 thru '19
50020	CJ all models '49 thru '86
50025	Grand Cherokee all models '93 thru '04
50026	Grand Cherokee '05 thru '19 & Dodge Durango '11 thru '19
50029	Grand Wagoneer & Pick-up '72 thru '91
50030	Wrangler all models '87 thru '17
50035	Liberty '02 thru '12 & Dodge Nitro '07 thru '11
50050	Patriot & Compass '07 thru '17

KIA
54050	Optima '01 thru '10
54060	Sedona '02 thru '14
54070	Sephia '94 thru '01, Spectra '00 thru '09, Sportage '05 thru '20
54077	Sorento '03 thru '13

LEXUS
	ES 300/330 - see TOYOTA Camry (92007, 92008)
	RX 300/330/350 - see TOYOTA Highlander (92095)

LINCOLN
	Navigator - see FORD Pick-up (36059)
59010	Rear-Wheel Drive Continental '70 thru '87, Mark Series '70 thru '92 & Town Car '81 thru '10

MAZDA
61010	GLC (rear wheel drive) '77 thru '83
61011	GLC (front wheel drive) '81 thru '85
61012	Mazda3 '04 thru '11
61015	323 & Protegé '90 thru '03
61016	MX-5 Miata '90 thru '14
61020	MPV all models '89 thru '98
61030	Pick-ups '72 thru '93
	Pick-ups '94 thru '09 - see Ford (36071)
61035	RX-7 all models '79 thru '85
61036	RX-7 all models '86 thru '91
61040	626 (rear-wheel drive) all models '79 thru '82
61041	626 & MX-6 (front-wheel drive) '83 thru '92
61042	626 '93 thru '01 & MX-6/Ford Probe '93 thru '02
61043	Mazda6 '03 thru '13

MERCEDES-BENZ
63012	123 Series Diesel '76 thru '85
63015	190 Series 4-cylinder gas models '84 thru '88
63020	230, 250 & 280 6-cylinder SOHC '68 thru '72
63025	280 123 Series gas models '77 thru '81
63030	350 & 450 all models '71 thru '80
63040	C-Class: C230/C240/C280/C320/C350 '01 thru '07

MERCURY
64200	Villager & Nissan Quest '93 thru '01
	All other titles, see FORD listing.

MG
66010	MGB Roadster & GT Coupe '62 thru '80
66015	MG Midget & Austin Healey Sprite Roadster '58 thru '80

MINI
67020	Mini '02 thru '13

MITSUBISHI
68020	Cordia, Tredia, Galant, Precis & Mirage '83 thru '93
68030	Eclipse, Eagle Talon & Plymouth Laser '90 thru '94
68031	Eclipse '95 thru '05 & Eagle Talon '95 thru '98
68035	Galant '94 thru '12
68040	Pick-up '83 thru '96 & Montero '83 thru '93

NISSAN
72010	300ZX all models incl. Turbo '84 thru '89
72011	350Z & Infiniti G35 all models '03 thru '08
72015	Altima all models '93 thru '06
72016	Altima '07 thru '12
72020	Maxima all models '85 thru '92
72021	Maxima all models '93 thru '08
72025	Murano '03 thru '14
72030	Pick-ups '80 thru '97 & Pathfinder '87 thru '95
72031	Frontier, Xterra & Pathfinder '96 thru '04
72032	Frontier & Xterra '05 thru '14
72037	Pathfinder '05 thru '14
72040	Pulsar all models '83 thru '86
72042	Roque all models '08 thru '20
72050	Sentra all models '82 thru '94
72051	Sentra & 200SX all models '95 thru '06
72060	Stanza all models '82 thru '90
72070	Titan pick-ups '04 thru '10 & Armada '05 thru '10
72080	Versa all models '07 thru '19

OLDSMOBILE
73015	Cutlass V6 & V8 gas models '74 thru '88
	For other OLDSMOBILE titles, see BUICK, CHEVROLET or GM listings.

PLYMOUTH
For PLYMOUTH titles, see DODGE.

PONTIAC
79008	Fiero all models '84 thru '88
79018	Firebird V8 models except Turbo '70 thru '81
79019	Firebird all models '82 thru '92
79025	G6 all models '05 thru '09
79040	Mid-size Rear-wheel Drive '70 thru '87
	Vibe '03 thru '10 - see TOYOTA Corolla (92037)
	For other PONTIAC titles, see BUICK, CHEVROLET or GM listings.

PORSCHE
80020	911 Coupe & Targa models '65 thru '89
80025	914 all 4-cylinder models '69 thru '76
80030	924 all models including Turbo '76 thru '82
80035	944 all models including Turbo '83 thru '89

RENAULT
	Alliance & Encore - see AMC (14025)

SAAB
84010	900 all models including Turbo '79 thru '88

SATURN
87010	Saturn all S-series models '91 thru '02
87020	Saturn Ion '03 thru '07- see GM (38017)
87040	Saturn L-series all models '00 thru '04
	Saturn VUE '02 thru '09

SUBARU
89002	1100, 1300, 1400 & 1600 '71 thru '79
89003	1600 & 1800 2WD & 4WD '80 thru '94
89080	Impreza '02 thru '11, WRX '02 thru '14 & WRX STI '04 thru '14
89100	Legacy all models '90 thru '99
89101	Legacy & Forester '00 thru '09
89102	Legacy '10 thru '16 & Forester '12 thru '16

SUZUKI
90010	Samurai/Sidekick & Geo Tracker '86 thru '01

TOYOTA
92005	Camry all models '83 thru '91
92006	Camry '92 thru '96 & Avalon '95 thru '96
92007	Camry, Avalon, Solara, Lexus ES 300 '97 thru '01
92008	Camry, Avalon, Lexus ES 300/330 '02 thru '06 & Solara '02 thru '08
92009	Camry, Avalon & Lexus ES 350 '07 thru '17
92015	Celica Rear-wheel Drive '71 thru '85
92020	Celica Front-wheel Drive '86 thru '99
92025	Celica Supra all models '79 thru '92
92030	Corolla all models '75 thru '79
92032	Corolla all rear-wheel drive models '80 thru '87
92035	Corolla front-wheel drive models '84 thru '92
92036	Corolla & Geo/Chevrolet Prizm '93 thru '02
92037	Corolla '03 thru '19, Matrix '03 thru '14, & Pontiac Vibe '03 thru '10
92040	Corolla Tercel all models '80 thru '82
92045	Corona all models '74 thru '82
92050	Cressida all models '78 thru '82
92055	Land Cruiser FJ40/43/45/55 '68 thru '82
92056	Land Cruiser FJ60/62/80/FZJ80 '80 thru '96
92060	Matrix '03 thru '11 & Pontiac Vibe '03 thru '10
92065	MR2 all models '85 thru '87
92070	Pick-ups all models '69 thru '78
92075	Pick-ups all models '79 thru '95
92076	Tacoma '95 thru '04, 4Runner '96 thru '02 & T100 '93 thru '98
92077	Tacoma all models '05 thru '18
92078	Tundra '00 thru '06, Sequoia '01 thru '07
92079	4Runner all models '03 thru '09
92080	Previa all models '91 thru '95
92081	Prius all models '01 thru '12
92082	RAV4 all models '96 thru '12
92085	Tercel all models '87 thru '94
92090	Sienna all models '98 thru '10
92095	Highlander '01 thru '19 & Lexus RX330/330/350 '99 thru '19
92179	Tundra '07 thru '19 & Sequoia '08 thru '19

TRIUMPH
94007	Spitfire all models '62 thru '81
94010	TR7 all models '75 thru '81

VW
96008	Beetle & Karmann Ghia '54 thru '79
96009	New Beetle '98 thru '10
96016	Rabbit, Jetta, Scirocco & Pick-up gas models '75 thru '92 & Convertible '80 thru '92
96017	Golf, GTI & Jetta '93 thru '98, Cabrio '95 thru '02
96018	Golf, GTI & Jetta '99 thru '05
96019	Jetta, Rabbit, GLI, GTI & Golf '05 thru '11
96020	Rabbit, Jetta, Pick-up diesel '77 thru '84
96021	Jetta '11 thru '18 & Golf '15 thru '19
96023	Passat '98 thru '05, Audi A4 '96 thru '01
96030	Transporter 1600 all models '68 thru '79
96035	Transporter 1700, 1800, 2000 '72 thru '79
96040	Type 3 1500 & 1600 '63 thru '73
96045	Vanagon Air-Cooled all models '80 thru '83

VOLVO
97010	120, 130 Series & 1800 Sports '61 thru '73
97015	140 Series all models '66 thru '74
97020	240 Series all models '76 thru '93
97040	740 & 760 Series all models '82 thru '88
97050	850 Series all models '93 thru '97

TECHBOOK MANUALS
10205	Automotive Computer Codes
10206	OBD-II & Electronic Engine Management
10210	Automotive Emissions Control Manual
10215	Fuel Injection Manual '78 thru '85
10225	Holley Carburetor Manual
10230	Rochester Carburetor Manual
10305	Chevrolet Engine Overhaul Manual
10320	Ford Engine Overhaul Manual
10330	GM and Ford Diesel Engine Repair Manual
10331	Duramax Diesel Engines '01 thru '19
10332	Cummins Diesel Engine Performance
10333	GM, Ford & Chrysler Engine Performance
10334	GM Engine Performance
10340	Small Engine Repair Manual, 5 HP & Less
10341	Small Engine Repair Manual, 5.5 HP to 20 HP
10345	Suspension, Steering & Driveline Manual
10355	Ford Automatic Transmission Overhaul
10360	GM Automatic Transmission Overhaul
10405	Automotive Body Repair & Painting
10410	Automotive Brake Manual
10411	Automotive Anti-lock Brake (ABS) Systems
10425	Automotive Heating & Air Conditioning
10435	Automotive Tools Manual
10445	Welding Manual

Over a 100 Haynes motorcycle manuals also available

10/22

Haynes North America, Inc. • (805) 498-6703 • www.haynes.com